Introductory Review on Sirtuins in Biology, Aging, and Disease

Introductory Review on Sirtuins in Biology, Aging, and Disease

Edited by

Leonard Guarente
MIT, Cambridge, MA, United States

Raul Mostoslavsky
The Massachusetts General Hospital Cancer Center,
Boston, MA, United States
Harvard Medical School, Boston, MA, United States
Broad Institute of Harvard and MIT, Cambridge, MA, United States

Aleksey Kazantsev
Massachusetts General Hospital and Harvard Medical School,
Cambridge, MA, United States

ACADEMIC PRESS
An imprint of Elsevier

Academic Press is an imprint of Elsevier
125 London Wall, London EC2Y 5AS, United Kingdom
525 B Street, Suite 1800, San Diego, CA 92101-4495, United States
50 Hampshire Street, 5th Floor, Cambridge, MA 02139, United States
The Boulevard, Langford Lane, Kidlington, Oxford OX5 1GB, United Kingdom

Notices

Knowledge and best practice in this field are constantly changing. As new research and experience broaden our
understanding, changes in research methods, professional practices, or medical treatment may become necessary.

Practitioners and researchers must always rely on their own experience and knowledge in evaluating and using any
information, methods, compounds, or experiments described herein. In using such information or methods they should
be mindful of their own safety and the safety of others, including parties for whom they have a professional
responsibility.

To the fullest extent of the law, neither the Publisher nor the authors, contributors, or editors, assume any liability for
any injury and/or damage to persons or property as a matter of products liability, negligence or otherwise, or from any
use or operation of any methods, products, instructions, or ideas contained in the material herein.

British Library Cataloguing-in-Publication Data
A catalogue record for this book is available from the British Library

Library of Congress Cataloging-in-Publication Data
A catalog record for this book is available from the Library of Congress

ISBN: 978-0-12-813499-3

For Information on all Academic Press publications
visit our website at https://www.elsevier.com/books-and-journals

  Working together
to grow libraries in
developing countries

www.elsevier.com • www.bookaid.org

Publisher: John Fedor
Acquisitions Editor: Tari K. Broderick
Editorial Project Manager: Tracy I. Tufaga
Production Project Manager: Poulouse Joseph
Cover Designer: Greg Harris

Typeset by MPS Limited, Chennai, India

We would like to dedicate this book to our families, for their constant support, and to the members of our laboratories and all the scientists in the field of sirtuins. Without their support and rigorous scientific work, this book would not have been possible.

Contents

List of Contributors

Maria Angulo-Ibanez
Stanford University School of Medicine, Stanford, CA, United States; Veterans Affairs Palo Alto Health Care System, Palo Alto, CA, United States

Johan Auwerx
École Polytechnique Fédérale de Lausanne, Lausanne, Switzerland

Katrin F. Chua
Stanford University School of Medicine, Stanford, CA, United States; Veterans Affairs Palo Alto Health Care System, Palo Alto, CA, United States

William Giblin
University of Michigan, Ann Arbor, MI, United States

David Gius
Northwestern University, Chicago, IL, United States

Leonard Guarente
MIT, Cambridge, MA, United States

Angela H. Guo
University of Michigan, Ann Arbor, MI, United States

Marcia C. Haigis
Harvard Medical School, Boston, MA, United States

Sylvana Hassanieh
The Massachusetts General Hospital Cancer Center, Boston, MA, United States; Harvard Medical School, Boston, MA, United States; Broad Institute of Harvard and MIT, Cambridge, MA, United States

Kathleen A. Hershberger
Duke University Medical Center, Durham, NC, United States

Matthew D. Hirschey
Duke University Medical Center, Durham, NC, United States

Shin-ichiro Imai
Washington University School of Medicine, St. Louis, MO, United States

Alice E. Kane
Harvard Medical School, Boston, MA, United States

Elena Katsyuba
École Polytechnique Fédérale de Lausanne, Lausanne, Switzerland

Aleksey G. Kazantsev
Massachusetts General Hospital and Harvard Medical School, Cambridge, MA, United States

Hening Lin
Howard Hughes Medical Institute, Cornell University, Ithaca, NY, United States

David B. Lombard
University of Michigan, Ann Arbor, MI, United States

Sébastien Moniot
University of Bayreuth, Bayreuth, Germany

Raul Mostoslavsky
The Massachusetts General Hospital Cancer Center, Boston, MA, United States; Harvard Medical School, Boston, MA, United States; Broad Institute of Harvard and MIT, Cambridge, MA, United States

Tiago F. Outeiro
University Medical Center Göttingen, Göttingen, Germany; Max Planck Institute for Experimental Medicine, Göttingen, Germany

David A. Sinclair
Harvard Medical School, Boston, MA, United States; The University of New South Wales, Sydney, NSW, Australia

Clemens Steegborn
University of Bayreuth, Bayreuth, Germany

Adam B. Stein
University of Michigan, Ann Arbor, MI, United States

Éva M. Szegő
University Medical Center Göttingen, Göttingen, Germany

Robert A.H. van de Ven
Harvard Medical School, Boston, MA, United States

Athanassios Vassilopoulos
Northwestern University, Chicago, IL, United States

Rui-Hong Wang
University of Macau, Macau SAR, China

Wen Yang
Harvard Medical School, Boston, MA, United States; Shanghai Jiao Tong University, School of Medicine, Shanghai, China

Mitsukuni Yoshida
Washington University School of Medicine, St. Louis, MO, United States

Weijie You
University of Bayreuth, Bayreuth, Germany

Biographies

Leonard Guarente is the Novartis Professor of Biology at the Massachusetts Institute of Technology, Director of the Glenn Center for the Science of Aging, and affiliate of the Koch Institute for Integrated Cancer Research.

He is best known for his research on life span extension in the budding yeast *Saccharomyces cerevisiae*, round worms (*Caenorhabditis elegans*), and mice. His lab identified sirtuins as key regulators of aging in yeast, roundworms, and mice. He also discovered that sirtuins are NAD + dependent deacylases, opening the door for therapeutic strategies to activate sirtuins by small molecules or by NAD + replenishment.

Thus he cofounded Elixir Pharmaceuticals, Elysium Health, and Galileo Biosciences to develop interventions to slow aging and extend healthspan. He was also cochair of the SAB of Sirtris Pharmaceuticals and consulted for GSK on sirtuin activators. He also consults for Segterra and Sibelius.

He serves on editorial boards of *Cell, Cell Metabolism, Trends in Genetics, EMBO Reports, Aging, Experimental Gerontology*, and *Clinical Epigenetics*. He wrote an autobiography *Ageless Quest: One Scientist's Search for Genes That Prolong Youth* (Cold Spring Harbor Press) 2003, and is an editor of the textbook *Molecular Biology of Aging* (Cold Spring Harbor Press) 2007.

Affiliation: Department of Biology, Massachusetts Institute of Technology, Cambridge, MA, USA

Raul Mostoslavsky received his M.D. from the University of Tucuman in Argentina and his Ph.D. from the Hebrew University in Jerusalem, Israel. His longstanding interest in basic science and epigenetics brought him to Harvard Medical School to pursue postdoctoral studies where he first became interested in sirtuins. In 2007, Dr. Mostoslavsky opened his own lab at the Massachusetts General Hospital Cancer Center, Harvard Medical School. He is currently the Laurel Schwartz Associate Professor of Oncology at the MGH Cancer Center, Harvard Medical School, the Kristin and Bob Higgins MGH Research Scholar, and an Associate Member of the Broad Institute.

His main research interests are on the crosstalk between epigenetics and metabolism, and the roles of chromatin sirtuins in health and disease, particularly cancer.

Affiliation: Department of Medicine, Massachusetts General Hospital Cancer Center, Harvard Medical School, Boston, MA, USA

Aleksey Kazantsev received his bachelor's degree in biology and chemistry from Moscow Pedagogical University. As a graduate student he was enrolled in the Genetics, Molecular Biology and Biotechnology Program at the University of North Carolina at Chapel Hill and completed his PhD thesis in 1997.

Dr. Kazantsev then joined David Housman's laboratory at the Massachusetts Institute of Technology, where he started work on cellular models of neurodegeneration. At MIT he developed a strong interest in drug discovery, and worked on the development of high-throughput drug-screening assays. Through involvement in collaborative work with Genzyme Corporation, he became coinventor on a few patent applications, disclosing novel neurodegenerative assays and therapeutic agents. He discovered an inhibitor of polyglutamine aggregation, which was neuroprotective in a Huntington's disease mouse model, and has published extensively in the field of Huntington's disease and drug discovery.

In 2002, Dr. Kazantsev was named assistant professor of neurology at Harvard Medical School and principal investigator at Massachusetts General Hospital, where he continued to pursue his interest in drug discovery for neurodegenerative diseases, setting up a high-throughput drug-screening facility and in collaboration with other investigators performing drug screens targeting various neurodegenerative disorders. Currently he is a co-PI on a collaborative research project with Novartis toward the development of a cure for Huntington's disease.

Affiliation: MassGeneral Institute for Neurodegenerative Disease, Charlestown, MA, USA

Introduction

In 1981, Amar Klar, James Hicks, and colleagues, working in Cold Spring Harbor Laboratories, performed a screen to identify factors that influenced yeast mating type. In a series of elegant experiments, they identified four loci that, when mutated, conferred derepression of the normal silent mating type (Klar et al., 1981). The loci were termed SIR (Silent Information Regulator), and the authors prophetically stated "The accessibility of the promoter would be established by the interactions of sites with the SIR products and could reflect differential chromatin structure." The study and function of these proteins may have remained confined to yeast biologists if it was not for two discoveries from the laboratory of Lenny Guarente that transformed the field: first, the discovery that extra activity of sirtuins in lower organisms could extend lifespan, and second the discovery that sirtuins from both yeast and mammals exhibit NAD-dependent deacetylase activity, suggesting that these proteins linked metabolism, epigenetics, and aging. Throughout the last decade sirtuins have emerged as proteins that modulate critical biological functions in cells, including mitochondrial metabolism and ATP generation, DNA repair, glucose and lipid metabolism, and preserve the integrity of organ systems, including liver, muscle, brain, kidney, pancreas, and endothelial cells. In this book, we attempt to provide a comprehensive overview of the current state of knowledge in the biology of sirtuins.

Lenny Guarente provides an historical perspective to introduce the field (Chapter 1: Sirtuins, NAD$^+$, Aging, and Disease: A Retrospective and Prospective Overview), particularly in the context of the roles of sirtuins in aging. In recent years, we have learned that NAD levels decline with age, and the pathways modulating NAD availability play key roles in maintaining the activity of SIRT1 and other sirtuins. Shin-Ichiro Imai and Johan Auwerx and their colleagues elegantly describe the different mechanisms governing NAD biology and SIRT1 function (Chapter 2: Regulation of Sirtuins by Systemic NAD+ Biosynthesis and Chapter 3: NAD+ Modulation: Biology and Therapy). Although originally defined as histone deacetylases, Hening Lin showed that sirtuins are broad-spectrum protein deacylases, capable of removing succinyl, glutaryl, and malonyl (SIRT5), or palmitoyl (SIRT6), modifications in hundreds of proteins. David Sinclair showed that small molecules could activate SIRT1 via an allosteric site, and in Chapter 4, The Enzymatic Activities of Sirtuins and Chapter 5, Structural and Mechanistic Insights in Sirtuin Catalysis and Pharmacological Modulation, Sinclair and Clemens Steenborn and their colleagues describe small-molecule activation of sirtuins from a structural standpoint. Three sirtuins, SIRT3, 4, and 5 reside in the mitochondria, and Matthew Hirschey and Marcia Haigis, and their colleagues provide details on their important functions in metabolism, energy production, and organelle integrity (Chapter 6: Pharmacological Approaches for Modulating Sirtuins and Chapter 7: Reactive Acyl-CoA Species and Deacylation by the Mitochondrial Sirtuins). SIRT6 and SIRT7, like SIRT1, function almost exclusively in the nucleus, and their roles in tumor suppression and metabolic regulation are summarized in Chapter 8, Mitochondrial Sirtuins: Coordinating Stress Responses Through Regulation of Mitochondrial Enzyme Networks and Chapter 9, Multitasking Roles of the Mammalian Deacetylase SIRT6 by Raul Mostoslavsky, Katrin Chua, and colleagues. Beyond their normal functions, sirtuins have emerged as critical protectors against numerous diseases, and the final chapters in the book (Chapters 10−12) by David Gius, David Lombard, Aleksey Kazantsev, and colleagues will cover their roles in cancer, cardiovascular disease, and neurodegenerative diseases.

Little did Klar and colleagues imagine that their initial screening for silencing factors could lead to the discovery of one of the most versatile and interesting families of proteins in biology. The study of sirtuins has provided us with a unique understanding on the crosstalk between metabolic homeostasis, chromatin dynamics, nutrient signaling, cellular proliferation, and genomic stability. Undoubtedly, there are still numerous questions for which we do not know the answers. How are the cellular and tissue specificity of the different sirtuins regulated? How is their deacylase function coordinated with the myriad of other protein modifications? How much redundancy exist, both among sirtuins and with other enzymes? Will modulation of sirtuins by small-molecule activators or NAD^+ replenishment be sufficient to delay, prevent, or treat diseases? We hope our book inspires the next generation of scientists to start unraveling the complexities posed by these questions.

Lenny Guarente, Raul Mostoslavsky and Aleksey Kazantsev

Editors

REFERENCE

1. Klar AJ, Strathern JN, Broach JR, Hicks JB. Regulation of transcription in expressed and unexpressed mating type cassettes of yeast. *Nature* 1981;**289**:239—44.

SIRTUINS, NAD$^+$, AGING, AND DISEASE: A RETROSPECTIVE AND PROSPECTIVE OVERVIEW

Leonard Guarente

MIT, Cambridge, MA, United States

Over the past 20 years, sirtuins have emerged as a major pathway regulating aging and age-related diseases in organisms ranging from yeast to mammals. When we started working on aging of yeast mother cells in 1991, our primary goals were to determine if the earlier claims that yeast mother cells really have a fixed life span (i.e., their cumulative number of replicative divisions to give daughter cells) were correct, and assuming this were the case, to try to identify what determined that life span. I personally was worried that it might not be possible to identify genes that regulated the process, because aging might be hopelessly complicated. I would have considered it a great success if we could identify any gene that regulated yeast aging, and assumed our findings would not have relevance for higher organisms. After years of slogging through genetic screens, students Brian Kennedy and Nic Austriaco identified a long-lived mutant that ultimately traced a path to sirtuins and indicated a link between aging, diet, metabolism, and epigenetics, as discussed below.

SIR2, SIR3, and SIR4 were first identified as yeast proteins that determined epigenetic silencing at repeated DNA sequences (telomeres, extra copies of mating type loci, and ribosomal DNA or rDNA). The SIR2/3/4 complex is required to silence at telomeres and extra mating type copies, while SIR2, but not SIR3 or SIR4, is required for silencing at the rDNA. SIR2 turned out to be universally conserved in all organisms ranging from bacteria to mammals, while SIR3 and SIR4 are found only in fungi. SIR2 homologs across all organisms have come to be called sirtuins. In mammals there are seven sirtuins termed SIRT1−7, where the T stands for "two." These seven proteins are not redundant, because they are found in different cellular compartments and display different tissue distributions. SIRT1, SIRT6, and SIRT7 are present in the nucleus, SIRT 3, SIRT4, and SIRT5 reside in the mitochondria (although encoded by nuclear genes), and SIRT2 is cytoplasmic, but has access to nuclear proteins during mitosis. SIRT1 has the broadest tissue expression pattern of all seven sirtuin genes, and SIRT1 knockout mice have the strongest phenotype.

We originally obtained several different yeast mutants that were long-lived, but were most intrigued by one of them because it caused sterility. This interesting mutant turned out to harbor a gain-of-function mutation in SIR4 that eliminated silencing at telomeres and mating type loci (thereby causing sterility), but reinforced silencing at the 150 or so tandem copies of ribosomal

Introductory Review on Sirtuins in Biology, Aging, and Disease. DOI: https://doi.org/10.1016/B978-0-12-813499-3.00001-0

DNA or rDNA repeats on chromosome 12. This was because in this mutant the SIR2/3/4 complex was no longer tethered at telomeres and mating type loci and all of the cell's SIR2/3/4 was redirected to the rDNA. Silencing at these repeated loci was known to be important to repress recombination, an event which would result in genome instability at that locus and loss of rDNA genes. This finding suggested that increasing silencing at the rDNA beyond that of wildtype cells in the SIR4 mutant is what increased their life span. Since only SIR2 (and not SIR3 or SIR4) was required for normal silencing at the rDNA, we reasoned that simply increasing SIR2 levels in cells, e.g., by inserting into yeast a second copy of the gene, would also increase rDNA silencing and extend the life span. Indeed, elevating SIR2 activity alone increased the life span of yeast mother cells, while knocking out SIR2 had the opposite effect.

A further analysis of silencing in the rDNA by postdoc David Sinclair led us to an understanding of an important cause of aging, which limited the life span of yeast mother cells. Recombination in the nine kilobase units of repeated rDNA could led to formation of rDNA circles, which would replicate each cell division along with the resident chromosomes of the cell. Remarkably, we found that the rDNA circles stayed exclusively confined to mother cells at cell division, rendering all of the daughter cells pristine and capable of enjoying a full life span. Therefore, if a mother cell sustained a recombination event at say generation five, in 10 generations, the copy number of rDNA circles in that cell would rise to 1000 in 10 generations and in 20 generations to 1,000,000. The burden imposed by all of these extra rDNA copies (titration of replication or transcription factors, unbalancing the RNA and protein subunits of the ribosome, etc.) would then lead to senescence of the mother cell. For a while, we thought that this molecular mechanism might be a universal feature of aging. Disappointingly, this particular mechanism did not seem to apply to higher organisms, since we did not observe the accumulation of rDNA in circular or any other form in aging mammalian tissue.

These findings reinforced my earlier view that yeast aging mechanisms and, in particular, the key role played by SIR2 in yeast aging would be idiosyncratic for that organism. However, contrary to these expectations, SIR2 orthologs were subsequently shown to extend the life span in *Caenorhabditis elegans*, *Drosophila*, mice, and perhaps humans. In mice, increased expression of SIRT1 in brain and SIRT6 globally extend the life span, as do compounds that activate SIRT1 or NAD$^+$ precursors (see below). It should be noted that there was a brief period of panic issuing from a highly visible 2011 report that sirtuins had nothing to do with aging. This erroneous report had the benefit of stimulating another round of studies on sirtuins and aging in yeast, *C. elegans*, *Drosophila*, and mice that resoundingly validated the earlier claims linking sirtuins and aging. Obviously, the mechanisms by which SIR2 orthologs counter aging in these other organisms must have evolved, which is now understandable with the knowledge that sirtuins are actually enzymes that alter protein modifications, and as such they should be able to evolve to have new protein substrates as dictated by evolutionary pressures.

Thus, one of the most important features of sirtuins is their unique biochemical activity, NAD$^+$-dependent deacylation of histones and other proteins. SIRT4 and SIRT6 can also catalyze ADP-ribosylation of proteins using NAD$^+$ as cosubstrate. SIRT5 actually removes longer-chain acyl groups from proteins, such as succinyl or malonyl groups, while SIRT6 can also remove still longer chains, such as myristyl groups. The finding that sirtuins are NAD$^+$-dependent deacetylases was indeed a fortuitous one. Postdoc Shin Imai and I searched for years for a robust biochemical activity for the purified yeast SIR2 and mammalian SIRT1 proteins in vitro. Since deacetylation at certain residues of histones H3 and H4 was associated with silencing, labs were searching for a deacetylase activity for SIR2. Indeed, Jim Broach's lab showed that overexpression of SIR2 in yeast cells led to a global deacetylation of H3 and H4. However, all attempts to demonstrate such an activity with purified SIR2 in vitro were not successful. We were intrigued by the demonstration by Roy Fry that SIR2 could transfer ADP ribose (albeit feebly) from NAD$^+$ to BSA. We set upon studying this reaction using histone substrates, and confirmed Fry's feeble ADP-ribose transfer. But in the course of conducting this work, we stumbled into the true activity of SIR2 and SIRT1— namely NAD$^+$-dependent protein deacetylation—which contrary to the feeble transfer of ADP-ribose, occurred in a robust fashion. In this unique enzymatic reaction, the acetyl group is removed from lysines in the amino terminal tails of H3 and H4, and the NAD$^+$ is concomitantly cleaved into *O*-acetyl ADP-ribose and nicotinamide. The reason nobody had found this activity previously, is that no sane person would have considered including NAD$^+$ in the reaction. To wit, deacetylation is an energetically downhill reaction, and, moreover, a different class of protein deacetylases, the HDACs, had been shown to have no cofactor requirements. So there was no earthly reason for us to associate NAD$^+$ with protein deacetylation, and indeed we actually thought we were studying ADP ribosylation when we discovered this unusual deacetylase activity.

The deacetylation of histones by sirtuins renders these proteins key mediators of epigenetics in yeast and higher organisms. The implications of this feature of sirtuins has probably not yet been fully fleshed out, in my opinion. For example, one of the defining traits of epigenetics, is that an epigenetic state, e.g., deacetylated histones and silenced genes, can be long-lived and may even be heritable. High fat diets lead to obesity in pregnant mothers and can result in poor glucose and lipid homeostasis and poor health in progeny. Since sirtuins promote metabolic health (see below), might this be due to down regulation of sirtuins in the mother leading to epigenetic changes that persist in the offspring? Might it be preventable by interventions that leverage what we have learned about sirtuin biology?

Beyond histones, mammalian nuclear sirtuins SIRT1, SIRT6, and SIRT7 deacetylate scores of other protein substrates to coordinate numerous physiological pathways that mediate mitochondrial function, DNA repair, cell survival, stress tolerance, metabolic strategies, and more. The cytosolic SIRT2 is important in numerous processes, including controlling the cell cycle via deacetylation of proteins during mitosis. The mitochondrial resident sirtuins SIRT3, SIRT4, and SIRT5 play crucial roles in driving catabolic processes, such as fatty acid oxidation, and also maintain optimal structural and functional integrity in that organelle. The role of SIRT3 is especially telling in determining the acetylation levels of proteins in the mitochondria, since it is the only HDAC there and there are no known histone acetyl transferases. Indeed, hundreds of mitochondrial proteins are acetylated, and SIRT3 can deacetylate a large fraction of these proteins. This may be an especially important buffer during calorie restriction (CR), which leads to a large increase in the concentration of mitochondrial acetyl-CoA, which drives the acetylation of mitochondrial proteins by mass action. The nuclear SIRT1 is also critical in mitochondrial biology, since it deacetylates and upregulates PGC-1-α. The latter protein is a coactivator that drives mitochondrial biogenesis, ATP production, and quality control. Meanwhile, SIRT6 downregulates glycolysis, the alternative cellular pathway that produces ATP. Therefore, it seems likely that the combined action of nuclear and mitochondrial sirtuins, which is focused on mitochondria, contributes to the regulation of aging and the diseases that occur later in life. As an example, numerous genetic diseases with features of premature aging are due to DNA repair defects in the nucleus. Remarkably, much of the pathogenesis in mouse models of these diseases appears to be due to mitochondrial deficits resulting from NAD$^+$ depletion and sirtuin inactivation. NAD$^+$ becomes depleted in these animals due to the chronic activation of poly ADP-ribose polymerases or PARPs, which consume NAD$^+$ in response to DNA damage by decorating the damage site with polyADP-ribose. Indeed, these mice are partially rescued by dietary treatments that boost NAD$^+$ levels in the animals (see below).

As mentioned above, the NAD$^+$ link to sirtuins is unique in that NAD$^+$ actively participates in the chemistry of deacetylation and is cleaved in each reaction cycle. Since the HDACs do not require any cofactors, one surmises that the NAD$^+$ requirement of sirtuins is playing a regulatory role. It was suggested that this NAD$^+$ requirement may link sirtuins to cellular metabolism, and that they are actively involved in mediating the extension of life span by CR. Previously it was thought by some that CR extended life by a passive mechanism, such as by slowing metabolism. Indeed, sirtuins are activated by a low-calorie diet, in part by elevated levels of NAD$^+$, which makes biological sense because sirtuins drive oxidative mitochondrial electron transport and this mechanism generates much more ATP per molecule of glucose than glycolysis. In support of a key role for sirtuins in CR, knockout mice that lack a sirtuin (e.g., SIRT1 or SIRT3) are defective in responding to many features of CR, and transgenic mice sporting elevated levels of sirtuins (e.g., SIRT1 or SIRT6) show partial overlap with CR phenotypes. How does a mitochondrial strategy of metabolism promote health and long life? It turns out that sirtuin activation not only enhances mitochondrial oxidative metabolism but also induces resistance to oxidative and other stressors, which may help explain the life extension by CR. A good example is SOD2, which is induced at the expression level by the SIRT1/PGC-1-α axis, and is further activated at the enzymatic level by SIRT3-mediated deacetylation.

CR is associated with many health benefits in animals, including resistance to many of the common diseases of aging. If CR works by raising the activities of sirtuins, it follows that other interventions that raise activity of sirtuins should also mediate these benefits, even on a normal diet. In many studies, manipulating sirtuins genetically or pharmacologically (see below) has protective effects in mice against diabetes, cardiovascular disease, cancer, kidney disease, fatty liver, neurodegenerative diseases, proinflammatory diseases, osteoporosis, hearing loss, etc. Many of the chapters in this book will describe studies that show how manipulation of sirtuins can affect specific organ systems, and frame disease areas for this kind of intervention to improve human health. What makes sirtuins particularly suitable as targets in human diseases is not only their proven benefit in preclinical studies, but also the advanced stage of the translational research that has occurred in this space. This area is now discussed below and in greater detail in other chapters of this book.

The finding that SIRT1 was an NAD$^+$-dependent deacetylase opened up the potential for screening for small molecules that could affect its specific activity. We started this approach at Elixir Pharmaceuticals, only to be discouraged by an influential advisor to our venture partner (VP) investors. Soon thereafter, Howich, Sinclair, and colleagues identified molecules of the polyphenol class that would activate SIRT1 in vitro by lowering the k_m for the binding of the protein substrate. The most famous compound in this category was resveratrol. In vivo studies of resveratrol show health benefits in mice and nonhuman primates, but in humans the results have been disappointing. The problem with this compound is that it is poorly available in people and also subject to oxidation. Another red herring that sowed panic presented itself when resveratrol was challenged as a bona fide SIRT1 activator because it exerted its effects by interacting with the fluorescent tag on the in vitro peptide substrate. However, further studies showed that resveratrol did the job by interacting with aromatic amino acids in peptide substrates that lacked any fluorescent tag, thus supporting the original claims that the compound is a bona fide SIRT1 activator. Because the fluorescent tag originally used is itself aromatic, it may have mimicked the activation mechanism in natural peptide substrates.

High-throughput screening by Sirtris Pharmaceuticals subsequently identified a variety of synthetic compounds that would activate SIRT1 and at a much lower concentration than resveratrol. These SIRT1-activating compounds, or SirT1 Activating Compound (STACs), also conferred metabolic and health benefits to mice, including extension of the life span. After Sirtris was bought and absorbed by Glaxo SmithKline (GSK), medicinal chemistry has continued and has now resulted in compounds that more closely approach the pharmacokinetics and bioavailability of a drug. It is anticipated that this new class of drug may be in the clinic within 1 year. Both the polyphenols and the synthetic GSK compounds bind to an allosteric site outside of the catalytic domain of SIRT1 to stimulate its activity. Single amino acid substitutions in this site abolish activation by over 100 of the STACs, including resveratrol.

A second way to activate sirtuins has emerged. For many years, NAD$^+$ was thought of as a ubiquitous cellular component in constant supply. This coenzyme has been known for more than a century and is important for certain metabolic redox enzymes, such as that catalyzed by the glycolytic enzyme, glyceraldehyde phosphate dehydrogenase. The reduced form of NAD$^+$, NADH, which is produced by shunting glucose metabolism through the mitochondrial TCA cycle, is what drives electron transport and ATP production by that organelle. Beyond these basic metabolic reactions, of course, NAD$^+$ is an essential cosubstrate for sirtuins and PARPs, which cleave the coenzyme.

Thus, in a more recent twist in the sirtuin story, NAD$^+$ boosting has emerged as a second important mechanism of activating SIRT1 and also the other six mammalian sirtuins. A key finding casting this spotlight on NAD$^+$ was a study by the Imai lab on SIRT1 transgenic overexpressing mice. These animals were more glucose-tolerant than wildtype, but the phenotype disappeared in old mice. Adding the NAD$^+$ precursor nicotinamide mononucleotide (NMN) to the food or drinking water of old mice restored the transgenic phenotype. Both NMN and nicotinamide riboside, which is phosphorylated by kinases in cells to make NMN, enter as immediate precursors that are converted into NAD$^+$ by NMN adenine transferases. Importantly, these NAD precursors enter after the rate-limiting step in NAD$^+$ synthesis, nicotinamide phosphoribosyl transferase, which makes NMN. The rescue of old animals by NMN shows first that NAD$^+$ levels decline in the course of normal aging, thereby inactivating SIRT1, which has now been verified by direct measurements in mice, *C. elegans*, and humans, and second that NAD$^+$ precursors can boost NAD$^+$ to youthful levels and reactivate SIRT1. Many studies have been carried out in the past several years, which show that NAD$^+$ replenishment can protect against kidney disease, liver disease, metabolic disease, adult stem cell loss, and extend the life span of mice.

These findings regarding STACs and NAD$^+$ replenishment pave a translational path which offers a golden opportunity to improve human health. It seems likely that interventions that activate sirtuins may serve a beneficial role in both a preventative setting as well as in disease settings. In the next decade it is likely that the full potential of sirtuin activating in preserving good health and treating human health maladies will begin to unfurl. Since these interventions appear to trigger at least some of the effects as CR, I anticipate that numerous fruitful applications will be discovered. If so, it certainly will have been a remarkable journey from the humble yeast-bearing agar of petri plates to the front lines of modern medicine. Stay tuned.

ACKNOWLEDGMENTS

I wish to thank all my colleagues in the sirtuins/NAD$^+$ field over the years, including students, postdocs, and colleagues in other labs for making this a most exciting and gratifying journey. Work in my lab has been funded by the NIH and by the Glenn Foundation for Medical Research.

FURTHER READING

Bonkowski MS, Sinclair DA. Slowing ageing by design: the rise of NAD$^+$ and sirtuin-activating compounds. *Nat Rev Mol Cell Biol* Nov 2016;**17**(11):679−90. Available from: https://doi.org/10.1038/nrm.2016.93. Epub 2016 Aug 24.

Fang EF, Scheibye-Knudsen M, Chua KF, Mattson MP, Croteau DL, Bohr VA. Nuclear DNA damage signalling to mitochondria in ageing. *Nat Rev Mol Cell Biol* May 2016;**17**(5):308−21. Available from: https://doi.org/10.1038/nrm.2016.14. Epub 2016 Mar 9.

Guarente L. Calorie restriction and sirtuins revisited. *Genes Dev* Oct 1 2013;**27**(19):2072−85. Available from: https://doi.org/10.1101/gad.227439.113.

Imai S, Guarente L. NAD + and sirtuins in aging and disease. *Trends Cell Biol* Aug 2014;**24**(8):464−71. Available from: https://doi.org/10.1016/j.tcb.2014.04.002. Epub 2014 Apr 29.

Kugel S, Mostoslavsky R. Chromatin and beyond: the multitasking roles for SIRT6. *Trends Biochem Sci* Feb 2014;**39**(2):72−81. Available from: https://doi.org/10.1016/j.tibs.2013.12.002. Epub 2014 Jan 14.

Qiu X, Brown KV, Moran Y, Chen D. Sirtuin regulation in calorie restriction. *Biochim Biophys Acta* Aug 2010;**1804**(8):1576−83. Available from: https://doi.org/10.1016/j.bbapap.2009.09.015. Epub 2009 Sep 24.

Satoh A, Imai S. Systemic regulation of mammalian ageing and longevity by brain sirtuins. *Nat Commun* Jun 26 2014;**5**:4211. Available from: https://doi.org/10.1038/ncomms5211.

Sinclair DA, Guarente L. Small-molecule allosteric activators of sirtuins. *Annu Rev Pharmacol Toxicol* 2014;**54**:363−80. Available from: https://doi.org/10.1146/annurev-pharmtox-010611-134657. Epub 2013 Oct 16.

van de Ven RAH, Santos D, Haigis MC. Mitochondrial sirtuins and molecular mechanisms of aging. *Trends Mol Med* Apr 2017;**23**(4):320−31. Available from: https://doi.org/10.1016/j.molmed.2017.02.005. Epub 2017 Mar 10.

Verdin E, Hirschey MD, Finley LW, Haigis MC. Sirtuin regulation of mitochondria: energy production, apoptosis, and signaling. *Trends Biochem Sci* Dec 2010;**35**(12):669−75. Available from: https://doi.org/10.1016/j.tibs.2010.07.003. Epub 2010 Sep 20.

Zhu Y, Yan Y, Principe DR, Zou X, Vassilopoulos A, Gius D. SIRT3 and SIRT4 are mitochondrial tumor suppressor proteins that connect mitochondrial metabolism and carcinogenesis. *Cancer Metab* Oct 20, 2014;**2**:15. Available from: https://doi.org/10.1186/2049-3002-2-15. eCollection 2014.

REGULATION OF SIRTUINS BY SYSTEMIC NAD$^+$ BIOSYNTHESIS

2

Mitsukuni Yoshida and Shin-ichiro Imai
Washington University School of Medicine, St. Louis, MO, United States

2.1 INTRODUCTION

Nicotinamide adenine dinucleotide (NAD$^+$) is a universal energy currency necessary for various cellular processes mediating metabolic homeostasis, damage response, immune reaction, and many others.[1-3] While NAD$^+$ has been well recognized for its importance as a coenzyme in redox reactions, its role as a cosubstrate has attracted significant attention over the past two decades. NAD$^+$ was originally discovered by Harden and Young as a low-molecular-weight substance extracted from yeast that promotes alcohol fermentation.[4] Since its discovery, NAD$^+$ and its reduced form NADH, as well as NADP$^+$ and NADPH, have been well studied as coenzymes for many redox reactions.[5] NAD$^+$ has also been identified as a cosubstrate for DNA ligase, poly-ADP-ribose polymerases (PARPs), and CD38/157 ADP-ribosyl cyclases.[6-8] In 2000, NAD$^+$ was identified as an essential cosubstrate for an evolutionarily conserved silent information regulator 2 (SIR2) family of protein deacetylases, also called sirtuins.[9] Yeast SIR2 protein and its mouse homolog, now called SIRT1, were demonstrated to deacetylate lysines 9 and 14 of histone H3 and lysine 16 of H4 in an NAD$^+$-dependent manner. This unprecedented enzymatic activity immediately suggested that sirtuins function as sensors of the cellular energy status represented by NAD$^+$, connecting between cellular metabolism and epigenetic regulation.

Since then, the biology of sirtuins has been rapidly evolving, demonstrating pleiotropic NAD$^+$-dependent functions of sirtuins in many critical biological processes, particularly in the regulation of aging and longevity, in diverse model organisms.[10-12] Although a challenge was raised for the importance of sirtuins in aging and longevity control, many studies have now firmly confirmed that sirtuins control the process of aging and promote longevity in yeast, worms, flies, and mice.[13-22] For example, in mammals, the brain-specific SIRT1-overexpressing (BRASTO) transgenic mice exhibit a significant delay in aging and lifespan extension in both male and female mice.[18] Additionally, whole-body SIRT6 transgenic mice show lifespan extension, although only males exhibit the phenotype.[19] Such an evolutionarily conserved function of sirtuins in aging and longevity control is mainly due to their importance in the regulation of physiological resilience in each organism. A good example of this case is dietary restriction. Dietary restriction is a well-studied dietary regimen that delays aging and extends lifespan in many diverse species including yeast, worms, flies, rodents, and primates.[23-27] Interestingly, sirtuins are critical in mediating physiological responses to dietary restriction, activating transcriptional programs that promote metabolic efficiency, and stimulating

Introductory Review on Sirtuins in Biology, Aging, and Disease. DOI: https://doi.org/10.1016/B978-0-12-813499-3.00002-2

mitochondrial oxidative metabolism, all of which augment physiological resilience throughout the body.[16,23,28–30] Therefore, it is conceivable that sirtuins have been evolved to maximize physiological resilience, particularly in life-threatening conditions, and promote survival.

Unfortunately, it seems that sirtuins cannot maintain their critical functions throughout the life course of an organism. It has now become clear that NAD$^+$ availability declines systemically in diverse organisms so that sirtuins cannot maintain their full activities, contributing to age-associated pathophysiologies in each organism.[31–34] For this reason, more studies have recently started focusing on the functional connection between NAD$^+$ biosynthesis and consumption and sirtuin functions. In this chapter, we will discuss how NAD$^+$ biosynthesis is regulated, how NAD$^+$ is consumed, and how sirtuin functions are regulated by NAD$^+$ availability, mainly in mammals. We will also discuss how this tight connection between NAD$^+$ and sirtuins impacts the regulation of aging and longevity and how sirtuin activities can be maintained to mitigate physiological functional decline.

2.2 NAD$^+$ BIOSYNTHESIS, CONSUMPTION, AND DEGRADATION

Given that NAD$^+$ is an essential cosubstrate for sirtuin activities, NAD$^+$ availability has been suggested to directly regulate sirtuin activities and impact their downstream signaling pathways. Several kinetic studies have demonstrated that K_m values of SIRT1–3 for NAD$^+$ are consistent with the range of NAD$^+$ changes in each subcellular compartment (Table 2.1). Indeed, another

Table 2.1 Biochemical Characteristics of Sirtuins and Subcellular NAD$^+$ Concentrations. Localization, Kinetics, and Dominant Activity Types of the Sirtuin Family Members and the NAD$^+$ Concentration in Subcellular Compartments are Listed[35–44]

Sirtuin	Localization	K_m (μM)	Subcellular NAD$^+$ Concentration (μM)[33]	Activity
SIRT1	Nucleus, cytoplasm	29[36] 94–96[37] 150–200[38]	109 (95%CI 87–136), 106 (92–122)	Deacetylation
SIRT2	Cytoplasm	18[36] 83[39] 100[38]	106 (92–122)	Deacetylation
SIRT3	Mitochondria	90[36] 880[40] 280[41]	230 (191–275)	Deacetylation
SIRT4	Mitochondria	35[42]	230 (191–275)	ADP-ribosylation Deacylation
SIRT5	Mitochondria	980[43]	230 (191–275)	Deacetylation Demalonylation Desuccinylation
SIRT6	Nucleus	26[44]	109 (87–136)	Deacetylation Defatty-acylation ADP-ribosylation
SIRT7	Nucleolus	Unknown		Deacetylation

recent study with a newly developed NAD$^+$ biosensor has reported that in cultured cells, free NAD$^+$ concentrations are 106, 109, and 230 μM in the cytoplasm, nucleus, and mitochondria, respectively (Table 2.1).[35] Interestingly, NAD$^+$ appears to be quickly equilibrated between the nucleus and cytoplasm, whereas mitochondrial NAD$^+$ levels are maintained separately. Therefore, any changes in NAD$^+$ availability due to the alteration in biosynthesis, consumption, or degradation likely have significant impacts on sirtuin activities in different subcellular compartments.

NAD$^+$ biosynthetic pathways: NAD$^+$ can be synthesized from five different substrates: tryptophan (Trp), nicotinamide (NAM), nicotinic acid (NA), nicotinamide ribose (NR), and nicotinamide mononucleotide (NMN).[2,45] Trp, NAM, and NA are precursors, whereas NR and NMN are intermediates. All five compounds of these are contained in different natural food sources.[46,47] De novo pathway starts from Trp, which is converted to nicotinic acid mononucleotide (NaMN) through five enzymatic steps and one nonenzymatic reaction (Fig. 2.1). NaMN is also synthesized from NA in the Preiss—Handler pathway (Fig. 2.1). NaMN is converted to deamido-NAD by nicotinamide/nicotinic acid mononucleotide adenylyltransferases (NMNATs) and then to NAD$^+$ by NAD$^+$ synthetase (Fig. 2.1). In invertebrates and many bacterial species, NAM needs to be converted to NA by nicotinamidase. In mammals, however, NAM is the major substrate for NAD$^+$ biosynthesis, and nicotinamide phosphoribosyltransferase (NAMPT) converts NAM and 5′-phosphoribosyl-pyrophophate to NMN (Fig. 2.1). NMN is then

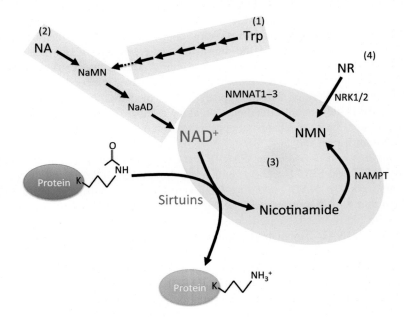

FIGURE 2.1

NAD$^+$ biosynthetic pathways. NAD$^+$ can be synthesized through: (1) de novo pathway, (2) Preiss—Handler pathway, (3) salvage pathway, or (4) from NR. De novo, Preiss—Handler, and salvage pathways use tryptophan (Trp), nicotinic acid (NA), and nicotinamide as starting substrates, respectively. NMN and NR are NAD$^+$ intermediates. In response to NAD$^+$ availability, sirtuins remove the acetyl group on the specific lysine residue from the target protein. Deacetylation alters the protein functions, initiating the downstream signaling cascade.

converted to NAD$^+$ by three different NMNATs, NMNAT1−3. NAMPT is the rate-limiting enzyme in this NAD$^+$ biosynthetic pathway and regulates the NAD$^+$ biosynthetic capacity of cells.[48] NAMPT is expressed ubiquitously, but the liver, kidney, and brown adipose tissue express high levels of NAMPT, the heart expresses an intermediate level, white adipose tissue, lung, spleen, testis, and skeletal muscle express low levels, and the brain and pancreas express barely detectable levels.[49] It has been demonstrated that NAMPT plays a critical role in regulating sirtuin activities in different cell types and tissues.[48,50,51]

Interestingly, mRNA and protein expression of *Nampt* is regulated in a circadian oscillation-dependent manner in peripheral tissues, such as liver and white adipose tissue.[52,53] The core circadian machinery mediated by BMAL1 and CLOCK promotes the transcription of *Nampt* through their binding to an E box on its promoter. Increased NAMPT expression leads to higher NAD$^+$ levels, stimulating SIRT1 activity. Reciprocally, increased SIRT1 activity suppresses the transcriptional activity of BMAL1 and CLOCK, creating a negative feedback loop. This dynamic interaction between BMAL/CLOCK and NAMPT/SIRT1 generates the circadian oscillation of NAD$^+$ in peripheral tissues, mediating the rhythmic regulation of many NAD$^+$-dependent enzymes including sirtuins. However, NAMPT expression and NAD$^+$ levels in peripheral tissues are reduced by a high-fat diet (HFD) and aging, impacting sirtuin activities.[32] Although the molecular mechanism underlying HFD- and aging-induced NAMPT suppression is yet to be elucidated, TNF-α, one of the major inflammatory cytokines, and oxidative stress have been found to suppress *Nampt* expression and NAD$^+$ levels in primary hepatocytes. Additionally, the activity of the circadian machinery also declines with age, particularly in the suprachiasmatic nucleus, likely resulting in reduced NAMPT and NAD$^+$ levels.[54] Contrarily, exercise training and dietary restriction increase *Nampt* expression and NAD$^+$ levels in skeletal muscle and adipose tissue, respectively, stimulating sirtuin activities.[55] This upregulation of *Nampt* appears to be mediated by AMPK. Thus, while the BMAL1/CLOCK/SIRT1-dependent circadian-regulatory feedback explains the regulation of basal, daily *Nampt* expression, other mechanisms appear to be required to induce *Nampt* in response to environmental and nutritional inputs.

NAD$^+$ consumption and degradation: PARP1 and CD38 knockout mice have revealed elevated levels of tissue NAD$^+$ and SIRT1 activity, indicating that PARPs, CD38, and the nuclear sirtuins all share the same pool of NAD$^+$ and that their activities are significantly influenced by each other.[56−58] Given that PARP activity and CD38 levels have been reported to significantly increase with age, it is quite likely that increased consumption and degradation also contribute to the systemic decline in tissue NAD$^+$ availability and sirtuin activities.[33,59] However, in contrast, it has been reported that PARP1 activity is inhibited by Deleted in breast cancer 1 (DBC1) in the aged liver.[60] DBC1 possesses a Nudix homology domain that binds to NAD$^+$. Binding of DBC1 to NAD$^+$ prevents it from binding to and inhibiting PARP1. With age, as NAD levels decline, DBC1 is freed to inhibit PARP1, promoting the accumulation of DNA damage. Therefore, the effect of PARP1 on NAD$^+$ availability might vary in a tissue-dependent manner.

Most recently, a new type of NAD$^+$-degrading enzyme, Sterile alpha and TIR motif-containing protein 1 (SARM1), has been identified.[61] Upon activation of SARM1 in response to axonal damage, the catalytic domain within its Toll/Interleukin-1 receptor domain shows a very potent NADase activity. Whereas the role of SARM1 has so far been implicated only in axonal degeneration, it would be interesting to see if SARM1 also contributes to the NAD$^+$ decline during aging and in other disease conditions.

2.3 THE ROLE OF NAD$^+$ BIOSYNTHESIS AND SIRTUINS IN TISSUE HOMEOSTASIS AND DISEASE CONDITIONS

2.3.1 LIVER

Liver is one of the tissues with the highest NAD$^+$ content (approximately 600−900 pmole/mg tissue), and intrahepatic NAD$^+$ levels oscillate in a circadian manner.[32] This NAD$^+$ oscillation drives the activities of SIRT1, SIRT3, and SIRT6, coordinating robust metabolic outputs (Fig. 2.2).[62,63] For example, SIRT1 regulates glucose metabolism, fatty acid oxidation, cholesterol transport, and peptide and cofactor biosynthesis in the liver. SIRT1 upregulates hepatic gluconeogenic genes and downregulates glycolytic genes through the deacetylation of peroxisome proliferator-activated receptor-gamma coactivator 1α (PGC-1α).[64] SIRT1 also promotes fatty acid oxidation and cholesterol transport through the activation of peroxisome proliferator-activated receptor α (PPARα) and the liver X receptor α (LXRα).[65,66] Under nutritional deprivation, hepatic SIRT1 activity is enhanced, likely due to the increase in NAD$^+$ biosynthesis, whereas during HFD feeding, hepatic NAMPT and NAD$^+$ levels are significantly reduced, likely due to elevated levels of an inflammatory cytokine, TNF-α, and a *Nampt*-targeting microRNA, mir-34a, so that SIRT1 activity

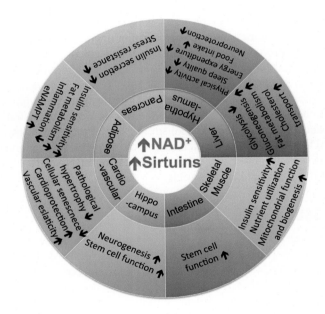

FIGURE 2.2

Pleiotropic effects of sirtuins and NAD$^+$ biosynthesis. Increasing the levels of sirtuins or their activity through the enhancement of NAD$^+$ biosynthesis has beneficial effects on tissue homeostasis and functions in pancreas, hypothalamus, liver, skeletal muscle, intestine, hippocampus, cardiovasculature, adipose tissue, and others.

decreases.[32,64,67] In response to the NAD$^+$ oscillation, SIRT3 generates circadian oscillation of the acetylation and activity of mitochondrial oxidative enzymes and respiratory chain components.[63] SIRT6 regulates a different subset of circadian genes in the liver through the oscillatory chromatin recruitment of sterol regulatory element-binding protein 1 (SREBP-1), producing the rhythmic regulation of genes related to fatty acid and cholesterol metabolism.[62]

Uniquely, liver is an organ with a high regenerative capacity. SIRT1 plays a critical role in the regenerative capacity of the liver in response to hepatectomy.[68] During the early stage of regeneration following hepatectomy, NAD$^+$ levels transiently decrease.[69] NAD$^+$ repletion by systemic administration of NR or hepatic NAMPT overexpression promotes liver regeneration and suppresses steatosis associated with regeneration.[69] In contrast, liver-specific deletion of *Sirt1* reveals a delay in cell cycle progression and hepatic lipid accumulation during regeneration.[68] Thus, NAD$^+$ biosynthesis and sirtuin functions are tightly connected to coordinate metabolic homeostasis in the liver.

2.3.2 SKELETAL MUSCLE

Skeletal muscle also shows high NAD$^+$ content (400–600 pmole/mg tissue).[32] Skeletal muscle plays a major role in determining insulin sensitivity (Fig. 2.2). Skeletal SIRT1 directly suppresses the transcription of the protein tyrosine phophatase-1B (PTP1B) gene.[70] PTP1B is a negative regulator of insulin signaling by suppressing insulin sensitivity through dephosphorylation of the insulin receptor.[71] HFD feeding decreases SIRT1 protein levels, subsequently increasing PTP1B levels and insulin resistance in skeletal muscle. Contrarily, fasting increases SIRT1 levels and decreases PTP1B expression, thereby promoting insulin sensitivity and glucose uptake. SIRT1 is also important to regulate mitochondrial biogenesis, oxidative metabolism, and antioxidant defense pathways through HIF-1α, PGC-1α, and FOXO1 (Fig. 2.2).[34,72] SIRT1 and SIRT3 activities are also involved in the mitochondrial unfolded protein response (UPR^{mt}) pathway and mitophagy.[33,73] These sirtuin functions are known to be controlled by NAD$^+$ availability. Indeed, administration of NMN significantly improves mitochondrial oxidative metabolism in skeletal muscle.[47] However, whether this is a direct effect of NAD$^+$ enhancement in skeletal muscle requires further investigation because it has been reported that skeletal muscle-specific overexpression of NAMPT increases NAD$^+$ levels but not mitochondrial functions in skeletal muscle.[74]

2.3.3 ADIPOSE TISSUE

Unlike the liver and skeletal muscle, the NAD$^+$ content in adipose tissue is relatively low (20–40 pmole/mg tissue).[32] However, a recent study has clearly demonstrated that NAMPT-mediated NAD$^+$ biosynthesis is critical for adipose tissue to control insulin sensitivity at the systemic level (Fig. 2.2).[75] Adipose tissue-specific *Nampt* knockout mice exhibit severe insulin resistance in adipose tissue, liver, and skeletal muscle and adipose tissue dysfunction including increased plasma free fatty acid levels, decreased plasma adiponectin levels, and increased adipose tissue inflammation.[75] Major downstream mediators appear to be SIRT1 and SIRT6.[76,77] Adipose tissue-specific knockout mice of each sirtuin show mild insulin resistance, decreased adiponectin levels, and increased inflammation under standard chow or low-fat diet conditions. Therefore, it is likely that their functions in adipose tissue are redundant or synergistic. SIRT1 deacetylates and activates FOXO1, thereby enhancing the expression of adiponectin.[78] On the other hand, both SIRT1 and SIRT6 affect NF-κB function and suppress proinflammatory genes including IL-6, TNF-α, and monocyte chemoattractant protein1 (MCP1).[76,79]

In adipose tissue, there is another interesting connection between sirtuin function and NAD$^+$ biosynthesis. SIRT1 promotes the secretion of extracellular NAMPT (eNAMPT). SIRT1 physically interacts with and deacetylates NAMPT specifically at lysine 53, enhancing its secretion and enzyme activity.[80] Remarkably, eNAMPT secreted from adipose tissue remotely regulates NAD$^+$ biosynthesis, SIRT1 activity, and neural activity in the hypothalamus.[80] Genetic modulation of plasma eNAMPT levels affects hypothalamic function, particularly physical activity levels during the dark time, in mice.[80] Another study has also reported the contribution of eNAMPT to cardiac NAD$^+$ biosynthesis.[81] Inhibition of monocyte-derived eNAMPT reduces cardiac SIRT1 activity, inducing cardiomyocyte apoptosis and dysfunction, consistent with the previously demonstrated role of cardiac SIRT1. These findings strongly suggest that the connection between NAMPT-mediated NAD$^+$ biosynthesis and sirtuin functions in adipose tissue and monocytes plays a critical role in regulating insulin sensitivity and modulating other tissue functions at a systemic level.

2.3.4 PANCREATIC β CELLS

In pancreatic β cells, SIRT1 enhances glucose-stimulated insulin secretion (Fig. 2.2). SIRT1 represses the transcription of uncoupling protein 2 (*Ucp2*) gene, which encodes a mitochondrial inner-membrane protein that uncouples respiration from ATP production. The suppression of *Ucp2* expression by SIRT1 enhances the coupling of glycolysis to ATP synthesis, resulting in the enhancement of glucose-stimulated insulin secretion.[82,83] SIRT1 also appears to control some unidentified step downstream of membrane depolarization.[83] Furthermore, SIRT1 provides the protection of pancreatic β cells against metabolic stress, such as glucolipotoxicity caused by a HFD, or inflammatory reactions caused by IL-1β and IFNγ through the suppression of iNOS expression and NF-κB activity.[84,85] SIRT1 also deacetylates FOXO1, enhancing the expression of NeuroD and MafA, which in turn offers β cell protection against oxidative stress.[86]

Pancreatic β cells appear to be very sensitive to changes in NAD$^+$ availability, likely because they have very low levels of intracellular NAMPT (iNAMPT). The insulin-secreting function of β cells decreases over age due to decreased NMN/NAD$^+$ availability.[31] Indeed, administration of NMN can improve glucose-stimulated insulin secretion in aged mice. Furthermore, adding eNAMPT to human islets also increases their insulin-secreting capability.[87] Therefore, intra- and extracellular NMN supply is critical for pancreatic β cells to maintain their insulin-secreting function.

2.3.5 HYPOTHALAMUS

The hypothalamus has a very unique feature in terms of the connection between sirtuin functions and systemic NAD$^+$ biosynthesis. As described above, NAD$^+$ biosynthesis, SIRT1 activity, and neural function in the hypothalamus are supported by eNAMPT secreted from adipose tissue.[80] Although how exactly eNAMPT can support hypothalamic NAD$^+$ levels requires further investigation, this finding strongly suggests that the NAD$^+$/SIRT1-driven interplay between the hypothalamus and adipose tissue significantly contributes to the systemic coordination of metabolic homeostasis in response to environmental and nutritional stimuli.

One good example is the regulation of adaptive response to food deprivation. During food deprivation, such as fasting and dietary restriction, mice increase their physical activity levels and maintain body temperature as an adaptive response to promote food-scavenging activity and survive through such a life-threatening condition. This adaptive response is regulated by SIRT1 in the hypothalamus, particularly in dorsomedial and lateral hypothalamic nuclei (DMH and LH, respectively).[88] Indeed, whereas whole-body *Sirt1* knockout mice show deficiencies in these adaptive responses and neural activation in the DMH and LH, BRASTO transgenic mice exhibit significant enhancement of these adaptive responses and neural activities in the DMH and LH.[88] Furthermore, eNAMPT secretion from adipose tissue is enhanced in response to fasting in a SIRT1-dependent manner, likely to support NAD$^+$ and SIRT1 activity levels in the hypothalamus under fasting.[80] Thus, SIRT1 and NAMPT together play a critical role in coordinating such an adaptive response to food deprivation.

Interestingly, BRASTO mice, both males and females, show significant delay in aging and lifespan extension.[18] In the DMH and LH, SIRT1 binds to and deacetylates a homeodomain transcription factor, NK2 homeobox 1 (Nkx2-1), promoting its transcriptional activity and upregulating one of the key downstream genes, orexin type 2 receptor (*Ox2r*). Elevated expression of *Ox2r* increases neural activities in the DMH and LH of BRASTO mice, mitigating the age-associated decline in physical activity, body temperature, oxygen consumption, and sleep quality (Fig. 2.2).[18] Consistent with the role of SIRT1 in the DMH and LH during food deprivation, OX2R is responsible for maintaining these functions during the process of aging. *Prdm13* is another downstream gene in the SIRT1/NKX2-1 signaling in the DMH, which is necessary for the maintenance of sleep quality and the control of adiposity.[89] *Prdm13* expression is significantly enhanced in the DMH of aged BRASTO mice. These findings indicate that the interplay between the hypothalamus and adipose tissue is mediated by a novel feedback loop regulated by SIRT1 and eNAMPT.

SIRT1 has also been reported to regulate energy homeostasis by modulating neuronal functions related to feeding behavior and energy expenditure in the arcuate nucleus (ARC) and ventromedial hypothalamus (VMH) (Fig. 2.2). In the ARC, neurons that are positive for neuropeptide Y (NPY) and agouti-related protein (AgRP) promote hunger, whereas neurons that are positive for proopiomelanocortin (POMC) promote satiety.[90] NPY/AgRP neuron-specific *Sirt1* knockout mice exhibit a marked reduction in food intake and gain less body weight compared to control mice.[91] They do not display any change in parameters related to energy expenditure. On the other hand, in POMC neurons, SIRT1 promotes energy expenditure in the peripheral tissue through sympathetic nerve stimulation.[92] SIRT1 deletion in POMC neurons causes leptin resistance and a defect in sympathetic nerve activity without altering feeding behavior, leading to reduced energy expenditure and excessive weight gain upon HFD. In the VMH, SIRT1 in SF1 neurons is necessary for maintaining insulin sensitivity in the skeletal muscle.[93] Loss of SIRT1 in SF1 neurons causes insulin resistance, making the mice prone to HFD-induced obesity, while its overexpression protects mice against HFD-induced obesity and metabolic dysregulation.

SIRT1 also governs the function of the suprachiasmatic nucleus (SCN), the central circadian pacemaker, by regulating the expression of circadian regulators (Fig. 2.2).[54] The cooperative binding of SIRT1 with PGC-1α and the nuclear receptor RORα enhances the expression of core clock regulators, *Bmal1* and *Clock*. Expression of these two core clock genes promotes the expression of circadian genes including *Per1/2*, *Cry1/2*, and *Nampt*, maintaining the amplitude and rhythmicity of circadian machinery. With age, SIRT1 protein levels decline in the SCN, resulting in the reduction of the core clock machinery and circadian gene expression. These changes cause a longer circadian period, contributing to age-associated decline in circadian functions.

2.3.6 HIPPOCAMPUS

The hippocampus is a brain region primarily associated with memory, learning, and emotion. It is also a unique site where neurogenesis is maintained in the adult brain. With age, hippocampal NAMPT and NAD⁺ levels and the number of neural stem cells significantly decline.[94] Loss of *Nampt* specifically in adult neural stem cells results in impairment of their ability for self-renewal, proliferation, and differentiation.[94] Among neuronal and glial lineages, the differentiation into oligodendrocyte precursor cells is specifically abrogated.[94] Gene expression profiles of lineage markers in *Nampt*-deficient neural stem cells can be recapitulated in neural stem cells deficient for both *Sirt1* and *Sirt2*, indicating that SIRT1 and SIRT2 likely have redundant functions downstream of NAMPT-mediated NAD⁺ biosynthesis. Importantly, long-term NMN administration maintains the number of neural stem cells in the dentate gyrus of aged mice.[94] Thus, it will be of great interest to examine whether NAMPT-mediated NAD⁺ biosynthesis plays an important role in the regulation of memory and learning.

2.3.7 SENSORY NEURONS

Neurons are susceptible to damage and degeneration in a variety of stress conditions. For example, upon nerve transection and chemotoxic stress, sensory neurons such as dorsal root ganglion neurons undergo rapid axonal degeneration, known as Wallerian degeneration.[95] It has been demonstrated that maintaining NAD⁺ levels by enhancing NAD⁺ biosynthesis or inhibiting NAD⁺ degradation leads to a significant delay in axonal degeneration.[96,97] SIRT1 appears to be required for this NAD⁺-mediated delay in axonal degeneration.[98] Recently, it has been reported that SARM1, the key regulator that triggers axonal degeneration, possesses NADase activity.[61] This NADase activity of SARM1 is necessary for axonal NAD⁺ depletion and axonal degeneration after injury.

Other studies have also demonstrated the neuroprotective role of NAD⁺ and sirtuins on special sensory neurons. Intense noise exposure induces cochlear NAD⁺ decline and hair cell degeneration.[99] Increased NAD⁺ biosynthesis, NAD⁺ repletion by NR, and overexpression of mitochondrial SIRT3 provide preventative effects against noise-induced hearing loss. Postexposure treatment by NR can also prevent hair cell degeneration in a SIRT3-dependent manner. Another type of sensory neuron, photoreceptor cells in the visual system, is highly dependent on NAMPT-mediated NAD⁺ biosynthesis.[100] NAMPT deficiency specifically in rod or cone photoreceptor neurons shows severe retinal dysfunction, which is dramatically ameliorated by NMN administration. Photoreceptor neurons are susceptible to dysfunction and degeneration by intensive light exposure, diabetes, and aging. Mitochondrial SIRT3 and SIRT5 play important roles in the NAD⁺-dependent regulation of retinal function. Interestingly, long-term NMN administration significantly improves photoreceptor functions in aged mice.[47] Thus, sirtuin functions and NAMPT-mediated NAD$^+$ biosynthesis tightly cooperate to regulate retinal homeostasis.

2.3.8 CARDIOVASCULAR SYSTEM

Cardiovascular disease is the leading cause of mortality worldwide, and age is its greatest risk factor. The cardiovascular system is constantly exposed to a variety of insults, such as mechanical, metabolic, and oxidative stresses. Pressure overload and aging induce the expression of SIRT1, perhaps as a compensatory mechanism to protect the heart from such stresses.[101] Indeed, increased levels of NAD$^+$ and SIRT1 provide protection from the oxidative stress caused by ischemic/reperfusion through the stimulation of autophagy (Fig. 2.2).[102] Moderate overexpression of SIRT1 also attenuates age-dependent cardiac changes including hypertrophy, cellular senescence, and apoptosis. Cardiac SIRT1 induces the expression of catalase, an antioxidant enzyme, by modulating FOXO1 function. SIRT1 also deacetylates NKX2-5, which is protective for cardiac function, and suppresses its transcriptional activity, suggesting that SIRT1 and NKX2-5 form a negative feedback loop to maintain cardiac function within an optimal range.[103,104] Consistently, an excessive overexpression of SIRT1 causes cardiomyopathy by increasing apoptosis, hypertrophy, and cardiac dysfunction.[101] SIRT3 also plays an important role in protecting the heart from cardiac hypertrophy, myocardial infarction, and heart failure.[105–108] These cardiac diseases cause mitochondrial dysfunction, and therefore, enhancing SIRT3 function is beneficial to ameliorate these disease conditions. Indeed, NMN administration improves cardiac function in these disease models, at least in part, through SIRT3.[109]

It has also been demonstrated that NAMPT-mediated NAD$^+$ biosynthesis and sirtuins together play a critical role in the regulation of endothelial cells, vascular smooth muscle cells, and endothelial progenitor cells (Fig. 2.2).[110] An age-dependent decline in SIRT1 levels is correlated with increased production of reactive oxygen species and increased levels of oxidative stress, leading to the impairment of endothelium-dependent dilation. Enhanced SIRT1 activity by NMN treatment upregulates the expression of manganese superoxide dismutase and mitochondrial antioxidant genes through the activation of PGC-1α, thereby improving vascular elasticity.

2.3.9 INTESTINE

The intestinal epithelium is comprised of cells lining the luminal surface of the gastrointestinal (GI) tract, functioning to absorb necessary nutrients and also form a barrier against harmful agents. Homeostasis of the GI tract is maintained by the continuous replenishment of intestinal epithelial cells differentiated from intestinal stem cells (ISCs). In response to dietary restriction, ISCs enhance their self-renewal capacity and expand their pool while reducing the differentiation and replenishment of intestinal epithelial cells in villi.[111] Interestingly, coordinated activation of SIRT1 and the mechanistic target of rapamycin complex 1 (mTORC1) is necessary for the ISC expansion mediated by dietary restriction. Genetic and pharmacological manipulations have demonstrated that deacetylation of S6 kinase 1 (S6K1) by SIRT1 promotes its phosphorylation by mTORC1 and drives protein synthesis, cell growth, and metabolism. Cyclic ADP-ribose (cADPR) secreted from neighboring Paneth cells upregulates *Nampt* expression in ISCs, enhancing NAD$^+$

biosynthesis and SIRT1 activity. These findings provide further support for the intimate interplay between NAMPT-mediated NAD^+ biosynthesis and sirtuin functions in different types of stem cells.

2.4 THE CONCEPT OF THE NAD WORLD: THE NAMPT/NAD$^+$/SIRT1-MEDIATED SYSTEMIC REGULATORY NETWORK FOR AGING AND LONGEVITY CONTROL IN MAMMALS

As discussed in the previous sections, NAMPT-mediated NAD^+ biosynthesis and sirtuins cooperate tightly to regulate physiological resilience, particularly in life-threatening conditions, and promote survival at a systemic level. This systemic, tight interplay between NAMPT-mediated NAD^+ biosynthesis and sirtuins, particularly SIRT1, forms a firm foundation for the comprehensive concept for mammalian aging and longevity control, named the "NAD World." The original concept of the NAD World was proposed in 2009.[112] In the original concept, one of the key predictions was that the decline in NAD^+ biosynthesis at a systemic level triggers a variety of pathophysiological changes, producing aging phenotypes through decreased activities of SIRT1 and other sirtuins. As described above, this prediction has been demonstrated to be correct over the past 6 years. Another important prediction was that when NAD^+ biosynthesis decreases, tissues and organs that have very low levels of iNAMPT and therefore rely on circulating NMN to maintain adequate levels of NAD^+ would become fragility points in the system. Pancreatic β cells and central neurons, both of which have very low levels of iNAMPT, have been predicted to be the most critical fragility points. This predicted susceptibility of both cell types to NAD^+ decline has also been confirmed, as described above.[18,31] Lastly, the concept proposed that the cascade of robustness breakdown triggered by a decrease in systemic NAD^+ biosynthesis is the central process of aging. Indeed, the decline in NAD^+ availability due to the decrease in its biosynthesis or the increase in its consumption and degradation results in a reduction in activities of SIRT1 and other sirtuins, leading to the acceleration of tissue dysfunctions and ultimately aging.[32,34]

Over the past 8 years, it has been demonstrated that SIRT1 and NAMPT together mediate critical intertissue communications, particularly among the hypothalamus, skeletal muscle, and adipose tissue. The hypothalamus functions as the high-order control center of aging, sending a signal to skeletal muscle through the sympathetic nervous system. Skeletal muscle appears to affect other tissue and organ functions by secreting various myokines. Adipose tissue remotely modulates hypothalamic function through eNAMPT, forming a NAD^+/SIRT1-mediated feedback loop. With all these developments, the concept has recently been reformulated as the NAD World 2.0, with an emphasis on the importance of intertissue communications among three key tissues for mammalian aging and longevity control (Fig. 2.3).[113] An important prediction from the concept of the NAD World 2.0 is that eNAMPT and NMN function as key systemic regulatory molecules that maintain physiological resilience over time. Indeed, long-term NMN administration shows remarkable anti-aging effects in mice[47]. The role of eNAMPT in mammalian aging and longevity is currently under investigation. This reformulated concept will provide a new foundation to understand the system architecture and dynamics for aging and longevity control in mammals and further promote the development of an effective anti-aging intervention that is translatable to humans.

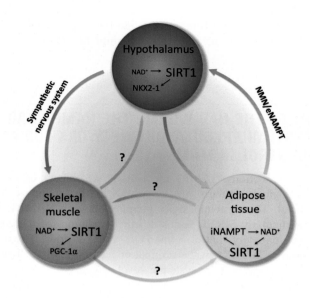

FIGURE 2.3

The concept of NAD World 2.0. Hypothalamus, skeletal muscle, and adipose tissue play as critical hubs for the systemic regulatory network of aging/longevity control in mammals. Hypothalamus acts as a control center, adipose tissue as a modulator, and skeletal muscle as a mediator. Each tissue communicates with and regulates one another to maintain physiological resilience through systemic NAD$^+$ biosynthesis and sirtuins.

2.5 CONCLUDING REMARKS

It is likely that NAMPT and sirtuins have coevolved together through different phyla. For example, vibriophage KVP40 possesses three genes encoding NAMPT, NMNAT, and a sirtuin family protein.[114] Furthermore, in bacterial species, the *Nampt* gene is carried on a plasmid that seems to be passed from one bacterial species to another.[115] Interestingly, the DNA sequences franking the *Nampt* gene on the plasmid have significant homology to the vibriophage genome.[116] Given that bacterial and mammalian NAMPT proteins show an unusually high homology, it might be possible that an ancestral species of mammals inherited the *Nampt* gene from some symbiotic bacterial species at some point in their evolution, which might have immediately given enormous benefits to activate sirtuins more efficiently and promote physiological resilience and survival.[48] Although this is purely speculation, it could provide food for thought to explain the tight connection between NAMPT-mediated NAD$^+$ biosynthesis and sirtuins. If this is indeed the case, some symbiotic relationship between mammalian hosts and their microbiota could still be maintained through NAMPT/NMN/NAD$^+$ and sirtuins in mammals. More examples of this intimate connection will surely be discovered in the coming years, and our understanding of this dynamic relationship will further promote the development of NAD$^+$-based preventive and therapeutic interventions against age-associated functional decline and diseases for humans.

ACKNOWLEDGMENTS

We apologize to those whose work is not cited due to space limitations. We thank members in the Imai lab for critical discussions and suggestions. S.I. is supported by grants from NIA (AG037457, AG047902), the American Federation for Aging Research, and the Tanaka Fund. The authors have no competing financial interest.

REFERENCES

1. Imai S, Guarente L. NAD$^+$ and sirtuins in aging and disease. *Trends Cell Biol* 2014;**24**:464−71. Available from: https://doi.org/10.1016/j.tcb.2014.04.002.
2. Canto C, Menzies KJ, Auwerx J. NAD(+) metabolism and the control of energy homeostasis: a balancing act between mitochondria and the nucleus. *Cell Metab* 2015;**22**:31−53. Available from: https://doi.org/10.1016/j.cmet.2015.05.023.
3. Yang Y, Sauve AA. NAD$^+$ metabolism: bioenergetics, signaling and manipulation for therapy. *Biochim Biophys Acta* 2016;**1864**:1787−800. Available from: https://doi.org/10.1016/j.bbapap.2016.06.014.
4. Harden A, Young WJ. The alcoholic ferment of yeast-juice. *Proc. Royal Soc. London. Series B, Contain. Papers Biol. Character* 1906;**77**:405−20.
5. Berger F, Ramirez-Hernandez MH, Ziegler M. The new life of a centenarian: signalling functions of NAD(P). *Trends Biochem Sci* 2004;**29**:111−18. Available from: https://doi.org/10.1016/j.tibs.2004.01.007.
6. Gellert M, Little JW, Oshinsky CK, Zimmerman SB. Joining of DNA strands by DNA ligase of *E. coli*. *Cold Spring Harb Symp Quant Biol* 1968;**33**:21−6.
7. Chambon P, Weill JD, Mandel P. Nicotinamide mononucleotide activation of new DNA-dependent polyadenylic acid synthesizing nuclear enzyme. *Biochem Biophys Res Commun* 1963;**11**:39−43.
8. Lee HC, Walseth TF, Bratt GT, Hayes RN, Clapper DL. Structural determination of a cyclic metabolite of NAD$^+$ with intracellular Ca2 + -mobilizing activity. *J Biol Chem* 1989;**264**:1608−15.
9. Imai S, Armstrong CM, Kaeberlein M, Guarente L. Transcriptional silencing and longevity protein Sir2 is an NAD-dependent histone deacetylase. *Nature* 2000;**403**:795−800. Available from: https://doi.org/10.1038/35001622.
10. Imai S, Guarente L. Ten years of NAD-dependent SIR2 family deacetylases: implications for metabolic diseases. *Trends Pharmacol Sci* 2010;**31**:212−20. Available from: https://doi.org/10.1016/j.tips.2010.02.003.
11. Satoh A, Imai S, Guarente L. The brain, sirtuins, and ageing. *Nat Rev Neurosci* 2017;**18**:362−74. Available from: https://doi.org/10.1038/nrn.2017.42.
12. van de Ven RAH, Santos D, Haigis MC. Mitochondrial sirtuins and molecular mechanisms of aging. *Trends Mol Med* 2017;**23**:320−31. Available from: https://doi.org/10.1016/j.molmed.2017.02.005.
13. Burnett C, et al. Absence of effects of Sir2 overexpression on lifespan in *C. elegans* and Drosophila. *Nature* 2011;**477**:482−5. doi:nature10296 [pii]10.1038/nature10296.
14. Kaeberlein M, McVey M, Guarente L. The SIR2/3/4 complex and SIR2 alone promote longevity in Saccharomyces cerevisiae by two different mechanisms. *Genes Dev* 1999;**13**:2570−80.
15. Tissenbaum HA, Guarente L. Increased dosage of a sir-2 gene extends lifespan in *Caenorhabditis elegans*. *Nature* 2001;**410**:227−30. Available from: https://doi.org/10.1038/35065638.
16. Rogina B, Helfand SL. Sir2 mediates longevity in the fly through a pathway related to calorie restriction. *Proc Natl Acad Sci U S A* 2004;**101**:15998−6003. Available from: https://doi.org/10.1073/pnas.0404184101.

17. Whitaker R, et al. Increased expression of Drosophila Sir2 extends life span in a dose-dependent manner. *Aging* 2013;**5**:682–91.

18. Satoh A, et al. Sirt1 extends life span and delays aging in mice through the regulation of Nk2 homeobox 1 in the DMH and LH. *Cell Metab* 2013;**18**:416–30. Available from: https://doi.org/10.1016/j.cmet.2013.07.013.

19. Kanfi Y, et al. The sirtuin SIRT6 regulates lifespan in male mice. *Nature* 2012;**483**:218–21. Available from: https://doi.org/10.1038/nature10815.

20. Stumpferl SW, et al. Natural genetic variation in yeast longevity. *Genome Res* 2012;**22**:1963–73. Available from: https://doi.org/10.1101/gr.136549.111.

21. Schmeisser K, et al. Role of sirtuins in lifespan regulation is linked to methylation of nicotinamide. *Nat Chem Biol* 2013;**9**:693–700. Available from: https://doi.org/10.1038/nchembio.1352.

22. Banerjee KK, et al. dSir2 in the adult fat body, but not in muscles, regulates life span in a diet-dependent manner. *Cell Rep* 2012;**2**:1485–91. Available from: https://doi.org/10.1016/j.celrep.2012.11.013.

23. Lin SJ, Defossez PA, Guarente L. Requirement of NAD and SIR2 for life-span extension by calorie restriction in *Saccharomyces cerevisiae*. *Science* 2000;**289**:2126–8.

24. Klass MR. Aging in the nematode Caenorhabditis elegans: major biological and environmental factors influencing life span. *Mech Ageing Dev* 1977;**6**:413–29.

25. Chapman T, Partridge L. Female fitness in Drosophila melanogaster: an interaction between the effect of nutrition and of encounter rate with males. *Proc Biol Sci* 1996;**263**:755–9. Available from: https://doi.org/10.1098/rspb.1996.0113.

26. McCay CM, Crowell MF, Maynard LA. The effect of retarded growth upon the length of life span and upon the ultimate body size: one figure. *J Nutr* 1935;**10**:63–79.

27. Colman RJ, et al. Caloric restriction delays disease onset and mortality in rhesus monkeys. *Science* 2009;**325**:201–4. Available from: https://doi.org/10.1126/science.1173635.

28. Anderson RM, Bitterman KJ, Wood JG, Medvedik O, Sinclair DA. Nicotinamide and PNC1 govern life-span extension by calorie restriction in Saccharomyces cerevisiae. *Nature* 2003;**423**:181–5. Available from: https://doi.org/10.1038/nature01578.

29. Boily G, et al. SirT1 regulates energy metabolism and response to caloric restriction in mice. *PLoS One* 2008;**3**:e1759. Available from: https://doi.org/10.1371/journal.pone.0001759.

30. Wang Y, Tissenbaum HA. Overlapping and distinct functions for a *Caenorhabditis elegans* SIR2 and DAF-16/FOXO. *Mech Ageing Dev* 2006;**127**:48–56. Available from: https://doi.org/10.1016/j.mad.2005.09.005.

31. Ramsey KM, Mills KF, Satoh A, Imai S. Age-associated loss of Sirt1-mediated enhancement of glucose-stimulated insulin secretion in beta cell-specific Sirt1-overexpressing (BESTO) mice. *Aging Cell* 2008;**7**:78–88. Available from: https://doi.org/10.1111/j.1474-9726.2007.00355.x.

32. Yoshino J, Mills KF, Yoon MJ, Imai S. Nicotinamide mononucleotide, a key NAD(+) intermediate, treats the pathophysiology of diet- and age-induced diabetes in mice. *Cell Metab* 2011;**14**:528–36. Available from: https://doi.org/10.1016/j.cmet.2011.08.014.

33. Mouchiroud L, et al. The NAD(+)/sirtuin pathway modulates longevity through activation of mitochondrial UPR and FOXO signaling. *Cell* 2013;**154**:430–41. Available from: https://doi.org/10.1016/j.cell.2013.06.016.

34. Gomes AP, et al. Declining NAD(+) induces a pseudohypoxic state disrupting nuclear-mitochondrial communication during aging. *Cell* 2013;**155**:1624–38. Available from: https://doi.org/10.1016/j.cell.2013.11.037.

35. Cambronne XA, et al. Biosensor reveals multiple sources for mitochondrial NAD(+). *Science* 2016;**352**:1474–7. Available from: https://doi.org/10.1126/science.aad5168.

111. Igarashi M, Guarente L. mTORC1 and SIRT1 cooperate to foster expansion of gut adult stem cells during calorie restriction. *Cell* 2016;**166**:436—50. Available from: https://doi.org/10.1016/j.cell.2016.05.044.

112. Imai S. The NAD world: a new systemic regulatory network for metabolism and aging--Sirt1, systemic NAD biosynthesis, and their importance. *Cell Biochem Biophys* 2009;**53**:65—74. Available from: https://doi.org/10.1007/s12013-008-9041-4.

113. Imai S. The NAD World 2.0: the importance of the inter-tissue communication mediated by NAMPT/NAD$^+$/SIRT1 in mammalian aging and longevity control. *NPJ Syst Biol Appl* 2016;**2**:16018. Available from: https://doi.org/10.1038/npjsba.2016.18.

114. Miller ES, et al. Complete genome sequence of the broad-host-range vibriophage KVP40: comparative genomics of a T4-related bacteriophage. *J Bacteriol* 2003;**185**:5220—33.

115. Martin PR, Shea RJ, Mulks MH. Identification of a plasmid-encoded gene from *Haemophilus ducreyi* which confers NAD independence. *J Bacteriol* 2001;**183**:1168—74. Available from: https://doi.org/10.1128/JB.183.4.1168-1174.2001.

116. Munson Jr RS, et al. Haemophilus ducreyi strain ATCC 27722 contains a genetic element with homology to the vibrio RS1 element that can replicate as a plasmid and confer NAD independence on haemophilus influenzae. *Infect Immun* 2004;**72**:1143—6.

NAD$^+$ MODULATION: BIOLOGY AND THERAPY

3

Elena Katsyuba and Johan Auwerx

École Polytechnique Fédérale de Lausanne, Lausanne, Switzerland

3.1 A BIT OF HISTORY

NAD$^+$ was the first cofactor ever described, and the history of its discovery is associated with several names of Nobel Prize laureates. In 1906 two British biochemists, Sir Arthur Harden and William John Young, made an interesting observation out of a relatively simple experiment: they separated yeast juice into two fractions based on their molecular weights through a gelatin filter. Once separated, each of these fractions was unable to perform fermentation and recombining them together again restored this capacity. Surprisingly, adding the low-molecular-weight fraction into yeast juice extracts could significantly accelerate the rate of the fermentation reaction, even after it was boiled, making them conclude that the low-molecular-weight yeast fraction contained some heat-resistant factor capable of promoting alcoholic fermentation.[1] Harden and Young presumed that the high-molecular-weight fraction of yeast juice contained the actual Eduard Buchner's "zymase," that had been described several years earlier, and gave the name of "cozymase" to the heat-resistant low-molecular-weight factor required for the "zymase" activity. In 1930, another Nobel laureate, Hans von Euler-Chelpin established that the "cozymase" is composed of a reducing sugar group, an adenine, and a phosphate.[2] Eight years later, in 1938, Otto Heinrich Warburg finally established the actual function of NAD$^+$ by discovering the capacity of the "cozymase" to transfer hydride from one molecule to another.[3]

3.2 HOW NAD$^+$ IS PRODUCED

NAD$^+$ can be synthetized de novo from the amino acid tryptophan. Otherwise, it can be produced via salvage pathways requiring a preformed pyridine ring. Three naturally occurring vitamins, nicotinamide (NAM), nicotinic acid (NA), and nicotinamide riboside (NR), can serve as precursor molecules for NAD$^+$ salvage.

Synthesis of NAD$^+$ from both NAM and NR is a two-step process, where during the first step both of these molecules are transformed into nicotinamide mononucleotide (NMN) (Fig. 3.1). The reactions leading to NMN production are however different: while NAM is converted into NMN by NAM phosphoribosyltransferase (NAMPRT), NR is phosphorylated by nicotinamide

Introductory Review on Sirtuins in Biology, Aging, and Disease. DOI: https://doi.org/10.1016/B978-0-12-813499-3.00003-4

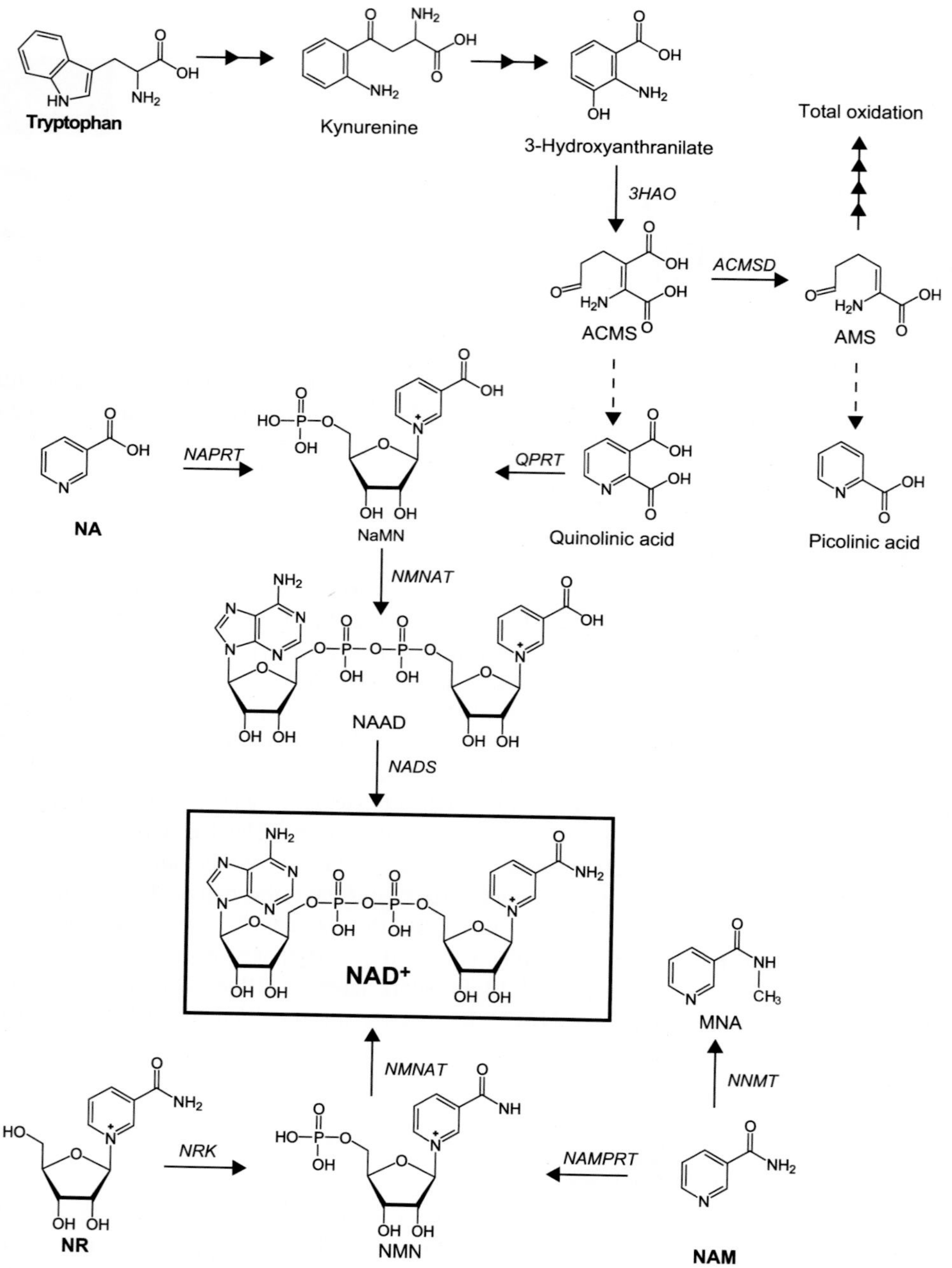

FIGURE 3.1

Different pathways leading to NAD^+ production in mammals.

NAD^+ can be synthesized de novo from the amino acid tryptophan via an eight-step process. ACMS represents a branching point of the pathway, since this intermediate can either undergo spontaneous cyclization, leading to formation of QA afterwards fusing with the Preiss—Handler pathway, or can be dissipated through decarboxylation by ACMSD. NAD^+ can also be produced from the precursor molecules, all naturally occurring vitamins, NAM, NR, or NA. *Solid arrows* represent enzymatic conversions; *dashed arrows* represent spontaneous reactions.

riboside kinase (NRK).[4] Two isoforms of NRK exist in mammals: NRK1, which is ubiquitously expressed and NRK2, which is mainly expressed in liver, skeletal muscle, heart, and BAT.[5] NMN is then transformed into NAD^+ by NMN adenylyltransferase (NMNAT) (Fig. 3.1). Three different isoforms of this enzyme exist and show differential tissue expression: NMNAT1 being ubiquitously expressed, with the highest activity detected in liver and kidney,[6] NMNAT2 expression limited to the brain[6−8] and NMNAT3 presence detected in multiple tissues, similarly to NMNAT1, but to a much lesser extent.[6,9]

Synthesis of NAD^+ from NA through the so-called Preiss−Handler pathway[10] consists of three steps: first NA gets transformed into NA mononucleotide (NAMN) by NA phosphoribosyltransferase (NAPRT). NAMN is then converted into NA adenine dinucleotide (NAAD) by NMNAT, which can use both NA and NAM as substrates. Finally, NAAD leads to NAD^+ in a reaction catalyzed by NAD synthetase (NADS) (Fig. 3.1). NADS activity was not detected in the brain, lung, and skeletal muscle, indicating that NAD^+ synthesis in these tissues is rather accomplished via routes other than the Preiss−Handler pathway.[6,11]

The de novo synthesis constitutes a more complex pathway including eight steps, during which an aromatic pyridine ring for the NAD^+ molecule is formed (Fig. 3.1). At first tryptophan is transformed into N-formylkynurenine. Two different enzymes can catalyze this reaction in mammals: tryptophan-2,3-dioxygenase (TDO) and indoleamine 2,3-dioxygenase (IDO). TDO seems to be the enzyme responsible for this transformation in liver, while in extrahepatic tissues IDO appears to accomplish this role.[12,13] The next step is catalyzed by kynurenine fomamidase (KFase) and leads to formation of kynurenine. Kynurenine is then transformed into 3-hydroxykynurenine by kynurenine 3-hydroxylase (K3H), which in turn is converted into 3-hydroxyanthranilate by kynureninase (Kyase). 3-Hydroxyanthranilate 3,4-dioxygenase (3HAO) catalyzes the next reaction transforming 3-hydroxyanthranilate into 2-amino 3-carboxymuconate 6-semialdehyde (ACMS), an unstable intermediate that can be decarboxylated by ACMS decarboxylase (ACMSD), leading to the formation of 2-aminomuconate-6-semialdehyde (AMS), which in turn can become picolinic acid or undergo total oxidation into CO_2 and H_2O (Fig. 3.1). A proportion of ACMS which does not become a substrate for ACMSD can undergo spontaneous cyclization into quinolinic acid (QA), which will then be transformed into NAMN by quinolinate phosphoribosyltransferase (QPRT) and lead to production of the NAD^+ molecule by fusing with the Preiss−Handler pathway (Fig. 3.1). Based on the tissue distribution of the enzymes involved in the de novo NAD^+ production, it is generally assumed that NAD^+ synthesis from tryptophan is limited to liver, kidney, and brain.[14,15] However, some studies detected no 3-HAO or QPRT in the brain, limiting the NAD^+ de novo synthesis pathway further to the liver and kidney.[11,16]

3.3 NAM, NA, NR, OR TRYPTOPHAN?

The fact that the NAD^+ molecule can be produced via four different pathways raises a logical question on the preferential source for NAD^+ biosynthesis. There is, however, no unanimity on that subject with the tissue-specific expression of various NAD^+ biosynthetic enzymes being a major determinant in balancing the contribution of the different synthesis routes. A large number of studies report that NA is a preferable precursor molecule for NAD^+ synthesis.[17−22] Several studies, on the contrary, present evidence that NAM has a higher capacity for NAD$^+$ production over NA.[6,23–25] It was also detected to be five times more abundant in human plasma than NA.[26] Moreover, it has been shown that NAM, but not NA, was capable of restoring a drop in NAD$^+$ levels, caused by streptozotocin-induced diabetes.[27] Due to the fact that NR was discovered several decades later than NAM and NA, the number of studies comparing all these three precursors together is limited. A recent report demonstrated that NR possessed a higher capacity for hepatic NAD$^+$ boosting compared to NA and NAM.[28] It is also important to mention that the conclusion that NAM is the main source for NAD$^+$ production reached earlier is somehow flawed, because with the employed experimental setup the "deamidated" (e.g., from NA) route of NAD$^+$ biosynthesis was compared with the "amidated" one, which includes both NR and NAM.[6] Hence, a possible contribution of NR should not be disregarded.

Often perceived as a less important pathway for NAD$^+$ production, de novo biosynthesis is the preferential source for NAD$^+$ synthesis in liver, the tissue where this pathway is basically active.[29] This might be due to the limited capacity of NA and NAM for NAD$^+$ production in hepatocytes, while tryptophan showed no such limitations.[29–31] When exposed to NA, NAM, or tryptophan, rat primary hepatocytes based their NAD$^+$ synthesis exclusively on tryptophan, despite the fact that they were still capable of taking up NA and NAM from the cell culture medium.[32]

3.4 COMPARTMENTALIZATION OF NAD$^+$ HOMEOSTASIS WITHIN THE CELL

The studies exploring compartmentalization of NAD$^+$ metabolism are strongly affected by the technical limitations in the NAD$^+$ quantification methods, which might explain the patchwork-like nature of the data available on this subject, as well as discrepancies in the values for the reported NAD$^+$ concentrations. Although there is no experimental evidence demonstrating NAD$^+$ shuttling between the nucleus and cytosol, it is generally assumed that these two pools are exchangeable (reviewed in Refs. [33,34]), especially after the NAD$^+$ concentrations were reported to be similar within the nucleus and cytosol.[35] The mitochondrial NAD$^+$ pool, in contrast, appears to be separated from the nucleocytosolic one, as mitochondria were shown to maintain their NAD$^+$ concentrations different from the rest of the cell.[35] The proportion of the mitochondrial pool, however, is tissue-specific, ranging from 30% to 40% in hepatocytes and astrocytes[36,37] up to 70% of total cellular NAD$^+$ stock in cardiac myocytes.[38] Interestingly, mitochondria are capable of maintaining their NAD$^+$ levels, even when the nucleocytosolic NAD$^+$ pool is depleted[38–40] and possess their own machinery for NAD$^+$ synthesis.[39,41,42]

It is established that NAD$^+$ cannot diffuse through the mitochondrial inner membrane.[43,44] In yeast two specific NAD$^+$ transporters (Ndt1 and Ndt2) exist, allowing an exchange through the inner mitochondrial membrane.[45] Loss of these transporters causes a drop in mitochondrial NAD$^+$ content, while their overexpression increases mitochondrial NAD$^+$ concentrations.[45,46] Although no similar transporters have been found in mammalian systems,[47] some reports indicate that NAD$^+$ can be somehow imported into the mitochondria: first, exogenous NAD$^+$ was shown to increase mitochondrial NAD$^+$ levels more than cytoplasmic ones;[48] second, NAD$^+$ produced in the cytosol was shown to also affect the mitochondrial stores.[35] Two NAD$^+$ precursor molecules, NMN and

NR, appear to be able to cross the mitochondrial inner membrane and influence mitochondrial NAD$^+$ levels,[49−51] but other, yet undiscovered, mechanisms for maintaining the mitochondrial NAD$^+$ pool might also exist.[35]

Compartmentalization of NAD$^+$ homeostasis is also achieved through differential subcellular distribution of the NAD$^+$ biosynthetic enzymes. For instance, all the enzymes involved in the de novo NAD$^+$ synthesis pathway are located within cytosol,[33] indicating that the de novo NAD$^+$ biosynthesis is most probably taking place there. Interestingly, the three reported isoforms of NMNAT, the enzyme common to all four pathways leading to NAD$^+$ production (Fig. 3.1), are all characterized by a specific localization within the cell: NMNAT1 is located in the nucleus,[52,53] NMNAT2 in the cytosol and Golgi apparatus,[53,54] and NMNAT3 is generally considered to be mitochondrial,[9,39,54] despite some reports questioning its location and even its existence.[55−57] The presence of intracellular NAMPRT was also detected in cytosol,[58,59] nucleus,[59] and mitochondria,[39] even if for mitochondria, similarly to NMNAT3, it is still a matter of debate.[40,41,50] Based on these reports, NAM can be a universal NAD$^+$ precursor in all three of these compartments. All the remaining enzymes involved in NAD$^+$ production—NADS, NRK1 and 2, NAPRT—until now have been only detected in cytosol.[19,50,60] Of note, since the seven mammalian sirtuins have different subcellular localizations,[61] the compartmentalization of NAD$^+$ metabolism may underpin the differential activation of specific members of sirtuin family.[62]

3.5 NAD$^+$ AS A REGULATOR OF SIRTUIN ACTIVITY

Sirtuins owe their name to the founder of the family the *silent information regulator 2* (SIR2), which was initially reported in yeast, *Saccharomyces cerevisiae*, as a gene regulating its mating types.[63] Twenty years later, SIR2 received much more attention due to the discovery of its ability to increase yeast lifespan.[64] A year after, the function of SIR2 as a nicotinamide adenine dinucleotide (NAD$^+$)-dependent histone deacetylase was reported.[65,66] In the deacetylation reaction catalyzed by the sirtuins, an acetyl group is removed from the substrate with NAD$^+$ as a cosubstrate, resulting in the deacetylated substrate and 2-*O*-acetyl-ADP-ribose (2-OAADPR), while NAD$^+$ is transformed into NAM (Fig. 3.2).

The unifying feature of all seven members of the mammalian sirtuin family is their highly conserved NAD$^+$-binding domain.[67] The K_m value for NAD$^+$, however, varies significantly between sirtuins (Table 3.1) and segregates them into two groups. For some sirtuins, like SIRT2, SIRT4, SIRT5, and SIRT6, NAD$^+$ might not be the rate-limiting substrate at least at a basal condition, since their K_m values are much below the physiological range of NAD$^+$ concentrations (which are in the 300−700 μM range).[33] The activity of SIRT1 and SIRT3, on the contrary, appears to be highly affected by NAD$^+$ availability within the cell (Table 3.1).

Since NAD$^+$ levels reflect the cellular energy status, it is very tempting to speculate that, by being dependent on NAD$^+$ for their activity, sirtuins act as metabolic sensors, which can modulate the metabolic responses of the cell accordingly. For instance, states of energetic deficit are associated with both increased NAD$^+$ levels and SIRT1 activity, as exemplified by caloric or glucose restriction,[68−71] fasting,[68,69,72] or exercise.[69,73,74] On the contrary, accumulation of fat within the liver, whether caused by a high-fat/high-fat high-sucrose feeding,[75−78] a methionine-choline-deficient diet,[76] or by orotic

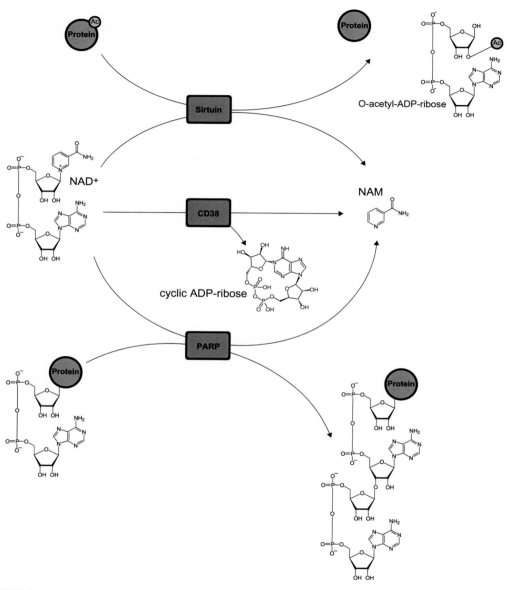

FIGURE 3.2

Enzymatic activities of NAD$^+$ consumers.

Three main classes of NAD$^+$-consuming enzymes are reported, including sirtuins, PARPs, and cADP-ribose synthases. Sirtuins use NAD$^+$ as a cosubstrate to perform deacetylation reaction, leading to the formation of O-acetyl-ADP-ribose and NAM. cADP-ribose synthases lead to cADP-ribose formation, also generating NAM as a product. Finally, PARPs also hydrolyze NAD$^+$ to generate NAM in a PARylation reaction, during which multiple ADP-ribose moieties are added in a sequential manner, resulting in linear or branched polymers consisting of multiple ADP-ribosyl groups attached to the target protein.

Table 3.1 K_m Values for NAD$^+$

Enzyme	K_m Value (μM)	References
SIRT1	94−888	1−3
SIRT2	~100	1
SIRT3	280−880	4, 5
SIRT4	35	6
SIRT5	26−200	7, 8
SIRT6	13	9
SIRT7	−	
PARP1	20−60; 50−97	10, 11
PARP2	130	11
CD38	15−25	12, 13

1. Smith BC, Hallows WC, Denu JM. A continuous microplate assay for sirtuins and nicotinamide-producing enzymes. Anal Biochem 2009;**394**:101−9.
2. Pacholec M, Bleasdale JE, Chrunyk B, et al. SRT1720, SRT2183, SRT1460, and resveratrol are not direct activators of SIRT1. J Biol Chem 2010;**285**:8340−51.
3. Gerhart-Hines Z, Dominy JE, Jr., Blattler SM, et al. The cAMP/PKA pathway rapidly activates SIRT1 to promote fatty acid oxidation independently of changes in NAD(+). Mol Cell 2011;**44**:851−63.
4. Jin L, Galonek H, Israelian K, et al. Biochemical characterization, localization, and tissue distribution of the longer form of mouse SIRT3. Protein Sci 2009;**18**:514−25.
5. Hirschey MD, Shimazu T, Jing E, et al. SIRT3 deficiency and mitochondrial protein hyperacetylation accelerate the development of the metabolic syndrome. Mol Cell 2011;**44**:177−90.
6. Laurent G, German NJ, Saha AK, et al. SIRT4 coordinates the balance between lipid synthesis and catabolism by repressing malonyl CoA decarboxylase. Mol Cell 2013;**50**:686−98.
7. Madsen AS, Andersen C, Daoud M, et al. Investigating the sensitivity of NAD + -dependent sirtuin deacylation activities to NADH. J Biol Chem 2016;**291**:7128−41.
8. Roessler C, Tuting C, Meleshin M, Steegborn C, Schutkowski M. A novel continuous assay for the deacylase sirtuin 5 and other deacetylases. J Med Chem 2015;**58**:7217−23.
9. Kugel S, Feldman JL, Klein MA, et al. Identification of and molecular basis for SIRT6 loss-of-function point mutations in cancer. Cell Rep 2015;**13**:479−88.
10. Mendoza-Alvarez H, Alvarez-Gonzalez R. Poly(ADP-ribose) polymerase is a catalytic dimer and the automodification reaction is intermolecular. J Biol Chem 1993;**268**:22575−80.
11. Ame JC, Rolli V, Schreiber V, et al. PARP-2, A novel mammalian DNA damage-dependent poly(ADP-ribose) polymerase. J Biol Chem 1999;**274**:17860−8.
12. Sauve AA, Munshi C, Lee HC, Schramm VL. The reaction mechanism for CD38. A single intermediate is responsible for cyclization, hydrolysis, and base-exchange chemistries. Biochemistry 1998;**37**:13239−49.
13. Cakir-Kiefer C, Muller-Steffner H, Oppenheimer N, Schuber F. Kinetic competence of the cADP-ribose-CD38 complex as an intermediate in the CD38/NAD + glycohydrolase-catalysed reactions: implication for CD38 signalling. Biochem J 2001;**358**:399−406.

acid[79] has been associated with a drop in hepatic NAD$^+$ content. A high-fat diet was reported to not exclusively affect the hepatic NAD$^+$ pool, but also reduce NAD$^+$ content in the muscle,[49] brown adipose tissue (BAT),[49] and white adipose tissue.[78] It is interesting to note that the initial response to high-fat feeding in the liver seems to be an increase in NAD$^+$-SIRT system components, representing a compensatory mechanism with the goal of fighting the energy overload. After prolonged exposure to a high-fat diet, the organism is no longer able to maintain its NAD$^+$ stocks, resulting in liver damage.[80]

Another important controller of NAD$^+$ content is the circadian clock. Hepatic NAD$^+$ levels were reported to oscillate in a daily manner[81,82] and seemed to be under tight control of the circadian clock machinery genes, since KO of circadian activators, *Bmal1* and *Clock*, reduced NAD$^+$ content, while KO of the core clock repressor genes, *Cry1* and *Cry2*, led to its increase.[82] These clock-driven oscillations of NAD$^+$ were reported to influence the activity of sirtuins, including SIRT1[81,82] and SIRT3.[83] The control over the NAD$^+$ pool by the core clock machinery appears to be effectuated via a controlled rhythmic expression of *Nampt*.[81,82]

3.6 NAD$^+$ BESIDES SIRTUINS

Long before NAD$^+$ was described as a cosubstrate for sirtuins and other enzyme families, it had been known as an electron carrier. NAD$^+$ and its reduced form, NADH, participate in many reactions requiring electron exchange. This includes glycolysis, gluconeogenesis, fatty acid oxidation, citric acid cycle, and oxidative phosphorylation. On top of that, a phosphate can be added to NAD$^+$ by NAD$^+$ kinases, transforming it into nicotinamide adenine dinucleotide phosphate (NADP$^+$). The NADP$^+$/NADPH redox couple has two major roles within the cell: it constitutes an important player in cell defense against oxidative stress, supporting glutathione regeneration and in several anabolic processes, including synthesis of fatty acids, cholesterol, and DNA.[84]

Besides sirtuins, other enzyme families are known to use NAD$^+$ as a cosubstrate to perform their respective functions, such as cyclic ADP-ribose (cADPR) synthases, like CD38 and CD157, and ADP-ribosyl transferases, including the mono- and poly- ADP ribose polymerases (PARPs). The enzymes performing mono-ADP-ribosylation reaction, sometimes called the MonoPARPs, consume NAD$^+$ at a much lower extent than PARPs and are significantly less studied.[85] PARP proteins, in their turn, constitute a family of 17 members in humans and 16 members in mice, and play a key role in DNA repair and maintenance of genomic integrity.[86] PARPs require NAD$^+$ to perform the PARylation reaction, a posttranscriptional protein modification during which a large polymer consisting of multiple ADP-ribosyl groups is attached to the target protein, which should already be MARylated (Fig. 3.2). The PARP-1 activity accounts for 90% of NAD$^+$ consumption by the whole family, while the remaining part of NAD$^+$ consumption is mostly accomplished by PARP-2.[86] PARP activity was shown to increase with aging,[87,88] probably due to increased DNA damage and this higher activity could be, at least partially, responsible for the depletion of NAD$^+$ stocks observed in different organisms with aging.[78,87–91] PARP activity is assumed not to be affected by the availability of NAD$^+$, since their K_m for NAD$^+$ lies in a low micromolar range (Table 3.1), which is significantly below the physiological subcellular NAD$^+$ concentrations[33] and, most importantly, is lower than the K_m of sirtuins (Table 3.1). PARP-1 also appears to be more effective in NAD$^+$ turnover than sirtuins,[86] making PARPs the principal NAD$^+$ consumers in the cell and the main competitors of sirtuins for the limited NAD$^+$ pool. Several conditions known to cause PARP overactivation, like aging or DNA damage, were reported to be associated with severe NAD$^+$ depletion, rendering it a rate-limiting substrate for sirtuins.[87,88,92] In line with this, pharmacological or genetic inhibition of PARPs was reported to increase the NAD$^+$ content and activate sirtuin signaling.[76,87,93,94] cADPR synthases use NAD$^+$ to produce cADPR (Fig. 3.2), which serves as a signaling molecule controlling cytoplasmic calcium fluxes. The activity of cADPR synthases therefore has an impact on multiple biological processes which depend on Ca^{2+} release, such as egg fertilization, muscle contraction, cell proliferation, immune responses, and glucose-triggered insulin release by β-cells.[95] The efficiency of cADPR production of CD38 is several hundredfold higher than that of CD157.[96] Initially detected as a surface molecule on thymocytes and T lymphocytes,[97] CD38 was considered for quite some time as a protein, the expression of which was limited to the immune system. Later, however, the expression of CD38 was detected in multiple organs and nowadays it is assumed to be ubiquitous.[95] CD38 is an important NAD$^+$ consumer with a low K_m, established to be ~15−25 μM[98,99] (Table 3.1) and its loss-of-function in mice raised NAD$^+$ levels in different tissues up to 30-fold.[100] Similarly to PARP-1, overactivation of CD38 upon aging was found to be responsible for the age-associated decrease in NAD$^+$ content.[101] By modulating NAD$^+$ levels, loss-of-function of CD38 was shown to promote sirtuin activity[102] and had protective effects against obesity.[103]

3.7 NAD$^+$ BOOSTING STRATEGIES

In order to boost NAD$^+$ content and through it promote sirtuin function two main strategies can be employed: either promote NAD$^+$ synthesis or reduce NAD$^+$ consumption by the enzymes competing with sirtuins for the NAD$^+$ pool. For someone wishing to choose the first option, the choice is still not limited to one unique strategy, since NAD$^+$ synthesis can be promoted either (1) by providing precursor molecules or (2) by increasing the activity or expression of the enzymes involved in NAD$^+$ production. Indeed, increasing NMNAT1, NMNAT3, or NAMPRT expression[104−108] or using P7C3—an activator of NAMPRT—were all reported as feasible approaches to increase NAD$^+$ levels.[109] Although of theoretical interest, the translational potential of these strategies seems more limited, given that enhancing the expression levels of proteins is not straightforward. It is also important to mention that NAMPRT activators should be used with caution, since NAMPRT overexpression has been previously reported in many different types of cancer, including ovarian, prostate, colorectal, breast and gastric carcinomas, myelomas, astrocytomas, and melanomas.[110]

Providing NAD$^+$ precursors has been extensively studied as an approach for NAD$^+$ boosting (see Refs. [111,112] for extensive reviews). Supplementation with tryptophan, NA, NAM, NMN, and NR were all reported to increase NAD$^+$ content (reviewed in Ref. [111]). One alternative method for promoting NAD$^+$ synthesis consists of the preservation of NAM for NAD$^+$ synthesis via the inhibition of its alternative consumption by nicotinamide methyl transferase (NNMT). NNMT catalyzes the addition of methyl group to NAM, transforming it into methylnicotinamide, which can no longer serve as an NAD$^+$ precursor (Fig. 3.1). Inhibition of NNMT hence increased NAD$^+$ levels in adipose tissue, which primarily relies on NAM for NAD$^+$ production, but not in the liver, where alternative pathways for NAD$^+$ synthesis can be employed.[113]

Overactivation of PARPs caused by DNA damage,[92] different diseases,[76,114−116] or during aging[87] was shown to lead to a severe depletion of NAD$^+$ content. Similarly, overexpression of CD38[117] is known to cause a reduction in NAD$^+$ content, indicating that pharmacological or genetic inhibition of these two enzymes might serve as an alternative NAD$^+$-boosting strategy.

Several studies have already demonstrated the efficiency of PARP[76,87,93,94,114,115] and CD38 inhibition[100−103] in preserving NAD$^+$ content for use by the sirtuins.

The existence of several strategies aiming to increase NAD$^+$ content raises questions about which particular one should be adopted. Many factors can influence the decision, including the tissue where the increase in NAD$^+$ content is expected or a particular disease state requiring a treatment. Potential side effects of different NAD$^+$-boosting techniques should also be taken into consideration. All PARP inhibitors, which are currently undergoing clinical trials or are even marketed (as is the case for Olaparib), possess genotoxic effects,[118] limiting their use for nononcologic indications. Administration of both NA and NAM has been associated with several side effects. For example, NA is known to cause severe flushing, which is due to the activation of the G-protein coupled receptor, GPR109A.[119] NAM does not activate GPR109A. However, at high doses, which has been used to treat diabetic patients, it was hepatotoxic.[120] NR has no reported adverse effects up to date and long-term administration of NR to mice (6−12 months) was well tolerated.[121−123] Furthermore, no side effects were reported after short-term administration of NR to humans.[28]

3.8 THERAPEUTIC POTENTIAL OF NAD$^+$ BOOSTING

A constantly increasing number of studies report beneficial effects of NAD$^+$ boosting in a wide range of disease states, probably laying the foundation for a future era of NAD$^+$ therapeutics. Since sirtuins have been known as important regulators of longevity and NAD$^+$ levels have been reported to decline with age, a number of studies have explored the beneficial effects of preservation of NAD$^+$ content in aging. Indeed, rising NAD$^+$ levels has been reported to increase lifespan in yeast, worms, and even mice[87,122,124−126] and protect from a wide range of aging-related complications, including muscular function and capacity to regenerate, insulin sensitivity, vision problems, vascular dysfunction, and alterations in mitochondrial activity.[87,90,106,122,127−129] Important benefits of different NAD$^+$-boosting strategies have been also observed in rare diseases associated with altered DNA repair and accelerated aging, such as xeroderma pigmentosum group A (XPA), ataxia-telangiectasia (A-T) and Cockayne syndrome group B (CSB).[116,130,131]

Restoring or increasing NAD$^+$ content was reported to improve exercise capacity and muscular function[49,94] in aged animals[122] and in animals with muscular dystrophies[114] and mitochondrial myopathies,[132,133] to provide cardiac protection against inschemia-reperfusion[134] and pressure overload,[135,136] to mitigate acute kidney injury (AKI),[91,137] and to induce myeloid differentiation, a function which is disturbed in congenital neutropenia.[138]

Beneficial effects of different NAD$^+$-boosting techniques were also reported in many metabolic diseases. Increasing NAD$^+$ levels prevented animals from developing obesity, improved glucose homeostasis in obese and diabetic animals,[49,77,78,93,103,113] and protected from nonalcoholic steatohepatitis (NASH)[75,139,140] and alcoholic steatohepatitis (ASH).[140]

Replenishing NAD$^+$ stocks also possesses multiple neuroprotective effects, e.g., against brain[141−146] and spinal cord injuries[147] caused by ischemia-reperfusion, Wallerian degeneration,[104,107,148,149] Parkinson's and Alzheimer's diseases,[70,123,150−154] vision and hearing problems,[106,155,156] as well as against peripheral neuropathies, caused by chemotherapy or diabetes.[77,157]

3.9 CONCLUSION

This chapter summarizes key features of our current knowledge in the field of NAD^+ metabolism. With its function as an electron carrier reported one century ago, the actual role of NAD^+ extends much further after the discovery of sirtuins as NAD^+-dependent deacylases. With a constantly increasing number of studies evidencing beneficial effects of manipulations of cellular NAD^+ content in disease models, we might be at the dawn of an era where NAD^+-based therapeutics are making their way to the clinic. An accurate dosing scheme, as well as meticulous safety assessment of different NAD^+ boosters would be certainly required prior to translation of the findings described in this chapter into the clinic.

In the meantime, many questions related to rather basic concepts of NAD^+ metabolism still remain unanswered and require further investigation. The technical advancements allowing more accurate quantification of NAD^+ content would help to clarify the subcellular distribution of the NAD^+ metabolome and synthesis machinery. Further elucidation of the relationship of NAD^+ with different members of the sirtuin family would shed light on the molecular mechanisms underlying the beneficial effects provided by the activation of the NAD^+−sirtuin signaling axis. Likewise, further work on other NAD^+-consuming enzymes will help to clarify their roles in disease pathogenesis and treatment.

REFERENCES

1. Harden A, Young WJ. The alcoholic ferment of yeast-juice. *P R Soc Lond B-Conta* 1906;**77**:405−20.
2. von Euler H, Myrbäck K. *Hoppe-Seylers Zeitschrift fur phyiologische Chemie* 1930;189.
3. Warburg O, Christian W. Pyridine, the hydrogen transfusing component of fermentative enzymes. *Helv Chim Acta* 1936;**19**:79−88.
4. Bieganowski P, Brenner C. Discoveries of nicotinamide riboside as a nutrient and conserved NRK genes establish a Preiss-Handler independent route to NAD^+ in fungi and humans. *Cell* 2004;**117**:495−502.
5. Bogan KL, Brenner C. Nicotinic acid, nicotinamide, and nicotinamide riboside: a molecular evaluation of NAD^+ precursor vitamins in human nutrition. *Annu Rev Nutr* 2008;**28**:115−30.
6. Mori V, Amici A, Mazzola F, et al. Metabolic profiling of alternative NAD biosynthetic routes in mouse tissues. *PLoS One* 2014;**9**:e113939.
7. Raffaelli N, Sorci L, Amici A, Emanuelli M, Mazzola F, Magni G. Identification of a novel human nicotinamide mononucleotide adenylyltransferase. *Biochem Biophys Res Commun* 2002;**297**:835−40.
8. Sood R, Bonner TI, Makalowska I, et al. Cloning and characterization of 13 novel transcripts and the human RGS8 gene from the 1q25 region encompassing the hereditary prostate cancer (HPC1) locus. *Genomics* 2001;**73**:211−22.
9. Zhang X, Kurnasov OV, Karthikeyan S, Grishin NV, Osterman AL, Zhang H. Structural characterization of a human cytosolic NMN/NaMN adenylyltransferase and implication in human NAD biosynthesis. *J Biol Chem* 2003;**278**:13503−11.
10. Preiss J, Handler P. Biosynthesis of diphosphopyridine nucleotide. I. Identification of intermediates. *J Biol Chem* 1958;**233**:488−92.
11. Shibata K, Hayakawa T, Iwai K. Tissue distribution of the enzymes concerned with the biosynthesis of Nad in rats. *Agr Biol Chem Tokyo* 1986;**50**:3037−41.

12. Yamazaki F, Kuroiwa T, Takikawa O, Kido R. Human indolylamine 2,3-dioxygenase. Its tissue distribution, and characterization of the placental enzyme. *Biochem J* 1985;**230**:635−8.
13. Kudo Y, Boyd CA. Human placental indoleamine 2,3-dioxygenase: cellular localization and characterization of an enzyme preventing fetal rejection. *Biochim Biophys Acta* 2000;**1500**:119−24.
14. Magni G, Orsomando G, Raffelli N, Ruggieri S. Enzymology of mammalian NAD metabolism in health and disease. *Front Biosci J Virtual Library* 2008;**13**:6135−54.
15. Magni G, Amici A, Emanuelli M, Orsomando G, Raffaelli N, Ruggieri S. Enzymology of NAD$^+$ homeostasis in man. *Cell Mol Life Sci* 2004;**61**:19−34.
16. Terakata M, Fukuwatari T, Kadota E, et al. The niacin required for optimum growth can be synthesized from L-tryptophan in growing mice lacking tryptophan-2,3-dioxygenase. *J Nutr* 2013;**143**:1046−51.
17. Hagino Y, Lan SJ, Ng CY, Henderson LM. Metabolism of pyridinium precursors of pyridine nucleotides in perfused rat liver. *J Biol Chem* 1968;**243**:4980−6.
18. Jackson TM, Rawling JM, Roebuck BD, Kirkland JB. Large supplements of nicotinic acid and nicotinamide increase tissue NAD$^+$ and poly(ADP-ribose) levels but do not affect diethylnitrosamine-induced altered hepatic foci in Fischer-344 rats. *J Nutr* 1995;**125**:1455−61.
19. Hara N, Yamada K, Shibata T, Osago H, Hashimoto T, Tsuchiya M. Elevation of cellular NAD levels by nicotinic acid and involvement of nicotinic acid phosphoribosyltransferase in human cells. *J Biol Chem* 2007;**282**:24574−82.
20. Ijichi H, Ichiyama A, Hayaishi O. Studies on the biosynthesis of nicotinamide adenine dinucleotide. 3. Comparative in vivo studies on nicotinic acid, nicotinamide, and quinolinic acid as precursors of nicotinamide adenine dinucleotide. *J Biol Chem* 1966;**241**:3701−7.
21. Williams GT, Lau KM, Coote JM, Johnstone AP. NAD metabolism and mitogen stimulation of human lymphocytes. *Exp Cell Res* 1985;**160**:419−26.
22. Lin LF, Henderson LM. Pyridinium precursors of pyridine nucleotides in perfused rat kidney and in the testis. *J Biol Chem* 1972;**247**:8023−30.
23. Collins PB, Chaykin S. Comparative metabolism of nicotinamide and nicotinic acid in mice. *Biochem J* 1971;**125**:117P.
24. Collins PB, Chaykin S. The management of nicotinamide and nicotinic acid in the mouse. *J Biol Chem* 1972;**247**:778−83.
25. Yang SJ, Choi JM, Kim L, et al. Nicotinamide improves glucose metabolism and affects the hepatic NAD-sirtuin pathway in a rodent model of obesity and type 2 diabetes. *J Nutr Biochem* 2014;**25**:66−72.
26. Jacobson EL, Dame AJ, Pyrek JS, Jacobson MK. Evaluating the role of niacin in human carcinogenesis. *Biochimie* 1995;**77**:394−8.
27. Ho CK, Hashim SA. Pyridine nucleotide depletion in pancreatic islets associated with streptozotocin-induced diabetes. *Diabetes* 1972;**21**:789−93.
28. Trammell SA, Schmidt MS, Weidemann BJ, et al. Nicotinamide riboside is uniquely and orally bioavailable in mice and humans. *Nat Commun* 2016;**7**:12948.
29. Bender DA, Magboul BI, Wynick D. Probable mechanisms of regulation of the utilization of dietary tryptophan, nicotinamide and nicotinic acid as precursors of nicotinamide nucleotides in the rat. *Br J Nutr* 1982;**48**:119−27.
30. McCreanor GM, Bender DA. The metabolism of high intakes of tryptophan, nicotinamide and nicotinic acid in the rat. *Br J Nutr* 1986;**56**:577−86.
31. Williams JN, Feigelson P, Elvehjem CA. Relation of tryptophan and niacin to pyridine nucleotides of tissue. *J Biol Chem* 1950;**187**:597−604.
32. Bender DA, Olufunwa R. Utilization of tryptophan, nicotinamide and nicotinic acid as precursors for nicotinamide nucleotide synthesis in isolated rat liver cells. *Br J Nutr* 1988;**59**:279−87.

33. Houtkooper RH, Canto C, Wanders RJ, Auwerx J. The secret life of NAD$^+$: an old metabolite controlling new metabolic signaling pathways. *Endocr Rev* 2010;**31**:194−223.
34. Menzies KJ, Zhang H, Katsyuba E, Auwerx J. Protein acetylation in metabolism - metabolites and cofactors. *Nat Rev Endocrinol* 2016;**12**:43−60.
35. Cambronne XA, Stewart ML, Kim D, et al. Biosensor reveals multiple sources for mitochondrial NAD (+). *Science* 2016;**352**:1474−7.
36. Tischler ME, Friedrichs D, Coll K, Williamson JR. Pyridine nucleotide distributions and enzyme mass action ratios in hepatocytes from fed and starved rats. *Arch Biochem Biophys* 1977;**184**:222−36.
37. Alano CC, Tran A, Tao R, Ying W, Karliner JS, Swanson RA. Differences among cell types in NAD(+) compartmentalization: a comparison of neurons, astrocytes, and cardiac myocytes. *J Neurosci Res* 2007;**85**:3378−85.
38. Di Lisa F, Menabo R, Canton M, Barile M, Bernardi P. Opening of the mitochondrial permeability transition pore causes depletion of mitochondrial and cytosolic NAD$^+$ and is a causative event in the death of myocytes in postischemic reperfusion of the heart. *J Biol Chem* 2001;**276**:2571−5.
39. Yang H, Yang T, Baur JA, et al. Nutrient-sensitive mitochondrial NAD$^+$ levels dictate cell survival. *Cell* 2007;**130**:1095−107.
40. Pittelli M, Formentini L, Faraco G, et al. Inhibition of nicotinamide phosphoribosyltransferase: cellular bioenergetics reveals a mitochondrial insensitive NAD pool. *J Biol Chem* 2010;**285**:34106−14.
41. Nakagawa T, Lomb DJ, Haigis MC, Guarente L. SIRT5 Deacetylates carbamoyl phosphate synthetase 1 and regulates the urea cycle. *Cell* 2009;**137**:560−70.
42. Barile M, Passarella S, Danese G, Quagliariello E. Rat liver mitochondria can synthesize nicotinamide adenine dinucleotide from nicotinamide mononucleotide and ATP via a putative matrix nicotinamide mononucleotide adenylyltransferase. *Biochem Mol Biol Int* 1996;**38**:297−306.
43. Di Lisa F, Ziegler M. Pathophysiological relevance of mitochondria in NAD(+) metabolism. *FEBS Lett* 2001;**492**:4−8.
44. van Roermund CW, Elgersma Y, Singh N, Wanders RJ, Tabak HF. The membrane of peroxisomes in *Saccharomyces cerevisiae* is impermeable to NAD(H) and acetyl-CoA under in vivo conditions. *EMBO J* 1995;**14**:3480−6.
45. Todisco S, Agrimi G, Castegna A, Palmieri F. Identification of the mitochondrial NAD$^+$ transporter in *Saccharomyces cerevisiae*. *J Biol Chem* 2006;**281**:1524−31.
46. Agrimi G, Brambilla L, Frascotti G, et al. Deletion or overexpression of mitochondrial NAD$^+$ carriers in *Saccharomyces cerevisiae* alters cellular NAD and ATP contents and affects mitochondrial metabolism and the rate of glycolysis. *Appl Environ Microbiol* 2011;**77**:2239−46.
47. VanLinden MR, Dolle C, Pettersen IK, et al. Subcellular distribution of NAD$^+$ between cytosol and mitochondria determines the metabolic profile of human cells. *J Biol Chem* 2015;**290**:27644−59.
48. Pittelli M, Felici R, Pitozzi V, et al. Pharmacological effects of exogenous NAD on mitochondrial bioenergetics, DNA repair, and apoptosis. *Mol Pharmacol* 2011;**80**:1136−46.
49. Canto C, Houtkooper RH, Pirinen E, et al. The NAD(+) precursor nicotinamide riboside enhances oxidative metabolism and protects against high-fat diet-induced obesity. *Cell Metab* 2012;**15**:838−47.
50. Nikiforov A, Dolle C, Niere M, Ziegler M. Pathways and subcellular compartmentation of NAD biosynthesis in human cells: from entry of extracellular precursors to mitochondrial NAD generation. *J Biol Chem* 2011;**286**:21767−78.
51. Brown K, Xie S, Qiu X, et al. SIRT3 reverses aging-associated degeneration. *Cell Rep* 2013;**3**:319−27.
52. Emanuelli M, Carnevali F, Saccucci F, et al. Molecular cloning, chromosomal localization, tissue mRNA levels, bacterial expression, and enzymatic properties of human NMN adenylyltransferase. *J Biol Chem* 2001;**276**:406−12.

53. Yalowitz JA, Xiao S, Biju MP, et al. Characterization of human brain nicotinamide 5′-mononucleotide adenylyltransferase-2 and expression in human pancreas. *Biochem J* 2004;**377**:317−26.

54. Berger F, Lau C, Dahlmann M, Ziegler M. Subcellular compartmentation and differential catalytic properties of the three human nicotinamide mononucleotide adenylyltransferase isoforms. *J Biol Chem* 2005;**280**:36334−41.

55. Felici R, Lapucci A, Ramazzotti M, Chiarugi A. Insight into molecular and functional properties of NMNAT3 reveals new hints of NAD homeostasis within human mitochondria. *PLoS One* 2013;**8**:e76938.

56. Yamamoto M, Hikosaka K, Mahmood A, et al. Nmnat3 is dispensable in mitochondrial NAD level maintenance in vivo. *PLoS One* 2016;**11**:e0147037.

57. Hikosaka K, Ikutani M, Shito M, et al. Deficiency of nicotinamide mononucleotide adenylyltransferase 3 (nmnat3) causes hemolytic anemia by altering the glycolytic flow in mature erythrocytes. *J Biol Chem* 2014;**289**:14796−811.

58. Rongvaux A, Shea RJ, Mulks MH, et al. Pre-B-cell colony-enhancing factor, whose expression is up-regulated in activated lymphocytes, is a nicotinamide phosphoribosyltransferase, a cytosolic enzyme involved in NAD biosynthesis. *Eur J Immunol* 2002;**32**:3225−34.

59. Kitani T, Okuno S, Fujisawa H. Growth phase-dependent changes in the subcellular localization of pre-B-cell colony-enhancing factor. *FEBS Lett* 2003;**544**:74−8.

60. Hara N, Yamada K, Terashima M, Osago H, Shimoyama M, Tsuchiya M. Molecular identification of human glutamine- and ammonia-dependent NAD synthetases. Carbon-nitrogen hydrolase domain confers glutamine dependency. *J Biol Chem* 2003;**278**:10914−21.

61. Houtkooper RH, Pirinen E, Auwerx J. Sirtuins as regulators of metabolism and healthspan. *Nat Rev Mol Cell Biol* 2012;**13**:225−38.

62. Yang T, Chan NY, Sauve AA. Syntheses of nicotinamide riboside and derivatives: effective agents for increasing nicotinamide adenine dinucleotide concentrations in mammalian cells. *J Med Chem* 2007;**50**:6458−61.

63. Klar AJ, Fogel S. Activation of mating type genes by transposition in *Saccharomyces cerevisiae*. *Proc Natl Acad Sci U S A* 1979;**76**:4539−43.

64. Kaeberlein M, McVey M, Guarente L. The SIR2/3/4 complex and SIR2 alone promote longevity in *Saccharomyces cerevisiae* by two different mechanisms. *Genes Dev* 1999;**13**:2570−80.

65. Imai S, Armstrong CM, Kaeberlein M, Guarente L. Transcriptional silencing and longevity protein Sir2 is an NAD-dependent histone deacetylase. *Nature* 2000;**403**:795−800.

66. Landry J, Sutton A, Tafrov ST, et al. The silencing protein SIR2 and its homologs are NAD-dependent protein deacetylases. *Proc Natl Acad Sci U S A* 2000;**97**:5807−11.

67. Frye RA. Phylogenetic classification of prokaryotic and eukaryotic Sir2-like proteins. *Biochem Biophys Res Commun* 2000;**273**:793−8.

68. Chen D, Bruno J, Easlon E, et al. Tissue-specific regulation of SIRT1 by calorie restriction. *Genes Dev* 2008;**22**:1753−7.

69. Cantó C, Jiang LQ, Deshmukh AS, et al. Interdependence of AMPK and SIRT1 for metabolic adaptation to fasting and exercise in skeletal muscle. *Cell Metab* 2010;**11**:213−19.

70. Qin W, Yang T, Ho L, et al. Neuronal SIRT1 activation as a novel mechanism underlying the prevention of Alzheimer disease amyloid neuropathology by calorie restriction. *J Biol Chem* 2006;**281**:21745−54.

71. Fulco M, Cen Y, Zhao P, et al. Glucose restriction inhibits skeletal myoblast differentiation by activating SIRT1 through AMPK-mediated regulation of Nampt. *Dev Cell* 2008;**14**:661−73.

72. Rodgers JT, Lerin C, Haas W, Gygi SP, Spiegelman BM, Puigserver P. Nutrient control of glucose homeostasis through a complex of PGC-1alpha and SIRT1. *Nature* 2005;**434**:113−18.

73. Canto C, Gerhart-Hines Z, Feige JN, et al. AMPK regulates energy expenditure by modulating NAD$^+$ metabolism and SIRT1 activity. *Nature* 2009;**458**:1056−60.

74. Costford SR, Bajpeyi S, Pasarica M, et al. Skeletal muscle NAMPT is induced by exercise in humans. *Am J Physiol Endocrinol Metab* 2010;**298**:E117−26.

75. Gariani K, Menzies KJ, Ryu D, et al. Eliciting the mitochondrial unfolded protein response by nicotinamide adenine dinucleotide repletion reverses fatty liver disease in mice. *Hepatology* 2016;**63**:1190−204.

76. Gariani K, Ryu D, Menzies KJ, et al. Inhibiting poly ADP-ribosylation increases fatty acid oxidation and protects against fatty liver disease. *J Hepatol* 2017;**66**:132−41.

77. Trammell SA, Weidemann BJ, Chadda A, et al. Nicotinamide riboside opposes type 2 diabetes and neuropathy in mice. *Sci Rep* 2016;**6**:26933.

78. Yoshino J, Mills KF, Yoon MJ, Imai S-I. Nicotinamide mononucleotide, a key NAD(+) intermediate, treats the pathophysiology of diet- and age-induced diabetes in mice. *Cell Metab* 2011;**14**:528−36.

79. Fukuwatari T, Morikawa Y, Sugimoto E, Shibata K. Effects of fatty liver induced by niacin-free diet with orotic acid on the metabolism of tryptophan to niacin in rats. *Biosci Biotechnol Biochem* 2002;**66**:1196−204.

80. Drew JE, Farquharson AJ, Horgan GW, Williams LM. Tissue-specific regulation of sirtuin and nicotinamide adenine dinucleotide biosynthetic pathways identified in C57Bl/6 mice in response to high-fat feeding. *J Nutr Biochem* 2016;**37**:20−9.

81. Nakahata Y, Sahar S, Astarita G, Kaluzova M, Sassone-Corsi P. Circadian control of the NAD^+ salvage pathway by CLOCK-SIRT1. *Science* 2009;**324**:654−7.

82. Ramsey KM, Yoshino J, Brace CS, et al. Circadian clock feedback cycle through NAMPT-mediated NAD^+ biosynthesis. *Science* 2009;**324**:651−4.

83. Peek CB, Affinati AH, Ramsey KM, et al. Circadian clock NAD^+ cycle drives mitochondrial oxidative metabolism in mice. *Science* 2013;**342**:1243417.

84. Ying W. NAD^+/NADH and NADP + /NADPH in cellular functions and cell death: regulation and biological consequences. *Antioxid Redox Signal* 2008;**10**:179−206.

85. Feijs KL, Forst AH, Verheugd P, Luscher B. Macrodomain-containing proteins: regulating new intracellular functions of mono(ADP-ribosyl)ation. *Nat Rev Mol Cell Biol* 2013;**14**:443−51.

86. Bai P, Canto C. The role of PARP-1 and PARP-2 enzymes in metabolic regulation and disease. *Cell Metab* 2012;**16**:290−5.

87. Mouchiroud L, Houtkooper RH, Moullan N, et al. The NAD(+)/sirtuin pathway modulates longevity through activation of mitochondrial UPR and FOXO signaling. *Cell* 2013;**154**:430−41.

88. Braidy N, Guillemin GJ, Mansour H, Chan-Ling T, Poljak A, Grant R. Age related changes in NAD^+ metabolism oxidative stress and Sirt1 activity in Wistar rats. *PLoS One* 2011;**6**:e19194.

89. Massudi H, Grant R, Braidy N, Guest J, Farnsworth B, Guillemin GJ. Age-associated changes in oxidative stress and NAD^+ metabolism in human tissue. *PLoS One* 2012;**7**:e42357.

90. Gomes AP, Price NL, Ling AJY, et al. Declining NAD(+) induces a pseudohypoxic state disrupting nuclear-mitochondrial communication during aging. *Cell* 2013;**155**:1624−38.

91. Guan Y, Wang SR, Huang XZ, et al. Nicotinamide mononucleotide, an NAD^+ precursor, rescues age-associated susceptibility to AKI in a sirtuin 1-dependent manner. *J Am Soc Nephrol JASN* 2017.

92. Berger NA. Poly(ADP-ribose) in the cellular response to DNA damage. *Radiat Res* 1985;**101**:4−15.

93. Bai P, Cantó C, Oudart H, et al. PARP-1 inhibition increases mitochondrial metabolism through SIRT1 activation. *Cell Metab* 2011;**13**:461−8.

94. Pirinen E, Canto C, Jo YS, et al. Pharmacological inhibition of poly(ADP-ribose) polymerases improves fitness and mitochondrial function in skeletal muscle. *Cell Metab* 2014;**19**:1034−41.

95. Quarona V, Zaccarello G, Chillemi A, et al. CD38 and CD157: a long journey from activation markers to multifunctional molecules. *Cytometry B Clin Cytom* 2013;**84**:207−17.

96. Hussain AMM, Lee HC, Chang CF. Functional expression of secreted mouse BST-1 in yeast. *Prot Exp Purif* 1998;**12**:133−7.

97. Bhan AK, Reinherz EL, Poppema S, Mccluskey RT, Schlossman SF. Location of T-cell and major histocompatibility complex antigens in the human thymus. *J Exp Med* 1980;**152**:771−82.

98. Sauve AA, Munshi C, Lee HC, Schramm VL. The reaction mechanism for CD38. A single intermediate is responsible for cyclization, hydrolysis, and base-exchange chemistries. *Biochemistry* 1998;**37**:13239−49.

99. Cakir-Kiefer C, Muller-Steffner H, Oppenheimer N, Schuber F. Kinetic competence of the cADP-ribose-CD38 complex as an intermediate in the CD38/NAD⁺ glycohydrolase-catalysed reactions: implication for CD38 signalling. *Biochem J* 2001;**358**:399−406.

100. Aksoy P, White TA, Thompson M, Chini EN. Regulation of intracellular levels of NAD: a novel role for CD38. *Biochem Biophys Res Commun* 2006;**345**:1386−92.

101. Camacho-Pereira J, Tarrago MG, Chini CC, et al. CD38 dictates age-related NAD decline and mitochondrial dysfunction through an SIRT3-dependent mechanism. *Cell Metab* 2016;**23**:1127−39.

102. Aksoy P, Escande C, White TA, et al. Regulation of SIRT 1 mediated NAD dependent deacetylation: a novel role for the multifunctional enzyme CD38. *Biochem Biophys Res Commun* 2006;**349**:353−9.

103. Barbosa MT, Soares SM, Novak CM, et al. The enzyme CD38 (a NAD glycohydrolase, EC 3.2.2.5) is necessary for the development of diet-induced obesity. *FASEB* 2007;**21**:3629−39.

104. Araki T, Sasaki Y, Milbrandt J. Increased nuclear NAD biosynthesis and SIRT1 activation prevent axonal degeneration. *Science* 2004;**305**:1010−13.

105. Hsu CP, Oka S, Shao D, Hariharan N, Sadoshima J. Nicotinamide phosphoribosyltransferase regulates cell survival through NAD⁺ synthesis in cardiac myocytes. *Circ Res* 2009;**105**:481−91.

106. Williams PA, Harder JM, Foxworth NE, et al. Vitamin B3 modulates mitochondrial vulnerability and prevents glaucoma in aged mice. *Science* 2017;**355**:756−60.

107. Sasaki Y, Araki T, Milbrandt J. Stimulation of nicotinamide adenine dinucleotide biosynthetic pathways delays axonal degeneration after axotomy. *J Neurosci* 2006;**26**:8484−91.

108. Tao R, Wei D, Gao H, Liu Y, DePinho RA, Dong XC. Hepatic FoxOs regulate lipid metabolism via modulation of expression of the nicotinamide phosphoribosyltransferase gene. *J Biol Chem* 2011;**286**:14681−90.

109. Wang G, Han T, Nijhawan D, et al. P7C3 neuroprotective chemicals function by activating the rate-limiting enzyme in NAD salvage. *Cell.* 2014;**158**:1324−34.

110. Garten A, Schuster S, Penke M, Gorski T, de Giorgis T, Kiess W. Physiological and pathophysiological roles of NAMPT and NAD metabolism. *Nat Rev Endocrinol* 2015;**11**:535−46.

111. Canto C, Menzies KJ, Auwerx J. NAD(+) metabolism and the control of energy homeostasis: a balancing act between mitochondria and the nucleus. *Cell Metab* 2015;**22**:31−53.

112. Imai S-I, Guarente L. *It takes two to tango: NAD⁺ and sirtuins in aging/longevity control.* NPJ Aging Mech Dis. 2016. p. 2.

113. Kraus D, Yang Q, Kong D, et al. Nicotinamide N-methyltransferase knockdown protects against diet-induced obesity. *Nature* 2014;**508**:258−62.

114. Ryu D, Zhang H, Ropelle ER, et al. NAD⁺ repletion improves muscle function in muscular dystrophy and counters global PARylation. *Sci Trans Med* 2016;**8**:361ra139.

115. Mukhopadhyay P, Horvath B, Rajesh M, et al. PARP inhibition protects against alcoholic and non-alcoholic steatohepatitis. *J Hepatol* 2017;**66**:589−600.

116. Fang EF, Scheibye-Knudsen M, Brace LE, et al. Defective mitophagy in XPA via PARP-1 hyperactivation and NAD(+)/SIRT1 reduction. *Cell* 2014;**157**:882−96.

117. Hu Y, Wang H, Wang Q, Deng H. Overexpression of CD38 decreases cellular NAD levels and alters the expression of proteins involved in energy metabolism and antioxidant defense. *J Proteome Res* 2014;**13**:786−95.

118. Ito S, Murphy CG, Doubrovina E, Jasin M, Moynahan ME. PARP inhibitors in clinical use induce genomic instability in normal human cells. *PLoS One* 2016;**11**:e0159341.
119. Benyo Z, Gille A, Bennett CL, Clausen BE, Offermanns S. Nicotinic acid-induced flushing is mediated by activation of epidermal langerhans cells. *Mol Pharmacol* 2006;**70**:1844−9.
120. Knip M, Douek IF, Moore WP, et al. Safety of high-dose nicotinamide: a review. *Diabetologia* 2000;**43**:1337−45.
121. Tummala KS, Gomes AL, Yilmaz M, et al. Inhibition of de novo NAD(+) synthesis by oncogenic URI causes liver tumorigenesis through DNA damage. *Cancer Cell* 2014;**26**:826−39.
122. Zhang H, Ryu D, Wu Y, et al. NAD(+) repletion improves mitochondrial and stem cell function and enhances life span in mice. *Science* 2016;**352**:1436−43.
123. Gong B, Pan Y, Vempati P, et al. Nicotinamide riboside restores cognition through an upregulation of proliferator-activated receptor-γ coactivator 1α regulated β-secretase 1 degradation and mitochondrial gene expression in Alzheimer's mouse models. *Neurobiol Aging* 2013;**34**:1581−8.
124. Belenky P, Racette FG, Bogan KL, McClure JM, Smith JS, Brenner C. Nicotinamide riboside promotes Sir2 silencing and extends lifespan via Nrk and Urh1/Pnp1/Meu1 pathways to NAD$^+$. *Cell* 2007;**129**:473−84.
125. Easlon E, Tsang F, Skinner C, Wang C, Lin SJ. The malate-aspartate NADH shuttle components are novel metabolic longevity regulators required for calorie restriction-mediated life span extension in yeast. *Genes Dev* 2008;**22**:931−44.
126. Lin SJ, Ford E, Haigis M, Liszt G, Guarente L. Calorie restriction extends yeast life span by lowering the level of NADH. *Genes Dev* 2004;**18**:12−16.
127. Mills KF, Yoshida S, Stein LR, et al. Long-term administration of nicotinamide mononucleotide mitigates age-associated physiological decline in mice. *Cell Metab* 2016;**24**:795−806.
128. Lin JB, Kubota S, Ban N, et al. NAMPT-mediated NAD(+) biosynthesis is essential for vision in mice. *Cell Rep* 2016;**17**:69−85.
129. de Picciotto NE, Gano LB, Johnson LC, et al. Nicotinamide mononucleotide supplementation reverses vascular dysfunction and oxidative stress with aging in mice. *Aging Cell* 2016;**15**:522−30.
130. Scheibye-Knudsen M, Mitchell SJ, Fang EF, et al. A high-fat diet and NAD(+) activate Sirt1 to rescue premature aging in cockayne syndrome. *Cell Metab* 2014;**20**:840−55.
131. Fang EF, Kassahun H, Croteau DL, et al. NAD$^+$ replenishment improves lifespan and healthspan in ataxia telangiectasia models via mitophagy and DNA repair. *Cell Metab* 2016;**24**:566−81.
132. Khan NA, Auranen M, Paetau I, et al. Effective treatment of mitochondrial myopathy by nicotinamide riboside, a vitamin B3. *EMBO Mol Med* 2014. n/a-n/a.
133. Cerutti R, Pirinen E, Lamperti C, et al. NAD(+)-dependent activation of Sirt1 corrects the phenotype in a mouse model of mitochondrial disease. *Cell Metab* 2014;**19**:1042−9.
134. Yamamoto T, Byun J, Zhai P, Ikeda Y, Oka S, Sadoshima J. Nicotinamide mononucleotide, an intermediate of NAD$^+$ synthesis, protects the heart from ischemia and reperfusion. *PLoS One* 2014;**9**:e98972.
135. Yano M, Akazawa H, Oka T, et al. Monocyte-derived extracellular Nampt-dependent biosynthesis of NAD(+) protects the heart against pressure overload. *Sci Rep* 2015;**5**:15857.
136. Lee CF, Chavez JD, Garcia-Menendez L, et al. Normalization of NAD$^+$ redox balance as a therapy for heart failure. *Circulation* 2016;**134**:883−94.
137. Tran MT, Zsengeller ZK, Berg AH, et al. PGC1alpha drives NAD biosynthesis linking oxidative metabolism to renal protection. *Nature* 2016;**531**:528−32.
138. Skokowa J, Lan D, Thakur BK, et al. NAMPT is essential for the G-CSF−induced myeloid differentiation via a NAD$^+$−sirtuin-1−dependent pathway. *Nat Med* 2009;**15**:151−8.
139. Gariani K, Ryu D, Menzies KJ, et al. Inhibiting poly ADP-ribosylation increases fatty acid oxidation and protects against fatty liver disease. *J Hepatol* 2016;**66**(1):132−41.

140. Mukhopadhyay P, Horvath B, Rajesh M, et al. PARP inhibition protects against alcoholic and nonalcoholic steatohepatitis. *J Hepatol* 2016;**66**(3):589−600.

141. Feng Y, Paul IA, LeBlanc MH. Nicotinamide reduces hypoxic ischemic brain injury in the newborn rat. *Brain Res Bull* 2006;**69**:117−22.

142. Kabra DG, Thiyagarajan M, Kaul CL, Sharma SS. Neuroprotective effect of 4-amino-1,8-napthalimide, a poly(ADP ribose) polymerase inhibitor in middle cerebral artery occlusion-induced focal cerebral ischemia in rat. *Brain Res Bull* 2004;**62**:425−33.

143. Kaundal RK, Shah KK, Sharma SS. Neuroprotective effects of NU1025, a PARP inhibitor in cerebral ischemia are mediated through reduction in NAD depletion and DNA fragmentation. *Life Sci* 2006;**79**:2293−302.

144. Klaidman L, Morales M, Kem S, Yang J, Chang ML, Adams Jr. JD. Nicotinamide offers multiple protective mechanisms in stroke as a precursor for NAD$^+$, as a PARP inhibitor and by partial restoration of mitochondrial function. *Pharmacology* 2003;**69**:150−7.

145. Sadanaga-Akiyoshi F, Yao H, Tanuma S, et al. Nicotinamide attenuates focal ischemic brain injury in rats: with special reference to changes in nicotinamide and NAD$^+$ levels in ischemic core and penumbra. *Neurochem Res* 2003;**28**:1227−34.

146. Zheng CB, Han J, Xia WL, Shi ST, Liu JR, Ying WH. NAD(+) administration decreases ischemic brain damage partially by blocking autophagy in a mouse model of brain ischemia. *Neurosci Lett* 2012;**512**:67−71.

147. Xie L, Wang Z, Li C, Yang K, Liang Y. Protective effect of nicotinamide adenine dinucleotide (NAD$^+$) against spinal cord ischemia-reperfusion injury via reducing oxidative stress-induced neuronal apoptosis. *J Clin Neurosci* 2017;**36**:114−19.

148. Wang J, Zhai Q, Chen Y, et al. A local mechanism mediates NAD-dependent protection of axon degeneration. *J Cell Biol* 2005;**170**:349−55.

149. Yin TC, Britt JK, De Jesus-Cortes H, et al. P7C3 neuroprotective chemicals block axonal degeneration and preserve function after traumatic brain injury. *Cell Rep.* 2014;**8**:1731−40.

150. Wang X, Hu X, Yang Y, Takata T, Sakurai T. Nicotinamide mononucleotide protects against beta-amyloid oligomer-induced cognitive impairment and neuronal death. *Brain Res* 2016;**1643**:1−9.

151. De Jesus-Cortes H, Xu P, Drawbridge J, et al. Neuroprotective efficacy of aminopropyl carbazoles in a mouse model of Parkinson disease. *Proc Natl Acad Sci U S A* 2012;**109**:17010−15.

152. Lehmann S, Loh SH, Martins LM. Enhancing NAD$^+$ salvage metabolism is neuroprotective in a PINK1 model of Parkinson's disease. *Biol Open* 2017;**6**:141−7.

153. Liu D, Pitta M, Jiang H, et al. Nicotinamide forestalls pathology and cognitive decline in Alzheimer mice: evidence for improved neuronal bioenergetics and autophagy procession. *Neurobiol Aging* 2013;**34**:1564−80.

154. Turunc Bayrakdar E, Uyanikgil Y, Kanit L, Koylu E, Yalcin A. Nicotinamide treatment reduces the levels of oxidative stress, apoptosis, and PARP-1 activity in Abeta(1-42)-induced rat model of Alzheimer's disease. *Free Radic Res* 2014;**48**:146−58.

155. Shindler KS, Ventura E, Rex TS, Elliott P, Rostami A. SIRT1 activation confers neuroprotection in experimental optic neuritis. *Invest Ophthalmol Vis Sci* 2007;**48**:3602−9.

156. Brown KD, Maqsood S, Huang JY, et al. Activation of SIRT3 by the NAD(+) precursor nicotinamide riboside protects from noise-induced hearing loss. *Cell Metab* 2014;**20**:1059−68.

157. Hamity MV, White SR, Walder RY, Schmidt MS, Brenner C, Hammond DL. Nicotinamide riboside, a form of vitamin B3 and NAD$^+$ precursor, relieves the nociceptive and aversive dimensions of paclitaxel-induced peripheral neuropathy in female rats. *Pain* 2017;**158**:962−72.

THE ENZYMATIC ACTIVITIES OF SIRTUINS

Hening Lin

Howard Hughes Medical Institute, Cornell University, Ithaca, NY, United States

Sirtuins are a class of nicotinamide adenine dinucleotide (NAD)-dependent enzymes that have generated a great deal of interest. Mammals have seven sirtuins, for which many interesting biological functions have been reported,[1−4] including the regulation of life span,[5,6] metabolism,[7−27] transcription,[9,24,28−40] genome stability,[41−45] and neuronal activity.[46−49] Small molecules that can regulate sirtuin activity are considered to have therapeutic potentials for several human diseases, including cancer, diabetes, and Parkinson's disease.[50,51] Controversies exist about the functions of sirtuins, especially concerning their effects on life span[52] and the effects of sirtuin activators.[53−59] However, there is no doubt that sirtuins play various important physiological functions, which are the topics of other chapters. The controversies to a large extent reflect that detailed molecular understanding about the function of sirtuins is still lacking, including their enzymatic activities and how they are regulated.

The founding member of this class of enzymes is yeast Sir2 (silencing information regulator 2), a protein known to be required for silencing genes in certain loci of the yeast genome.[60] It was reported that Sir2 is important for calorie restriction-induced life span extension.[61−63] Shortly after that, it was discovered that Sir2 is an enzyme that catalyzes the hydrolysis of acetyl lysine in histones using NAD as a co-substrate (Fig. 4.1).[64] Several other sirtuins were also reported to have this activity.[33,65−70]

The mechanism of the NAD-dependent deacetylation has been well studied (Fig. 4.1).[65,67,71−77] In general, the acyl peptide binds to sirtuin first, which is followed by NAD binding. The amide bond in the acyl peptide substrate then attacks the anomeric position of the ribose in NAD, displacing nicotinamide, forming an ADP-ribosyl-substrate covalent intermediate I, which is often referred to as the "alkylamidate" intermediate. This intermediate then is converted to a cyclic intermediate II, which is facilitated by a conserved His residue that serves as a general base catalyst. Recent evidence suggests that the cyclic intermediate can also be converted to another alkylamidate intermediate III, which is then hydrolyzed to produce the 2′-O-acyl ADP-ribose and the deacylated lysine product.[78] The 2′-O-acyl ADP-ribose can nonenzymatically isomerize to 3′-O-acyl ADP-ribose.

Among the seven mammalian sirtuins, SIRT1−3 have robust deacetylase activity in vitro, but SIRT4−7 have only weak deacetylase activity in vitro.[21,43,79−81] Why do SIRT4−7 have weak deacetylase activity? Is the weak deacetylase activity physiologically relevant or is other

Introductory Review on Sirtuins in Biology, Aging, and Disease. DOI: https://doi.org/10.1016/B978-0-12-813499-3.00004-6

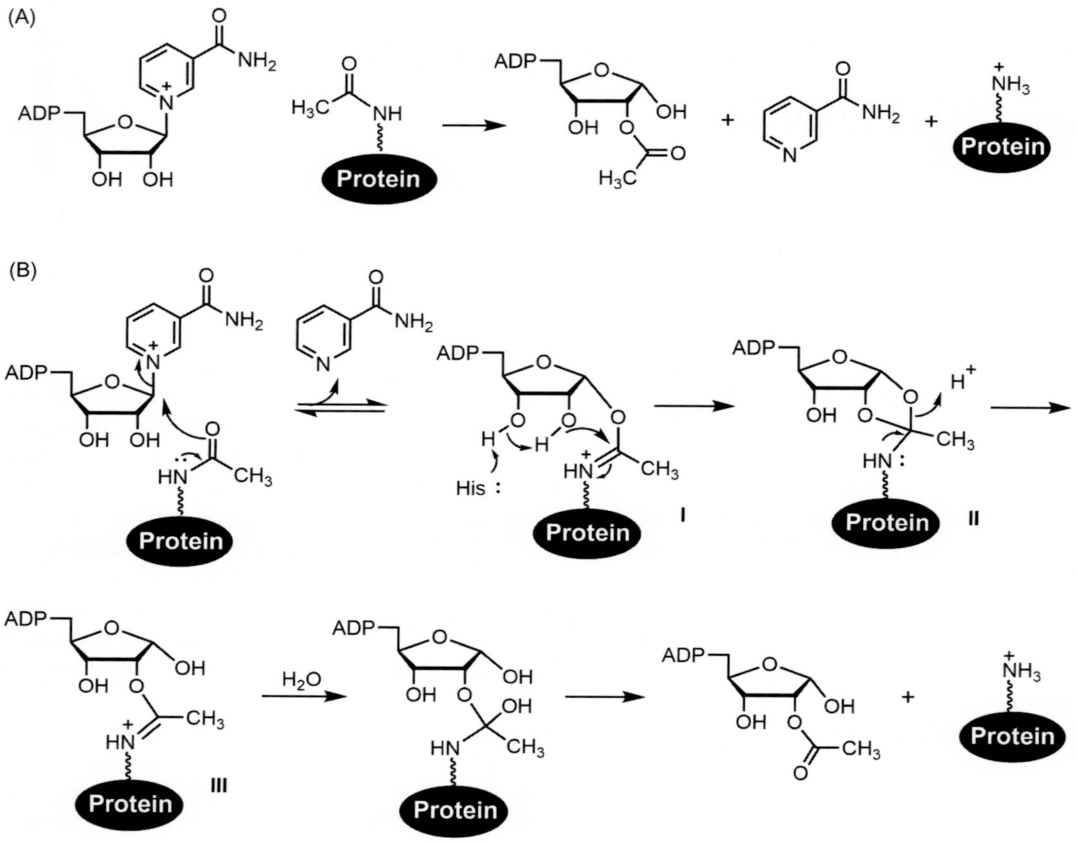

FIGURE 4.1

Sirtuin-catalyzed deacetylation reaction (A) and its enzymatic reaction mechanism (B).

more efficient activity responsible for their physiological function? In the past few years, significant progress has been made in this regard and several novel activities have been reported. Below I will summarize the different activities that have been reported for different sirtuins and comment on their physiological relevance. In particular, some sirtuins are capable of removing multiple acyl groups that are structurally very different. Thus, different acyl lysine substrates may compete for the same sirtuin enzyme, which may produce interesting physiological and pharmacological effects.

This chapter does not focus on the acyl lysine modifications that are more similar to acetyl lysine, such as propionyl, butyryl, and crotonyl lysines,[82,83] although they have been reported to be substrates for sirtuins with deacetylase activities.[84,85] Because they are similar to acetyl lysine and are likely put on by the same acyltransferases on the same sites and recognized by the same reader proteins,[86−93] it is hard to differentiate their function from that of acetyl lysine. Similarly, it is also hard to analyze the contribution of the corresponding deacylation activity to the biological functions of the sirtuins. Of course, this does not mean that these modifications or their regulation by sirtuins are not important.

4.1 DESUCCINYLATION, DEMALONYLATION, AND DEGLUTARYLATION CATALYZED BY SIRT5

Human SIRT5 is the first sirtuin that has taught us that sirtuins can remove acyl groups that are structurally very different from the acetyl group.[94] SIRT5 was reported to have weak but detectable deacetylase activity.[95] The new and more efficient desuccinylation and demalonylation activity was discovered based on the crystal structure of SIRT5 in complex with a thioacetyl H3 K9 peptide, a mechanism-based inhibitor synthesized for SIRT1, SIRT2, and SIRT3.[73,96] The structure contained a bound buffer molecule, 2-(cyclohexylamino)ethanesulfonate or CHES, not far from the thioacetyl group (Fig. 4.2A). The negatively charged sulfonate group interacts with an Arg residue and a Tyr residue (Arg105 and Tyr102) in SIRT5 through electrostatic interaction and hydrogen bondings. The sulfonate is close to the thioacetyl group. Based on this structure, it is hypothesized that peptides with negatively charged acyl lysine modifications (Fig. 4.2) could bind SIRT5 better and could be better substrates for SIRT5. Both succinyl-CoA and malonyl-CoA are common metabolites in cells, like acetyl-CoA, which is the acetyl donor for protein lysine acetylation. Thus, proteins in cells might be succinylated and malonylated and SIRT5's function was to remove succinyl and malonyl groups. This hypothesis was quickly confirmed by testing SIRT5's activity on synthetic succinyl and malonyl peptides. On the same H3 K9 peptide sequence, the catalytic efficiencies of SIRT5 for desuccinylation and demalonylation were more than 550-fold better than that for deacetylation.[94] The crystal structure of SIRT5 in complex with a succinyl peptide showed that the carboxylate of succinyl indeed interacted with Arg105 and Tyr102 of SIRT5 (Fig. 4.2B), as predicted from the structure of SIRT5 in complex with CHES and the thioacetyl peptide. Mutating these two residues increased the K_m values for succinyl peptide significantly, confirming that these two residues were important for recognizing negatively charged acyl groups. Several studies demonstrated that protein lysine succinylation and malonylation occurred in both bacterial and mammalian cells.[94,97−99] Later, SIRT5 was also demonstrated to be able to remove glutaryl from proteins.[100]

Various proteomic studies using antibodies that can recognize succinyl lysine have led to the identification of close to 1000 desuccinylation substrates of SIRT5.[27,101,102] Furthermore, SIRT5 was shown to regulate the activity and succinylation level of several proteins, including carbamoyl synthetase 1 (CPS1),[94] succinate dehydrogenase, pyruvate dehydrogenase (PDH),[27] 3-hydroxy-3-methylglutaryl-CoA synthase 2 (HMGCS2),[101] pyruvate kinase M2 (PKM2),[103,104] isocitrate dehydrogenase 2 (IDH2), glucose 6-phosphate dehydrogenase (G6PD),[105] and mitochondrial trifunctional enzyme ECHA.[102] Demalonylation substrates have also been identified by proteomics for SIRT5.[106,107]

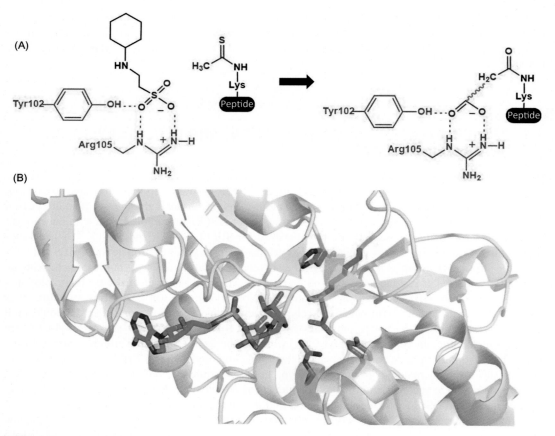

FIGURE 4.2

SIRT5 selectively recognizes negatively charged acyl lysine modifications. (A) The structure of SIRT5 in complex with CHES and a thioacetyl lysine peptide led to the hypothesis that SIRT5 may recognize negatively charged acyl groups, such as malonyl and succinyl. (B) Structure of SIRT5 in complex with a succinyl peptide (PDB 3RIY). The Tyr102 and Arg105 that are important for recognizing the negatively charged acyl groups are shown, as well as the catalytic His residue and the NAD cosubstrate.

Thus, studying the enzymatic activity of SIRT5 not only led to the discovery of more efficient biochemical activity of SIRT5, but also led to the identification of previously unknown protein posttranslational modifications. It also pointed out a new research direction for the sirtuins with weak deacetylase activities. Other sirtuins with weak deacetylase activities may also be able to remove other acyl lysine modifications.

4.2 HYDROLYSIS OF LONG-CHAIN FATTY ACYL GROUPS (DEFATTY-ACYLATION) BY SEVERAL SIRTUINS

The first sirtuin reported to have efficient defatty-acylation activity was PfSir2A, a sirtuin from the malaria parasite, *Plasmodium falciparum*.[108] PfSir2A has been reported to have both deacetylase activity and ADP-ribosyltransferase activity.[109] The reported deacetylase activity was rather weak.[110] Inspired by the SIRT5 studies, several peptides with different acyl lysine modifications were tested to find out whether PfSir2A could hydrolyze other acyl lysine modifications.[111] This study revealed that PfSir2A could hydrolyze long-chain fatty acyl groups more efficiently. Using a ^{32}P-NAD assay, it was shown that *P. falciparum* proteins contained fatty acyl lysine modifications,[111] but the exact physiological substrate for this activity of PfSir2A is still unknown.

Mammalian SIRT6 was the first sirtuin for which a physiological defatty-acylation substrate was identified. SIRT6 was reported to have deacetylase activity specifically on histone H3 K9 and K56.[42,43,112] SIRT6 knockout was shown to increase histone H3 K9 and K56 acetylation in vivo. However, the catalytic efficiency in vitro on a H3 K9 acetyl peptide, is still very low (k_{cat}/K_m is about $5 \text{ s}^{-1} \text{ M}^{-1}$, slightly weaker than the activity of SIRT5 on the same acetyl peptide), raising the possibility that SIRT6 may also have other enzymatic activities. Indeed, similar to PfSir2A, SIRT6 can hydrolyze long-chain fatty acyl groups several hundred-fold more efficiently than hydrolyzing acetyl groups.[113] The structure of SIRT6 in complex with a myristoyl lysine peptide revealed a large hydrophobic pocket that accommodated the myristoyl group, providing a nice structural explanation for SIRT6's preference for long-chain fatty acyl groups.[113]

TNFα was the first physiological substrate protein found to be regulated by the defatty-acylation activity of SIRT6. TNFα is a cytokine molecule that is important for immune responses.[114] It was one of the two mammalian proteins that were reported to be myristoylated on lysine residues.[115,116] It was reported that SIRT6 could regulate TNFα protein synthesis without affecting TNFα gene transcription.[117] However, the detailed molecular mechanism was not clear. TNFα is a type II membrane protein with a single transmembrane domain. After translation, TNFα is trafficked to the plasma membrane via secretory pathways.[114] At the plasma membrane, TNFα can also be cleaved by proteases to release the extracellular domain that can bind to the receptors on other cells.[114] Using a chemical probe for detecting long-chain fatty acylation,[118] it was demonstrated that SIRT6 regulated the fatty acylation level on both transfected and endogenous TNFα. Lysine fatty acylation affected the secretion level of TNFα (the percentage of TNFα secreted to the extracellular environment). This is because lysine fatty-acylated TNFα somehow is targeted more to the lysosome for degradation instead of going to the plasma membrane for secretion.[119] The effects of SIRT6 knockout on TNFα secretion were more pronounced when palmitic acid was added to the cell culture medium, suggesting that protein lysine fatty acylation could be regulated by the metabolic state of the cells.[113]

Protein lysine fatty acylation has been known for more than 20 years.[115,116] However, the original reports had not been followed up by further studies, while a lot of studies were carried out for glycine myristoylation, cysteine palmitoylation, and isoprenylation. The discovery of SIRT6 as a protein lysine defatty-acylase now provides a useful handle to further investigate the occurrence and biological functions of protein lysine fatty acylation. Indeed, recently, a Ras small GTPase, R-Ras2, has been reported to be regulated by lysine fatty acylation and SIRT6-catalyzed defatty-acylation. The Ras family of proteins regulates various cell signaling processes and mutation of Ras is a major cause of human tumors. Lysine fatty acylation promotes the plasma membrane localization and activation of R-Ras2. Activated R-Ras2 in turn activates PI3K/AKT and promotes cell proliferation. SIRT6, by defatty-acylation of R-Ras2, serves to suppress cell proliferation,[120] which may be important for the reported tumor suppressor role of SIRT6.[121]

Other sirtuins are also found to be able to hydrolyze long-chain fatty acyl groups. For example, SIRT7 can hydrolyze myristoyl lysine in a nucleic acid-dependent manner, although the physiological substrates for this activity have not been identified. Interestingly, it was reported that even SIRT1, SIRT2, and SIRT3 have efficient defatty-acylase activity in vitro.[122,123] The demyristoylase activity of SIRT2 is even slightly higher than its deacetylase activity based on the k_{cat}/K_m values.[124]

4.3 THE VARIOUS DEACYLATION ACTIVITIES REPORTED FOR SIRT4

Compared to other mammalian sirtuins, the enzymatic activity of SIRT4 is still not well understood. This is partially because the recombinant protein is more difficult to obtain. Its deacetylase activity is very weak and in many cases cannot be detected in vitro. However, it was demonstrated to exhibit weak deacetylase activity on certain peptide sequences. For example, it was shown to deacetylate malonyl-CoA decarboxylase (MCD) and regulate lipid metabolism.[22]

The weak deacetylase activity of SIRT4 has attracted people to look for other activities. In 2014, it was reported that SIRT4 can remove lipoyl- and biotinyl-lysine modifications better than its deacetylation activity.[125] SIRT4 could remove the lipoyl group (Fig. 4.3) from the E2 component dihydrolipoyllysine acetyltransferase of pyruvate dehydrogenase (PDH), diminishing PDH activity. However, the in vitro catalytic efficiency of SIRT4 on lipoyl and biotinyl lysine is still

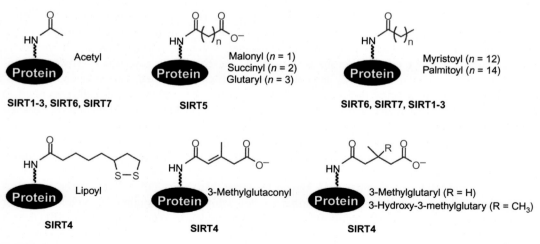

FIGURE 4.3

Different acyl lysine modifications that can be removed by sirtuins.

low compared to other more efficient sirtuin activities.[125] In 2017, it was reported that SIRT4 removes methylglutaryl (MG), hydroxymethylglutaryl (HMG), and 3-methylglutaconyl (MGc; Fig. 4.3) from protein lysine residues more efficiently, although the detailed k_{cat} and K_m values were not reported.[126] The acyl-CoA metabolites leading to these posttranslational modifications are intermediates in leucine metabolism. SIRT4 plays an important role in activating methylcrotonyl-CoA carboxylase A and B in the leucine oxidation pathway by deacylating them. This in turn regulates insulin secretion and SIRT4 KO mice have elevated basal and stimulated insulin secretion and eventually develop glucose intolerance and insulin resistance.[126]

4.4 **THE ADP-RIBOSYLTRANSFERASE ACTIVITY OF SIRTUINS**

Sirtuins had been reported to have ADP-ribosyltransferase activity,[21,45,81,109,127] which transfers ADP-ribosyl groups from NAD to proteins.[127–129] Before yeast Sir2's deacetylase activity was discovered, it was reported that it could ADP-ribosylate BSA and histones in vitro.[127] This activity did not gain major recognition since the deacetylase activity was discovered to be a more efficient activity.[64,66] However, for sirtuins with no or very weak deacetylase activity, the ADP-ribosyltransferase activity was often postulated to be important. For example, SIRT4 was thought to regulate glutamate dehydrogenase activity by ADP-ribosylating it,[21] and SIRT6 was reported to ADP-ribosylate poly(ADP-ribose) polymerase 1 or PARP1.[45] Quantitative measurement showed that the reported ADP-ribosyltransferase activity of sirtuins is very weak and to a large extent depends on the deacetylation activity.[80,130] Thus, the physiological relevance for the ADP-ribosyltransferase activity of sirtuins remains debatable.

Under extreme conditions, the ADP-ribosyltransferase activity of certain sirtuins could be relevant. The *Salmonella* CobB is known to deacetylate acetyl-CoA synthetase and activates its activity.[131] As with other sirtuins, it has weak ADP-ribosyltransferase activity.[132] This enzyme was named "CobB" because its overexpression complements the deletion of CobT, which is a phosphoribosyltransferase/ADP-ribosyltransferase required for the biosynthesis of vitamin B12.[133] However, CobB has to be overexpressed to high levels for this activity to be relevant and this activity is only relevant when CobT is not present.[133] CobT deletion leads to B12 deficiency, but CobB deletion does not.[133]

Interestingly, a sirtuin variant called SirTM that only exists in certain pathogenic bacteria or fungi is reported to only have ADP-ribosyltransferase activity, but not deacetylase activity.[134] These sirtuins are found in the genome together with a macro domain protein. The SirTM and macro domain proteins are typically found within the same operon or as a fusion within a single open reading frame. SirTM proteins can efficiently ADP-ribosylate another protein, GcvH-L, encoded in the same operon on acidic Asp and Glu residues, and can be reversed by the macro domain protein. In SirTM, the conserved catalytic His residue is replaced with Gln, which is important for the ADP-ribosyltransferase activity, as mutating it to His eliminated the ADP-ribosylation activity. The ADP-ribosylation of GcvH-L is hypothesized to regulate the interaction of GcvH-L with other proteins in response to oxidative stress.[134]

4.5 SPECIAL CONSIDERATION FOR SIRTUINS THAT CAN RECOGNIZE STRUCTURALLY VERY DIFFERENT ACYL GROUPS

As discussed above, several sirtuins can catalyze the hydrolysis of structurally very different acyl groups. For example, SIRT1, SIRT2, SIRT3, and SIRT6 can all hydrolyze acetyl and myristoyl lysine. This situation has raised several interesting questions that will be discussed below.

How can one sirtuin recognize both a short acetyl lysine and a very long fatty acyl lysine? A possible structural explanation is the following. The pocket that accommodates the myristoyl group is lined up by many hydrophobic residues, such as leucine, isoleucine, and phenylalanine. These residues may be relatively flexible. When acetyl lysine binds, these residues can move around to form a seemingly smaller hydrophobic pocket that was observed in reported crystal structures of sirtuins with acetyl peptides bound. However, when a long-chain fatty acyl peptide binds, these resides will move away slightly to accommodate the long-chain fatty acyl groups. However, direct support for this explanation requires comparing two structures of the same sirtuin, one in complex with acetyl peptide and another in complex with myristoyl peptide, which are currently not available (typically only one such structure is available).

The dual specificity for both substrates can also be explained by kinetics. The long-chain fatty acyl groups form a more extensive hydrophobic interaction with the sirtuin. Thus, a long-chain fatty acyl peptide would have a much higher binding affinity and a much lower K_m value on these sirtuins. However, the product, the acyl ADP-ribose, will also have a higher binding affinity and its dissociation may be rate-limiting, which will significantly decrease the k_{cat} value. In contrast, the acetyl peptide will have a higher K_m value, but also higher k_{cat} value, making the catalytic efficiency (k_{cat}/K_m) similar to that of long-chain fatty acyl peptides.

How can the functional contributions of different deacylation activities of the same sirtuin be differentiated? As mentioned at the beginning of the chapter, it is very difficult to differentiate the functional contributions of sirtuins' activities in removing acetyl, priopionyl, butyryl, and crotonyl groups. However, differentiating the functional contribution of sirtuins' deacetylation activity and demyristoylation/depalmitoylation activity is feasible. This is due to several reasons. First, the acetylation and long-chain fatty acylation can occur on different substrate proteins. For example, for SIRT6 substrates, acetylation occurs on histones, but myristoylation/palmitoylation occurs on TNFα. In other words, each substrate is preferentially only regulated by one of the modifications, but not both. Thus, the functional contribution can be assigned to only one of the modifications and one of the sirtuin activities. Second, acetylation and long-chain fatty acylation are likely added by different enzymes. Although lysine myristoyl/palmitoyl transferases are not known yet, currently there is no evidence to suggest that known acetyltransferases could transfer myristoyl/palmitoyl. Thus, myristoyl/palmitoyl groups are likely added via a different system. Finally, it is possible to obtain mutant sirtuins with only one of these activities. For example, recently it was shown that SIRT6 G60A mutant retains demyristoylation activity but essentially abolishes deacetylation activity in vitro and in cells.[135] This mutant is thus very useful for dissecting the functional contributions of different activities of SIRT6. Similar mutants for other sirtuins will also be useful if they can be obtained.

The competition between acetyl substrates and long-chain fatty acyl substrates for the same sirtuin. The existence of two very different substrates also creates a situation where the two substrates may compete for the limiting enzyme. Under this situation, the substrate with a much higher binding affinity (a much lower K_m value) will be preferentially bound and turned over by the enzyme. In other words, the substrate with a much lower K_m value will inhibit the turnover of the substrate with a much higher K_m value.[136] For sirtuins that catalyze both deacetylation and demyristoylation, typically the myristoyl peptide has a much lower K_m value than the acetyl peptide. Thus, if both are present with limiting sirtuin concentrations, the myristoyl peptide will be preferentially bound by the sirtuin and turned over. It is likely that such a competing situation could exist in vivo. For example, one can imagine that a high-fat diet that increases protein lysine fatty acylation may decrease the deacetylation activity of sirtuins, leading to increased protein acetylation.[137]

The possibility to selectively inhibit one of the enzymatic activities. The fact that sirtuins can accept both a short acetyl lysine and a long fatty acyl lysine as substrates also provides opportunities to selectively perturb only one of the activities. For an inhibitor that is competitive with the acetyl/myristoyl substrate, the inhibitor will inhibit the turnover of the substrate with a much higher K_m value (the acetyl peptide) more effectively because the inhibitor can more effectively displace it from the enzyme active site. Thus, it will be easier to inhibit the deacetylase activity than the demyristoylase activity of SIRT1/SIRT2/SIRT3 by an inhibitor that is competitive with the acyl peptide. In other words, it is possible to design inhibitors that only inhibit the deacetylase activity of SIRT1/SIRT2/SIRT3 without significantly affecting their demyristoylase activity.[138]

On the other hand, it is possible to have small molecules that can activate one of the activities. For example, it is reported that in vitro, high concentrations of free fatty acids can activate the deacetylase activity of SIRT6 but at the same time inhibit the demyristoylase activity of SIRT6.[122] Whether this is physiologically relevant is not clear now because cellular free fatty acids typically exist in much lower concentrations. However, this study suggests that it is possible to selectively activate one of the activities of SIRT6.

The situation also suggests that small-molecule inhibitors of sirtuin enzymes may produce effects that are different from knockout or knockdown of the sirtuin.[138] These phenomena are very interesting and provide unique opportunities for developing small-molecule modulators of sirtuins. For example, it may be possible to selectively inhibit one of activity to avoid the toxicity associated with complete knockout of the sirtuin.

4.6 FACTORS THAT AFFECT SIRTUIN SUBSTRATE SPECIFICITY

After summarizing the different activities of sirtuins toward different acyl lysine substrates, it is also helpful to discuss factors that determine sirtuin substrate specificity.[139]

1. *The lysine residue.* All the deacylation activity of sirtuins requires the lysine residue. So far, there has been no report that sirtuins can deacylate other residues, such as palmitoyl cysteine. In the reported sirtuin structures, there is a hydrophobic pocket that accommodates the four methylene groups from the lysine side chain. Down in the pocket, the amide bond is positioned

to react with NAD. Other residues, such as palmitoyl cysteine, are likely too short to reach the NAD cosubstrate and the conserved catalytic histidine residue.

2. *The acyl group.* As discussed above, the acyl group is important for determining the substrate specificity of different sirtuins. This is why SIRT5 preferentially hydrolyzes negatively charged acyl groups over others and SIRT6 has much better activity towards long-chain fatty acyl groups in vitro.

3. *The substrate peptide backbone.* Crystal structures of sirtuins with peptide substrates bound reveal that the substrate peptide and the sirtuin form a small β sheet-like structure with the substrate peptide forming one of the β strands. As with other β sheet structures, the major interactions between the β strands are main chain (backbone) hydrogen-bonding interactions. Typically, the interaction can have 5−8 main chain hydrogen bonds involving the main chain amide groups from the peptide substrate.[94,140]

4. *The substrate peptide side chains (i.e., peptide sequence).* Available crystal structures of sirtuins with peptide substrate bound did not reveal any specific pocket for the peptide side chains. However, the substrate side chains sometimes do form hydrogen bonding with the sirtuin. Typically only 1−3 hydrogen bonds involve the side chains of the substrate. Thus, the peptide sequence contributes only a small portion to the substrate binding. This is why, when a good acyl group is present, the sirtuins typically do not show much sequence selectivity. However, when the acyl group is not optimal, the peptide sequence may become important. This may explain why SIRT6 was found to be a H3 K9- and K56-specific deacetylase,[42,43,112] and SIRT7 was found to be a H3 K18-specific deacetylase,[40] while SIRT4 was found to be able to specifically deacetylate MCD.[22]

5. *The three-dimensional structures of the substrate proteins in vivo.* Although in vitro, the sirtuins do not display much sequence selectivity, in cells, sirtuins typically display more selectivity toward different protein substrates.[139] For example, although both SIRT1 and SIRT2 are present in the nucleus and the cytosol, in many cases they have unique substrates.[139] This is likely because the three-dimensional structures of the substrate proteins determine the binding affinity and specificity to different sirtuins. The high binding affinity and specificity may explain why sirtuins and their corresponding substrates can often be coimmunoprecipitated.

6. *Interacting partners in vivo.* In cells, sirtuins could bind to different molecules, which could influence the substrate specificity. For example, when given a nucleosome substrate, the histone deacetylation activity of SIRT6 becomes much more efficient than when a histone peptide or naked histone protein is used.[141] Acetyl peptides do not bind to SIRT6 well (K_m values are very high), which contributes to the low deacetylation activity of SIRT6 on histone peptides. The binding affinity of nucleosome substrate to SIRT6 is likely much higher, leading to increased deacetylation activity. Thus, the chromatin contact can promote the deacetylation activity of SIRT6.

7. *Subcellular localization.* Subcellular localization of different sirtuins plays an important role in determining the substrate specificity of sirtuins. This is most obvious if one compares SIRT1 and SIRT3, both are deacetylases, but with distinct subcellular localizations.[139] The subcellular localization determines that most SIRT1 substrates are nuclear or cytosolic, while most SIRT3 substrates are mitochondrial.

4.7 SUMMARY AND CONCLUDING REMARKS

In the last few decades, research on sirtuins has significantly expanded the enzymatic activity of sirtuins from deacetylation to a variety of deacylation activities. We now have a much better understanding of how substrates are recognized by sirtuins. The expanded enzymatic activities in turn have helped to discover previously unknown protein post-translational modifications (PTMs), and facilitated the investigation of the biological functions of these PTMs and sirtuins. Although for several PTMs, it is still difficult to differentiate their functions from that of deacetylation because of their structural similarity and shared modifying enzymes or readers, it has been possible to investigate the functions of the PTMs that are more different from acetylation (e.g., succinylation and myristoylation). This, to a large extent, utilizes the ability of sirtuins to regulate the levels of these PTMs in cells. Interestingly, several sirtuins can efficiently hydrolyze both acetyl and long-chain fatty acyl groups. This dual specificity may raise interesting possibilities about substrate competition and selective modulation of one of the activities. It is amazing how much has been learned from this fascinating type of enzyme and there is still a lot of interesting biochemistry to be learned from sirtuins.

ACKNOWLEDGMENTS

I would like to thank all the researchers in the sirtuin field whose work made this manuscript possible and many of which I could not cite. Work in my laboratory on sirtuins and protein lysine acylation is supported by R01GM086703, R01GM098596, R01CA163255, R01DK107868 from NIH, and support from HHMI and Cornell University.

REFERENCES

1. Finkel T, Deng C-X, Mostoslavsky R. Recent progress in the biology and physiology of sirtuins. *Nature* 2009;**460**:587−91.
2. Houtkooper RH, Pirinen E, Auwerx J. Sirtuins as regulators of metabolism and healthspan. *Nat Rev Mol Cell Biol* 2012;**13**:225−38.
3. Haigis MC, Sinclair DA. Mammalian sirtuins: biological insights and disease relevance. *Annu Rev Pathol* 2010;**5**:253−95.
4. Imai S-I, Guarente L. Ten years of NAD-dependent SIR2 family deacetylases: implications for metabolic diseases. *Trends Pharmacol Sci* 2010;**31**:212−20.
5. Cohen HY, Miller C, Bitterman KJ, Wall NR, Hekking B, Kessler B, et al. Calorie restriction promotes mammalian cell survival by inducing the SIRT1 deacetylase. *Science* 2004;**305**:390−2.
6. Kanfi Y, Naiman S, Amir G, Peshti V, Zinman G, Nahum L, et al. The sirtuin SIRT6 regulates lifespan in male mice. *Nature* 2012;**483**:218−21.
7. Lin H, Su X, He B. Protein lysine acylation and cysteine succination by intermediates of energy metabolism. *ACS Chem Biol* 2012;**7**:947−60.
8. Verdin E, Hirschey MD, Finley LWS, Haigis MC. Sirtuin regulation of mitochondria: energy production, apoptosis, and signaling. *Trends Biochem Sci* 2010;**35**:669−75.
9. Picard F, Kurtev M, Chung N, Topark-Ngarm A, Senawong T, Machado de Oliveira R, et al. Sirt1 promotes fat mobilization in white adipocytes by repressing PPAR-γ. *Nature* 2004;**429**:771−6.

10. Rodgers JT, Lerin C, Haas W, Gygi SP, Spiegelman BM, Puigserver P. Nutrient control of glucose homeostasis through a complex of PGC-1α and SIRT1. *Nature* 2005;**434**:113−18.

11. Hallows WC, Yu W, Denu JM. Regulation of glycolytic enzyme phosphoglycerate mutase-1 by Sirt1 protein-mediated deacetylation. *J Biol Chem* 2012;**287**:3850−8.

12. Jiang W, Wang S, Xiao M, Lin Y, Zhou L, Lei Q, et al. Acetylation regulates gluconeogenesis by promoting PEPCK1 degradation via recruiting the UBR5 ubiquitin ligase. *Mol Cell* 2011;**43**:33−44.

13. Ahn B-H, Kim H-S, Song S, Lee IH, Liu J, Vassilopoulos A, et al. A role for the mitochondrial deacetylase Sirt3 in regulating energy homeostasis. *Proc Natl Acad Sci U S A* 2008;**105**:14447−52.

14. Hallows WC, Lee S, Denu JM. Sirtuins deacetylate and activate mammalian acetyl-CoA synthetases. *Proc Natl Acad Sci U S A* 2006;**103**:10230−5.

15. Schwer B, Bunkenborg J, Verdin RO, Andersen JS, Verdin E. Reversible lysine acetylation controls the activity of the mitochondrial enzyme acetyl-CoA synthetase 2. *Proc Natl Acad Sci U S A* 2006;**103**:10224−9.

16. Hirschey MD, Shimazu T, Capra JA, Pollard KS, Verdin E. SIRT1 and SIRT3 deacetylate homologous substrates: AceCS1,2 and HMGCS1,2. *Aging* 2011;**3**:635−42.

17. Kim H-S, Patel K, Muldoon-Jacobs K, Bisht KS, Aykin-Burns N, Pennington JD, et al. SIRT3 is a mitochondria-localized tumor suppressor required for maintenance of mitochondrial integrity and metabolism during stress. *Cancer Cell* 2010;**17**:41−52.

18. Hirschey MD, Shimazu T, Goetzman E, Jing E, Schwer B, Lombard DB, et al. SIRT3 regulates mitochondrial fatty-acid oxidation by reversible enzyme deacetylation. *Nature* 2010;**464**:121−5.

19. Shimazu T, Hirschey MD, Hua L, Dittenhafer-Reed KE, Schwer B, Lombard DB, et al. SIRT3 deacetylates mitochondrial 3-hydroxy-3-methylglutaryl CoA synthase 2 and regulates ketone body production. *Cell Metab* 2010;**12**:654−61.

20. Someya S, Yu W, Hallows WC, Xu J, Vann JM, Leeuwenburgh C, et al. Sirt3 mediates reduction of oxidative damage and prevention of age-related hearing loss under caloric restriction. *Cell* 2010;**143**:802−12.

21. Haigis MC, Mostoslavsky R, Haigis KM, Fahie K, Christodoulou DC, Murphy AJ, et al. SIRT4 Inhibits glutamate dehydrogenase and opposes the effects of calorie restriction in pancreatic β cells. *Cell* 2006;**126**:941−54.

22. Laurent G, German NJ, Saha AK, de Boer VCJ, Davies M, Koves TR, et al. SIRT4 coordinates the balance between lipid synthesis and catabolism by repressing malonyl CoA decarboxylase. *Mol Cell* 2013;**50**:686−98.

23. Nakagawa T, Lomb DJ, Haigis MC, Guarente L. SIRT5 deacetylates carbamoyl phosphate synthetase 1 and regulates the urea cycle. *Cell* 2009;**137**:560−70.

24. Zhong L, D'Urso A, Toiber D, Sebastian C, Henry RE, Vadysirisack DD, et al. The histone deacetylase Sirt6 regulates glucose homeostasis via Hif1α. *Cell* 2010;**140**:280−93.

25. Xiao C, Kim H-S, Lahusen T, Wang R-H, Xu X, Gavrilova O, et al. SIRT6 deficiency results in severe hypoglycemia by enhancing both basal and insulin-stimulated glucose uptake in mice. *J Biol Chem* 2010;**285**:36776−84.

26. Kim H-S, Xiao C, Wang R-H, Lahusen T, Xu X, Vassilopoulos A, et al. Hepatic-specific disruption of SIRT6 in mice results in fatty liver formation due to enhanced glycolysis and triglyceride synthesis. *Cell Metab* 2010;**12**:224−36.

27. Park J, Chen Y, Tishkoff DX, Peng C, Tan M, Dai L, et al. SIRT5-mediated lysine desuccinylation impacts diverse metabolic pathways. *Mol Cell* 2013;**50**:919−30.

28. Brunet A, Sweeney LB, Sturgill JF, Chua KF, Greer PL, Lin Y, et al. Stress-dependent regulation of FOXO transcription factors by the SIRT1 deacetylase. *Science* 2004;**303**:2011−15.

29. Motta MC, Divecha N, Lemieux M, Kamel C, Chen D, Gu W, et al. Mammalian SIRT1 represses forkhead transcription factors. *Cell* 2004;**116**:551−63.

30. Qiao L, Shao J. SIRT1 regulates adiponectin gene expression through Foxo1-C/Enhancer-binding -protein α transcriptional complex. *J Biol Chem* 2006;**281**:39915−24.

31. Vaquero A, Scher M, Lee D, Erdjument-Bromage H, Tempst P, Reinberg D. Human SirT1 interacts with histone H1 and promotes formation of facultative heterochromatin. *Mol Cell* 2004;**16**:93−105.

32. Xie M, Liu M, He C-S. SIRT1 regulates endothelial Notch signaling in lung cancer. *PLoS One* 2012;**7**: e45331.

33. Vaziri H, Dessain SK, Eaton EN, Imai S-I, Frye RA, Pandita TK, et al. hSIR2(SIRT1) functions as an NAD-dependent p53 deacetylase. *Cell* 2001;**107**:149−59.

34. van der Horst A, Tertoolen LGJ, de Vries-Smits LMM, Frye RA, Medema RH, Burgering BMT. FOXO4 is acetylated upon peroxide stress and deacetylated by the longevity protein hSir2 SIRT1. *J Biol Chem* 2004;**279**:28873−9.

35. Yeung F, Hoberg JE, Ramsey CS, Keller MD, Jones DR, Frye RA, et al. Modulation of NF-[kappa]B-dependent transcription and cell survival by the SIRT1 deacetylase. *EMBO J* 2004;**23**:2369−80.

36. Yang Y, Hou H, Haller EM, Nicosia SV, Bai W. Suppression of FOXO1 activity by FHL2 through SIRT1-mediated deacetylation. *EMBO J* 2005;**24**:1021−32.

37. Wang C, Chen L, Hou X, Li Z, Kabra N, Ma Y, et al. Interactions between E2F1 and SirT1 regulate apoptotic response to DNA damage. *Nat Cell Biol* 2006;**8**:1025−31.

38. Kawahara TLA, Michishita E, Adler AS, Damian M, Berber E, Lin M, et al. SIRT6 links histone H3 lysine 9 deacetylation to NF-κB-dependent gene expression and organismal life span. *Cell* 2009;**136**:62−74.

39. Ford E, Voit R, Liszt G, Magin C, Grummt I, Guarente L. Mammalian Sir2 homolog SIRT7 is an activator of RNA polymerase I transcription. *Genes Dev* 2006;**20**:1075−80.

40. Barber MF, Michishita-Kioi E, Xi Y, Tasselli L, Kioi M, Moqtaderi Z, et al. SIRT7 links H3K18 deacetylation to maintenance of oncogenic transformation. *Nature* 2012;**487**:114−18.

41. Mostoslavsky R, Chua KF, Lombard DB, Pang WW, Fischer MR, Gellon L, et al. Genomic instability and aging-like phenotype in the absence of mammalian SIRT6. *Cell* 2006;**124**:315−29.

42. Yang B, Zwaans BMM, Eckersdorff M, Lombard DB. The sirtuin SIRT6 deacetylates H3 K56Ac in vivo to promote genomic stability. *Cell Cycle* 2009;**8**:2662−3.

43. Michishita E, McCord RA, Berber E, Kioi M, Padilla-Nash H, Damian M, et al. SIRT6 is a histone H3 lysine 9 deacetylase that modulates telomeric chromatin. *Nature* 2008;**452**:492−6.

44. Kaidi A, Weinert BT, Choudhary C, Jackson SP. Human SIRT6 promotes DNA end resection through CtIP deacetylation. *Science* 2010;**329**:1348−53.

45. Mao Z, Hine C, Tian X, Van Meter M, Au M, Vaidya A, et al. SIRT6 promotes DNA repair under stress by activating PARP1. *Science* 2011;**332**:1443−6.

46. Donmez G, Wang D, Cohen DE, Guarente L. SIRT1 suppresses β-amyloid production by activating the α-secretase gene ADAM10. *Cell* 2010;**142**:320−32.

47. Libert S, Pointer K, Bell EL, Das A, Cohen DE, Asara JM, et al. SIRT1 activates MAO-A in the brain to mediate anxiety and exploratory drive. *Cell* 2011;**147**:1459−72.

48. Outeiro TF, Kontopoulos E, Altmann SM, Kufareva I, Strathearn KE, Amore AM, et al. Sirtuin 2 inhibitors rescue α-synuclein-mediated toxicity in models of parkinson's disease. *Science* 2007;**317**:516−19.

49. Chopra V, Quinti L, Kim J, Vollor L, Narayanan KL, Edgerly C, et al. The Sirtuin 2 inhibitor AK-7 is neuroprotective in huntington's disease mouse models. *Cell Rep* 2012;**2**:1492−7.

50. Sanchez-Fidalgo S, Villegas I, Sanchez-Hidalgo M, Alarcon de la Lastra C. Sirtuin modulators: mechanisms and potential clinical implications. *Curr Med Chem* 2012;**19**:2414−41.

51. Villalba JM, Alcaín FJ. Sirtuin activators and inhibitors. *BioFactors* 2012;**38**:349−59.

52. Burnett C, Valentini S, Cabreiro F, Goss M, Somogyvari M, Piper MD, et al. Absence of effects of Sir2 overexpression on lifespan in *C. elegans* and *Drosophila*. *Nature* 2011;**477**:482−5.

53. Milne JC, Lambert PD, Schenk S, Carney DP, Smith JJ, Gagne DJ, et al. Small molecule activators of SIRT1 as therapeutics for the treatment of type 2 diabetes. *Nature* 2007;**450**:712−16.

54. Borra MT, Smith BC, Denu JM. Mechanism of human SIRT1 activation by resveratrol. *J Biol Chem* 2005;**280**:17187−95.

55. Howitz KT, Bitterman KJ, Cohen HY, Lamming DW, Lavu S, Wood JG, et al. Small molecule activators of sirtuins extend *Saccharomyces cerevisiae* lifespan. *Nature* 2003;**425**:191−6.

56. Baur JA, Pearson KJ, Price NL, Jamieson HA, Lerin C, Kalra A, et al. Resveratrol improves health and survival of mice on a high-calorie diet. *Nature* 2006;**444**:337−42.

57. Pacholec M, Bleasdale JE, Chrunyk B, Cunningham D, Flynn D, Garofalo RS, et al. SRT1720, SRT2183, SRT1460, and resveratrol are not direct activators of SIRT1. *J Biol Chem* 2010;**285**:8340−51.

58. Dai H, Kustigian L, Carney D, Case A, Considine T, Hubbard BP, et al. SIRT1 activation by small molecules: kinetic and biophysical evidence for direct interaction of enzyme and activator. *J Biol Chem* 2010;**285**:32695−703.

59. Hubbard BP, Gomes AP, Dai H, Li J, Case AW, Considine T, et al. Evidence for a common mechanism of SIRT1 regulation by allosteric activators. *Science* 2013;**339**:1216−19.

60. Ivy JM, Klar AJ, Hicks JB. Cloning and characterization of four SIR genes of *Saccharomyces cerevisiae*. *Mol Cell Biol* 1986;**6**:688−702.

61. Kaeberlein M, McVey M, Guarente L. The SIR2/3/4 complex and SIR2 alone promote longevity in *Saccharomyces cerevisiae* by two different mechanisms. *Genes Dev* 1999;**13**:2570−80.

62. Lin S-J, Defossez P-A, Guarente L. Requirement of NAD and SIR2 for life-span extension by calorie restriction in *Saccharomyces cerevisiae*. *Science* 2000;**289**:2126−8.

63. Anderson RM, Bitterman KJ, Wood JG, Medvedik O, Sinclair DA. Nicotinamide and PNC1 govern lifespan extension by calorie restriction in *Saccharomyces cerevisiae*. *Nature* 2003;**423**:181−5.

64. Imai S-I, Armstrong CM, Kaeberlein M, Guarente L. Transcriptional silencing and longevity protein Sir2 is an NAD-dependent histone deacetylase. *Nature* 2000;**403**:795−800.

65. Sauve AA, Celic I, Avalos J, Deng H, Boeke JD, Schramm VL. Chemistry of gene silencing: the mechanism of NAD$^+$-dependent deacetylation reactions. *Biochemistry* 2001;**40**:15456−63.

66. Tanner KG, Landry J, Sternglanz R, Denu JM. Silent information regulator 2 family of NAD- dependent histone/protein deacetylases generates a unique product, 1-*O*-acetyl-ADP-ribose. *Proc Natl Acad Sci U S A* 2000;**97**:14178−82.

67. Jackson MD, Denu JM. Structural identification of 2′- and 3′-*O*-acetyl-ADP-ribose as novel metabolites derived from the Sir2 family of beta -NAD$^+$-dependent histone/protein deacetylases. *J Biol Chem* 2002;**277**:18535−44.

68. Zhao K, Chai X, Marmorstein R. Structure and substrate binding properties of cobB, a Sir2 homolog protein deacetylase from *Escherichia coli*. *J Mol Biol* 2004;**337**:731−41.

69. North BJ, Marshall BL, Borra MT, Denu JM, Verdin E. The human Sir2 ortholog, SIRT2, is an NAD$^+$-dependent tubulin deacetylase. *Mol Cell* 2003;**11**:437−44.

70. Schwer B, North BJ, Frye RA, Ott M, Verdin E. The human silent information regulator (Sir)2 homologue hSIRT3 is a mitochondrial nicotinamide adenine dinucleotide−dependent deacetylase. *J Cell Biol* 2002;**158**:647−57.

71. Sauve AA, Schramm VL. Sir2 regulation by nicotinamide results from switching between base exchange and deacetylation chemistry. *Biochemistry* 2003;**42**:9249−56.

72. Jackson MD, Schmidt MT, Oppenheimer NJ, Denu JM. Mechanism of nicotinamide inhibition and transglycosidation by Sir2 histone/protein deacetylases. *J Biol Chem* 2003;**278**:50985−98.

73. Smith BC, Denu JM. Mechanism-based inhibition of Sir2 deacetylases by thioacetyl-lysine peptide. *Biochemistry* 2007;**46**:14478−86.
74. Hawse WF, Hoff KG, Fatkins DG, Daines A, Zubkova OV, Schramm VL, et al. Structural insights into intermediate steps in the Sir2 deacetylation reaction. *Structure (London, England: 1993)* 2008;**16**:1368−77.
75. Hoff KG, Avalos JL, Sens K, Wolberger C. Insights into the sirtuin mechanism from ternary complexes containing NAD$^+$ and acetylated peptide. *Structure (London, England: 1993)* 2006;**14**:1231−40.
76. Zhao K, Chai X, Marmorstein R. Structure of the yeast Hst2 protein deacetylase in ternary complex with 2′-O-acetyl ADP ribose and histone peptide. *Structure (London, England: 1993)* 2003;**11**:1403−11.
77. Sauve AA, Wolberger C, Schramm VL, Boeke JD. The biochemistry of sirtuins. *Annu Rev Biochem* 2006;**75**:435−65.
78. Wang Y, Fung YM, Zhang W, He B, Chung MW, Jin J, et al. Deacylation mechanism by SIRT2 revealed in the 1′-SH-2′-O-myristoyl intermediate structure. *Cell Chem Biol* 2017;**24**:339−45.
79. Michishita E, Park JY, Burneskis JM, Barrett JC, Horikawa I. Evolutionarily conserved and nonconserved cellular localizations and functions of human SIRT proteins. *Mol Biol Cell* 2005;**16**:4623−35.
80. Du J, Jiang H, Lin H. Investigating the ADP-ribosyltransferase activity of sirtuins with NAD analogs and ^{32}P-NAD. *Biochemistry* 2009;**48**:2878−90.
81. Liszt G, Ford E, Kurtev M, Guarente L. Mouse Sir2 homolog SIRT6 is a nuclear ADP-ribosyltransferase. *J Biol Chem* 2005;**280**:21313−20.
82. Chen Y, Sprung R, Tang Y, Ball H, Sangras B, Kim SC, et al. Lysine propionylation and butyrylation are novel post-translational modifications in histones. *Mol Cell Proteomics* 2007;**6**:812−19.
83. Tan M, Luo H, Lee S, Jin F, Yang JS, Montellier E, et al. Identification of 67 histone marks and histone lysine crotonylation as a new type of histone modification. *Cell* 2011;**146**:1016−28.
84. Smith BC, Denu JM. Acetyl-lysine analog peptides as mechanistic probes of protein deacetylases. *J Biol Chem* 2007;**282**:37256−65.
85. Bao X, Wang Y, Li X, Li XM, Liu Z, Yang T, et al. Identification of 'erasers' for lysine crotonylated histone marks using a chemical proteomics approach. *Elife* 2014;**3**. Available from: http://dx.doi.org/10.7554/eLife.02999.
86. Li Y, Sabari BR, Panchenko T, Wen H, Zhao D, Guan H, et al. Molecular coupling of histone crotonylation and active transcription by AF9 YEATS domain. *Mol Cell* 2016;**62**:181−93.
87. Kaczmarska Z, Ortega E, Goudarzi A, Huang H, Kim S, Marquez JA, et al. Structure of p300 in complex with acyl-CoA variants. *Nat Chem Biol* 2017;**13**:21−9.
88. Vollmuth F, Geyer M. Interaction of propionylated and butyrylated histone H3 lysine marks with Brd4 bromodomains. *Angew Chem Int Ed Engl* 2010;**49**:6768−72.
89. Flynn EM, Huang OW, Poy F, Oppikofer M, Bellon SF, Tang Y, et al. A subset of human bromodomains recognizes butyryllysine and crotonyllysine histone peptide modifications. *Structure (London, England : 1993)* 2015;**23**:1801−14.
90. Zhang Q, Zeng L, Zhao C, Ju Y, Konuma T, Zhou MM. Structural insights into histone crotonyl-lysine recognition by the AF9 YEATS domain. *Structure (London, England : 1993)* 2016;**24**:1606−12.
91. Andrews FH, Shinsky SA, Shanle EK, Bridgers JB, Gest A, Tsun IK, et al. The Taf14 YEATS domain is a reader of histone crotonylation. *Nat Chem Biol* 2016;**12**:396−8.
92. Zhao D, Guan H, Zhao S, Mi W, Wen H, Li Y, et al. YEATS2 is a selective histone crotonylation reader. *Cell Res* 2016;**26**:629−32.
93. Xiong X, Panchenko T, Yang S, Zhao S, Yan P, Zhang W, et al. Selective recognition of histone crotonylation by double PHD fingers of MOZ and DPF2. *Nat Chem Biol* 2016;**12**:1111−18.
94. Du J, Zhou Y, Su X, Yu J, Khan S, Jiang H, et al. Sirt5 is an NAD-dependent protein lysine demalonylase and desuccinylase. *Science* 2011;**334**:806−9.

95. Schuetz A, Min J, Antoshenko T, Wang C-L, Allali-Hassani A, Dong A, et al. Structural basis of inhibition of the human NAD$^+$-dependent deacetylase SIRT5 by suramin. *Structure (London, England: 1993)* 2007;**15**:377—89.
96. He B, Du J, Lin H. Thiosuccinyl peptides as Sirt5-specific inhibitors. *J Am Chem Soc* 2012;**134**:1922—5.
97. Zhang Z, Tan M, Xie Z, Dai L, Chen Y, Zhao Y. Identification of lysine succinylation as a new post-translational modification. *Nat Chem Biol* 2011;**7**:58—63.
98. Peng C, Lu Z, Xie Z, Cheng Z, Chen Y, Tan M, et al. The first identification of lysine malonylation substrates and its regulatory enzyme. *Mol Cell Proteomics* 2011;**10**. Available from: https://doi.org/10.1074/mcp.M1111.012658.
99. Xie Z, Dai J, Dai L, Tan M, Cheng Z, Wu Y, et al. Lysine succinylation and lysine malonylation in histones. *Mol Cell Proteomics* 2012;**11**:100—7.
100. Tan M, Peng C, Anderson KA, Chhoy P, Xie Z, Dai L, et al. Lysine glutarylation is a protein posttranslational modification regulated by SIRT5. *Cell Metab* 2014;**19**:605—17.
101. Rardin MJ, He W, Nishida Y, Newman JC, Carrico C, Danielson SR, et al. SIRT5 regulates the mitochondrial lysine succinylome and metabolic networks. *Cell Metab* 2013;**18**:920—33.
102. Sadhukhan S, Liu X, Ryu D, Nelson OD, Stupinski JA, Li Z, et al. Metabolomics-assisted proteomics identifies succinylation and SIRT5 as important regulators of cardiac function. *Proc Natl Acad Sci U S A* 2016;**113**:4320—5.
103. Wang F, Wang K, Xu W, Zhao S, Ye D, Wang Y, et al. SIRT5 desuccinylates and activates pyruvate kinase M2 to block macrophage IL-1beta production and to prevent DSS-induced colitis in mice. *Cell Rep* 2017;**19**:2331—44.
104. Xiangyun Y, Xiaomin N, Linping G, Yunhua X, Ziming L, Yongfeng Y, et al. Desuccinylation of pyruvate kinase M2 by SIRT5 contributes to antioxidant response and tumor growth. *Oncotarget* 2017;**8**:6984—93.
105. Zhou L, Wang F, Sun R, Chen X, Zhang M, Xu Q, et al. SIRT5 promotes IDH2 desuccinylation and G6PD deglutarylation to enhance cellular antioxidant defense. *EMBO Rep* 2016;**17**:811—22.
106. Nishida Y, Rardin MJ, Carrico C, He W, Sahu AK, Gut P, et al. SIRT5 regulates both cytosolic and mitochondrial protein malonylation with glycolysis as a major target. *Mol Cell* 2015;**59**:321—32.
107. Colak G, Pougovkina O, Dai L, Tan M, Te Brinke H, Huang H, et al. Proteomic and biochemical studies of lysine malonylation suggest its malonic aciduria-associated regulatory role in mitochondrial function and fatty acid oxidation. *Mol Cell Proteomics* 2015;**14**:3056—71.
108. Frye RA. Phylogenetic classification of prokaryotic and eukaryotic Sir2-like proteins. *Biochem Biophys Res Commun* 2000;**273**:793—8.
109. Merrick CJ, Duraisingh MT. *Plasmodium falciparum* Sir2: an unusual sirtuin with dual histone deacetylase and ADP-ribosyltransferase activity. *Eukaryot Cell* 2007;**6**:2081—91.
110. French JB, Cen Y, Sauve AA. *Plasmodium falciparum* Sir2 is an NAD$^+$-dependent deacetylase and an acetyllysine-dependent and acetyllysine-independent NAD$^+$ glycohydrolase. *Biochemistry* 2008;**47**:10227—39.
111. Zhu AY, Zhou Y, Khan S, Deitsch KW, Hao Q, Lin H. *Plasmodium falciparum* Sir2A preferentially hydrolyzes medium and long chain fatty acyl lysine. *ACS Chem Biol* 2012;**7**:155—9.
112. Michishita E, McCord RA, Boxer LD, Barber MF, Hong T, Gozani O, et al. Cell cycle-dependent deacetylation of telomeric histone H3 lysine K56 by human SIRT6. *Cell Cycle* 2009;**8**:2664—6.
113. Jiang H, Khan S, Wang Y, Charron G, He B, Sebastian C, et al. SIRT6 regulates TNF-α secretion through hydrolysis of long-chain fatty acyl lysine. *Nature* 2013;**496**:110—13.
114. Chu W-M. Tumor necrosis factor. *Cancer Lett* 2013;**328**:222—5.

115. Stevenson FT, Bursten SL, Locksley RM, Lovett DH. Myristyl acylation of the tumor necrosis factor alpha precursor on specific lysine residues. *J Exp Med* 1992;**176**:1053−62.

116. Stevenson FT, Bursten SL, Fanton C, Locksley RM, Lovett DH. The 31-kDa precursor of interleukin 1 alpha is myristoylated on specific lysines within the 16-kDa N-terminal propiece. *Proc Natl Acad Sci U S A* 1993;**90**:7245−9.

117. Van Gool F, Galli M, Gueydan C, Kruys V, Prevot P-P, Bedalov A, et al. Intracellular NAD levels regulate tumor necrosis factor protein synthesis in a sirtuin-dependent manner. *Nat Med* 2009;**15**:206−10.

118. Charron G, Zhang MM, Yount JS, Wilson J, Raghavan AS, Shamir E, et al. Robust fluorescent detection of protein fatty-acylation with chemical reporters. *J Am Chem Soc* 2009;**131**:4967−75.

119. Jiang H, Zhang X, Lin H. Lysine fatty acylation promotes lysosomal targeting of TNF-alpha. *Sci Rep* 2016;**6**:24371.

120. Zhang X, Spiegelman NA, Nelson OD, Jing H, Lin H. SIRT6 regulates Ras-related protein R-Ras2 by lysine defatty-acylation. *Elife* 2017;**6** pii: e25158.

121. Sebastián C, Zwaans BMM, Silberman DM, Gymrek M, Goren A, Zhong L, et al. The histone deacetylase SIRT6 is a tumor suppressor that controls cancer metabolism. *Cell* 2012;**151**:1185−99.

122. Feldman JL, Baeza J, Denu JM. Activation of the protein deacetylase SIRT6 by long-chain fatty acids and widespread deacylation by mammalian sirtuins. *J Biol Chem* 2013;**288**:31350−6.

123. He B, Hu J, Zhang X, Lin H. Thiomyristoyl peptides as cell-permeable Sirt6 inhibitors. *Org Biomol Chem* 2014;**12**:7498−502.

124. Teng Y-B, Jing H, Aramsangtienchai P, He B, Khan S, Hu J, et al. Efficient demyristoylase activity of SIRT2 revealed by kinetic and structural studies. *Sci Rep* 2015;**5**. Available from: https://doi.org/10.1038/srep08529.

125. Mathias RA, Greco TM, Oberstein A, Budayeva HG, Chakrabarti R, Rowland EA, et al. Sirtuin 4 is a lipoamidase regulating pyruvate dehydrogenase complex activity. *Cell* 2014;**159**:1615−25.

126. Anderson KA, Huynh FK, Fisher-Wellman K, Stuart JD, Peterson BS, Douros JD, et al. SIRT4 is a lysine deacylase that controls leucine metabolism and insulin secretion. *Cell Metab* 2017;**25** 838−855. e815.

127. Tanny JC, Dowd GJ, Huang J, Hilz H, Moazed D. An enzymatic activity in the yeast Sir2 protein that is essential for gene silencing. *Cell* 1999;**99**:735−45.

128. Hassa PO, Haenni SS, Elser M, Hottiger MO. Nuclear ADP-ribosylation reactions in mammalian cells: where are we today and where are we going? *Microbiol Mol Biol Rev* 2006;**70**:789−829.

129. Lin H. Nicotinamide adenine dinucleotide: beyond a redox coenzyme. *Org Biomol Chem* 2007;**5**:2541−54.

130. Kowieski TM, Lee S, Denu JM. Acetylation-dependent ADP-ribosylation by *Trypanosoma brucei* Sir2. *J Biol Chem* 2008;**283**:5317−26.

131. Starai VJ, Celic I, Cole RN, Boeke JD, Escalante-Semerena JC. Sir2-dependent activation of acetyl-CoA synthetase by deacetylation of active lysine. *Science* 2002;**298**:2390−2.

132. Frye RA. Characterization of five human cDNAs with homology to the yeast Sir2 gene: Sir2-like proteins (sirtuins) metabolize NAD and may have protein ADP-ribosyltransferase activity. *Biochem Biophys Res Commun* 1999;**260**:273−9.

133. Tsang AW, Escalante-Semerena JC. CobB, a new member of the SIR2 family of eucaryotic regulatory proteins, is required to compensate for the lack of nicotinate mononucleotide:5,6-dimethylbenzimidazole phosphoribosyltransferase activity in cobT mutants during cobalamin biosynthesis in Salmonella typhimurium LT2. *J Biol Chem* 1998;**273**:31788−94.

134. Rack JGM, Morra R, Barkauskaite E, Kraehenbuehl R, Ariza A, Qu Y, et al. Identification of a class of protein ADP-ribosylating sirtuins in microbial pathogens. *Mol Cell* 2015;**59**:309−20.

135. Zhang X, Khan S, Jiang H, Antonyak MA, Chen X, Spiegelman NA, et al. Identifying the functional contribution of the defatty-acylase activity of SIRT6. *Nat Chem Biol* 2016;**12**:614−20.
136. Lin H, Fischbach MA, Liu DR, Walsh CT. In vitro characterization of salmochelin and enterobactin trilactone hydrolases IroD, IroE, and Fes. *J Am Chem Soc* 2005;**127**:11075−84.
137. Kendrick AA, Choudhury M, Rahman SM, McCurdy CE, Friederich M, Van Hove JL, et al. Fatty liver is associated with reduced SIRT3 activity and mitochondrial protein hyperacetylation. *Biochem J* 2011;**433**:505−14.
138. Jing H, Hu J, He B, Yashira LNA, Stupinski J, Weiser K, et al. A SIRT2-selective inhibitor promotes c-Myc oncoprotein degradation and exhibits broad anticancer activity. *Cancer Cell* 2016;**29**:297−310.
139. Bheda P, Jing H, Wolberger C, Lin H. The substrate specificity of sirtuins. *Annu Rev Biochem* 2016;**85**:405−29.
140. Cosgrove MS, Bever K, Avalos JL, Muhammad S, Zhang X, Wolberger C. The structural basis of sirtuin substrate affinity. *Biochemistry* 2006;**45**:7511−21.
141. Gil R, Barth S, Kanfi Y, Cohen HY. SIRT6 exhibits nucleosome-dependent deacetylase activity. *Nucleic Acids Res* 2013;**41**:8537−45.

STRUCTURAL AND MECHANISTIC INSIGHTS IN SIRTUIN CATALYSIS AND PHARMACOLOGICAL MODULATION

5

Sébastien Moniot, Weijie You and Clemens Steegborn

University of Bayreuth, Bayreuth, Germany

5.1 SIRTUIN FUNCTION, STRUCTURE, AND CATALYTIC MECHANISM

Sirtuins are an evolutionary conserved family of NAD^+-dependent protein lysine deacylase enzymes. They regulate the activity, localization, or stability of their target proteins and thereby control, e.g., metabolic pathways and stress responses.[1] Sirtuins have been implicated in aging processes and aging-related dysfunctions, and the seven human isoforms, Sirt1−7, are considered therapeutic targets for diseases such as cancer, metabolic syndrome, and neurodegenerative disorders.[2,3]

Sirt1−7 differ in their subcellular localization and in their acyl specificities. Sirt1, 6, 7 are mainly localized in the nucleus, Sirt3−5 in mitochondria, and Sirt2 in the cytosol.[1] Sirt1−3 are preferentially deacetylases, Sirt5 prefers dicarboxylate modifications (succinylation, glutarylation) and Sirt6 long-chain acyls (myristoylation), and efficient deacylation substrates remain to be identified for Sirt4 and Sirt7.[4] Despite their differing acyl selectivities, sirtuins share a common molecular architecture and catalytic mechanism.[4−6] Their conserved catalytic core comprises ~ 275 amino acids and is flanked by N- and C-terminal extensions of various sequences and lengths. The extensions contribute to proper cellular localization, regulation of sirtuin activity, and/or interactions with regulatory proteins.[1,7] High-resolution structures of several sirtuins, including human Sirt1, 2, 3, 5, and 6, in apo- or in (co)substrate/inhibitor-bound forms have been reported and have contributed to reveal their catalytic mechanism. The sirtuin catalytic core has a bilobal fold composed of a large α/β Rossmann-fold domain typical for nucleotide-binding proteins and of a smaller, more structurally diverse zinc-binding domain, both connected by several loops (Fig. 5.1A).[5] In contrast to other protein deacetylase families, the zinc atom has a structural role and is not involved in catalysis. The catalytic site is located in a cleft between the two domains, which undergo large conformational changes upon substrate and cosubstrate binding.[5,8] The acylated substrate lysine inserts into a narrow active site channel, and the substrate main chain inserts into a mixed β-sheet composed of strands from both, the zinc-binding and the Rossmann-fold domain. The cosubstrate NAD^+ binds to subpockets A (adenosine moiety), B (diphosphoribose), and C (nicotinamide), positioning the nicotinamide ribose next to the substrate acyl (Fig. 5.1A). The first catalytic step is formation of a $1'$-O-alkylimidate intermediate between NAD^+ and the acyl group, which is

Introductory Review on Sirtuins in Biology, Aging, and Disease. DOI: https://doi.org/10.1016/B978-0-12-813499-3.00005-8
© 2018 Elsevier Inc. All rights reserved.

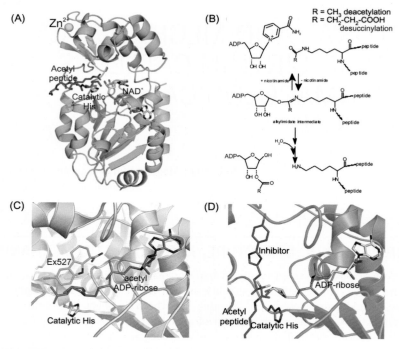

FIGURE 5.1 Sirtuin structure, catalysis, and inhibition.

(A) Overall structure of Sirt3 in complex with acetyl peptide substrate (*magenta*) and a nonhydrolyzable NAD^+ analog (PDB entry 4FTV). The conserved catalytic His in the active site is indicated. (B) Mechanism of sirtuin-dependent deacylations. (C) Crystal structure of Sirt3 in complex with the inhibitor Ex527 (*green*) and the coproduct acetyl-ADP-ribose (*orange*; PDB entry 4BVH). (D) Crystal structure of Sirt2 in complex with a potent oxadiazole-based inhibitor (*blue*) and ADP-ribose (*gray*; PDB entry 5MAR).

accompanied by nicotinamide release (Fig. 5.1B).[4,6] A conserved histidine residue activates the ribose 2'-OH to attack the imidate carbon to form a bicyclic intermediate, followed by rearrangement into a 2'-O-alkylimidate intermediate, which is finally hydrolyzed to produce deacylated polypeptide and 2'-O-acetyl-ADP-ribose.[4,6] During catalysis, the cosubstrate binding loop gets ordered upon (co)substrate binding and changes to a more closed conformation upon acyl transfer, with a suggested role of this conformational change in expelling the first reaction product, nicotinamide.[5,8] This unique mechanism couples each deacylation reaction to the consumption of one NAD^+ molecule, rendering sirtuins metabolic sensors.[6]

5.2 SIRTUIN INHIBITORS

Due to their (patho)physiological functions sirtuins are considered attractive therapeutic targets, which has stimulated considerable efforts to develop small-molecule sirtuin modulators. Although various sirtuin inhibitors have been described, only a few of them show high potency, isoform selectivity, and favorable pharmacological properties, and their binding sites and mechanisms are often unknown.[3,7,9] Most studies were focusing on Sirt1 and 2, and inhibitors were only tested on these or few additional isoforms. The ongoing advances in the characterization of acyl substrates for Sirt4−7, however, will enable full isoform selectivity tests and the discovery of novel small-molecule modulators for these isoforms.[4] Comprehensive descriptions of reported compound classes are available,[3,9] and we will focus here on selected inhibitors illustrating several approaches for sirtuin inhibition.

Many sirtuin inhibitors target the binding site for polypeptide and/or nucleotide (co)substrate in a competitive inhibition strategy. For example, the huge diarylurea suramine, which inhibits most sirtuins with low micromolar potency, blocks both, substrate- and NAD^+-binding site.[10] Suramin also affects other cellular targets and has several unfavorable pharmacological properties, however, and it appears difficult to improve.[11] Smaller molecules targeting exclusively the NAD^+-binding site are also likely to lack isoform selectivity due to the high degree of conservation of this site in sirtuins and possibly even other NAD^+-dependent enzymes. The available structural data indeed indicate that the most promising compounds either target parts of the NAD^+ site plus surrounding, isoform-specific areas and/or the polypeptide and acyl-Lys binding channels, which also feature differences between isoforms.

A straightforward approach to block the acyl-Lys peptide site is the use of modified peptides. Replacing the acyl oxygen with sulfur results in thioacyl-peptides, which form a relatively stable thioalkylimidate intermediate and thereby trap an inactive sirtuin/intermediate complex. This approach has yielded several low-μM inhibitors, some of them with substantial isoform selectivity.[12,13] Since sirtuins exhibit only weak sequence preferences around the modified Lys,[14] isoform selectivity is mostly achieved by the acyl modification.[13,15] The unique succinyl substrate specificity of Sirt5, e.g., was exploited for potent and highly isoform-specific inhibition using a 3-methyl-3-phenyl-succinyl-CPS1 peptide (IC$_{50}$ 4.3 μM).[15] However, peptides are often incapable of crossing biological membranes and rapidly degraded in vivo. Although the replacement of N- and C-terminal extensions to a thioacyl-Lys by anilino and benzyloxycarbonyl groups, respectively, produced encouraging results on compound stability and potency,[13] further development will be required to obtain acyl-peptide/acyl-Lys-based inhibitors suitable for in vivo applications.

During the catalytic cycle, NAD^+ binds with its nicotinamide group in the so-called C-pocket, which distorts the glycosidic bond and thus facilitates intermediate formation and nicotinamide release. At this step, nicotinamide can rebind and reform NAD^+ in the reverse reaction, causing pan sirtuin inhibition with IC$_{50}$ values ranging from 50 to 200 μM, well within the range of physiological nicotinamide concentrations.[16,17] Sirtuin isoform sensitivities to nicotinamide can also depend on the catalyzed reaction. Sirt5 desuccinylation activity, e.g., is affected by nicotinamide (IC$_{50}$ of 21 μM), while its deacetylation activity is insensitive due to a Sirt5-specific Arg residue responsible for substrate recognition.[17] Such molecular differences neighboring the C-pocket should enable the development of more potent and selective compounds. A starting point for such compounds could be Ex527, which targets the C-pocket and is used as a potent Sirt1 inhibitor (IC$_{50}$ ~0.1 μM) in many physiological studies, although it also inhibits Sirt2 and Sirt3 (IC$_{50}$ of 19.6 and 22.4 μM, respectively).[18,19] Crystal structures revealed that Ex527 binds to the C-pocket and a neighboring hydrophobic cavity (Fig. 5.1C), the so-called extended C-site (ECS).[19] The compound can form a ternary complex with sirtuin/NAD^+, forcing the cosubstrate in a nonproductive conformation with nicotinamide outside the C-pocket, but a stable inhibitory complex is obtained only after formation of the reaction intermediate or the coproduct 2′-O-acetyl-ADP-ribose (Fig. 5.1C).[19] The compound thus exploits the unique NAD^+-dependent catalytic mechanism of sirtuins, but the binding site is almost identical in various isoforms and the moderate selectivity in fact due to kinetic isoform differences.[19] Molecular differences neighboring the C-site (see above) should enable, however, to develop it into a more specific compound through extensions that reach such isoform-specific areas. In fact, several subcavities exist next to the C-site and are exploited by potent and/or specific inhibitors. They are designated ECS I-III:[7] ECSI is a kinked C-site extension occupied by part of Ex527; ECSII part of the acyl-Lys channel and exploited, e.g., by ELT-11c, the most potent sirtuin inhibitor reported so far (IC_{50} ~3 nM), which lacks, however, any isoform selectivity;[20] and ECSIII an extra cavity at the back of the active site occupied, e.g., by bromo-resveratrol.[21]

Interestingly, in particular for Sirt2, various potent and relatively selective inhibitors were reported (e.g., Refs. 22−24). Sirt2 is an efficient deacetylase but also shows significant activity against longer lysine modification such as myristoylation due to a large hydrophobic extension of the acetyl-Lys channel that can accommodate such substrates.[8,25] Several crystal structures of Sirt2 complexes with potent and selective inhibitors revealed that they utilize this pocket as well as ECSI and II but not the C-pocket itself. SirReal compounds (IC_{50} 0.4 μM) as well as thienopyrimidinone compounds (<1 μM IC_{50} against Sirt2) were shown to bind to an open Sirt2 conformation, trapping it in a "locked open" state.[22,24] Consistent with them occupying the ECSII acyl-binding channel but not the C-pocket, these compounds are competitive against the peptide substrate but not against NAD^+. Binding to the same hydrophobic Sirt2 pocket, but to the protein's closed conformation, was shown for an oxadiazole-based class of rather small inhibitors (Fig. 5.1D).[23] Assuming an unprecedented orientation, they exploit, in addition, the ECSI site, resulting in potent, highly Sirt2-specific inhibition that is uncompetitive with substrate and cosubstrate. These compounds seem particularly attractive for further development due to their limited size yet already pronounced potency and specificity.

5.3 SIRTUIN ACTIVATORS

Sirtuins show beneficial effects on lifespan and healthspan. For example, Sirt1 activity protects against the negative effects of a high-fat diet, and Sirt6 extends the lifespan in male mice and suppresses aging phenotypes and cancer growth.[26] Sirtuin-activating small molecules are thus considered attractive therapeutics for metabolic disorders and aging-related diseases.[1,26]

Sirtuin-activating compounds (STACs) were initially described for Sirt1. The first activator identified, the plant polyphenol resveratrol that is found, e.g., in red wine, extends yeast lifespan in a sirtuin-dependent manner and promotes human cell survival by stimulating Sirt1.[26] Direct Sirt1 activation was initially debated, since resveratrol appeared to increase peptide substrate affinity but only in the presence of the artificial fluorophore label of the FdL substrate used.[27] Moreover, indirect mechanisms were suggested to cause the in vivo effects of resveratrol on sirtuins as the compound affects a variety of cellular targets.[28] However, direct sirtuin binding and activation by resveratrol and synthetic STACs was validated later through biophysical methods and crystal structure analyses, and testing the resveratrol effect on ~6500 mammalian acetylation sites in peptide microarrays identified physiological substrate sites whose deacetylation can be activated.[29–31] Interestingly, this study revealed a dramatic influence of the substrate sequence on resveratrol effects, ranging from strong activation to significant inhibition,[31] and the relevance of the substrate sequence was confirmed independently.[29] The molecular basis of resveratrol activation was first studied with Sirt5, whose deacetylase activity against FdL and a substrate protein could also be stimulated.[32] A crystal structure of Sirt5 in complex with FdL peptide and resveratrol revealed activator binding between two loops and through extensive hydrophobic interactions to the artificial substrate fluorophore (Fig. 5.2A).[32] However, the fluorophore occupies the sirtuin channel accommodating substrate residues C-terminal from the acyl-Lys in regular substrates, and the direct activator/substrate contact thus provides a rationale for the resveratrol effects against physiological

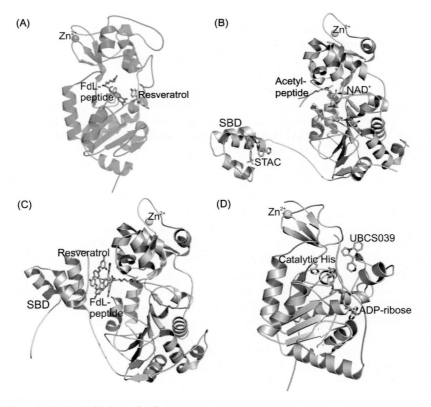

FIGURE 5.2 Pharmacological sirtuin activation.

(A) Crystal structure of Sirt5 in complex with FdL peptide (*blue*) and resveratrol (*gray*; PDB entry 4HDA). (B) Crystal structure of Sirt1 in complex with a STAC (*cyan*; PDB entry 4ZZH). Acetyl peptide (*beige*) and NAD$^+$ (*yellow*) were modeled to indicate their binding sites. (C) Crystal structure of Sirt1 in complex with FdL peptide (*beige*) and three resveratrol molecules (*cyan*) around its fluorophore (PDB entry 4BTR). (D) Crystal structure of Sirt6 in complex with UBCS039 (*cyan*) and the product fragment ADP-ribose (*yellow*; PDB entry 5MF6).

substrates and for the influence of the substrate sequence. Interestingly, the resveratrol effect on Sirt5 also depends on the acyl modification, since resveratrol activated Sirt5-dependent FdL deacetylation but inhibited Sirt5-dependent FdL desuccinylation.[32]

Resveratrol effects on Sirt5 and on Sirt3, which can be inhibited by this compound, are weak and likely not relevant for resveratrol's in vivo effects, in contrast to Sirt1 activation, which appears essential for several physiological resveratrol effects.[26,29] However, resveratrol potency against Sirt1 is also moderate, and its effects on other targets and limited bioavailability render it less suited as a pharmacological Sirt1 activator.[28] Tests with resveratrol derivatives identified, e.g., similar effects for the more soluble piceatannol, and 4′-bromo-resveratrol as a potent Sirt1/3 inhibitor, but no significantly improved sirtuin activators.[21,32] Thus, compound screens were employed for the identification of unrelated Sirt1 activators and yielded a variety of potent substances. Although resveratrol and these compounds represent several differing chemotypes, biochemical and structural studies suggest that they all employ a common activation mechanism.[29,30] Sirt1/STAC complex structures revealed a Sirt1-specific STAC binding domain (SBD) N-terminal from the catalytic core, which provides a rather hydrophobic and flat depression as a STAC docking site, leaving the second activator surface solvent accessible (Fig. 5.2B).[30] However, in the presence of FdL substrate, resveratrol was found to interact with the SBD but to rotate with this domain on top of the active site (Fig. 5.2C), leading to a closed conformation and enabling the same activator/FdL fluorophore interactions previously observed in the Sirt5/FdL/resveratrol complex (Fig. 5.2A).[32,33] The flexibility of the linker between SBD and catalytic core and of a salt bridge between Glu230 and Arg446 that can form in this closed Sirt1 conformation were indeed essential for Sirt1 activation also with regular substrates.[30] Thus, resveratrol and synthetic STACs are assumed to activate Sirt1 against artificial as well as natural substrates through binding to the SBD and formation of the closed conformation, resulting in direct activator/substrate contacts that increase substrate affinity and rationalize the relevance of the substrate sequence for activation (see above). Although some mechanistic details still remain to be clarified, the first synthetic STACs have entered clinical trials for evaluation as therapeutic Sirt1 activators against inflammatory and metabolic disorders.[26]

Activator development for the other isoforms, which lack Sirt1's SBD, has lagged behind, but promising compounds are now emerging. 1,4-Dihydropyridine (DHP)-based compounds, e.g., which can activate Sirt1 against regular substrates, showed activating effects on Sirt2 and Sirt3 in the FdL assay, but such effects remain to be confirmed with physiological substrates.[34] The isoform besides Sirt1 for which potent activation against physiological substrates is well established is Sirt6. Activation of Sirt6-dependent deacetylation can be achieved with fatty acids, and competition experiments implicated the enzyme's long acyl-binding channel as a binding site.[35] This activating effect required several hundred µM fatty acid concentrations, however, and only more recently, pyrrolo[1,2-a]quinoxaline-derived compounds were discovered as first potent Sirt6 activators.[36] These compounds increase Sirt6-dependent deacetylation of peptide substrates, histone proteins, and complete nucleosomes. Crystal structures of Sirt6/activator complexes revealed that the pyrrolo[1,2-a]quinoxalines bind to the acyl channel exit (Fig. 5.2D).[36] The structures together with first structure−activity relationship analyses indicate a key role in binding for the compounds' pyridine substituents, which occupy a Sirt6 pocket that branches off from the acyl channel. Consistent with the observed binding mode, the compounds activate Sirt6 only against acetylated substrate and instead show competition against substrate carrying the longer myristoyl modification that requires these Sirt6 regions for binding, indicating the exciting possibility to develop acyl-specific Sirt6 modulators. Interestingly, the compounds show no effects against Sirt1, 2, and 3, as expected from the Sirt6-specific binding site, but stimulate Sirt5's desuccinylase activity.[36] The molecular basis of this Sirt5 effect remains to be characterized, but it shows that the Sirt6 activators and the structures of Sirt6/activator complexes will now have to be exploited for further development to obtain highly specific Sirt6 modulators.

5.4 OUTLOOK

Several promising sirtuin inhibitors have been identified, in particular for Sirt1 and 2, and the structural and mechanistic insights in sirtuin activities and inhibition obtained in recent years should enable efficient further development of these compounds and of compounds for additional sirtuin isoforms. Similarly, small-molecule activation is now established and structurally characterized for Sirt1 and Sirt6, supporting further activator development for these and additional isoforms. The first sirtuin modulators have already entered clinical trials, illustrating the exciting development of the field of sirtuin modulation.

REFERENCES

1. Morris BJ. Seven sirtuins for seven deadly diseases of aging. *Free Radic Biol Med* 2013;**56**:133−71.
2. Sanchez-Fidalgo S, Villegas I, Sanchez-Hidalgo M, de la Lastra CA. Sirtuin modulators: mechanisms and potential clinical implications. *Curr Med Chem* 2012;**19**:2414−41.
3. Yoon YK, Oon CE. Sirtuin inhibitors: an overview from medicinal chemistry perspective. *Anticancer Agents Med Chem* 2016;**16**:1003−16.
4. Schutkowski M, Fischer F, Roessler C, Steegborn C. New assays and approaches for discovery and design of Sirtuin modulators. *Exp Opin Drug Disc* 2014;**9**:183−99.
5. Moniot S, Weyand M, Steegborn C. Structures, substrates, and regulators of Mammalian sirtuins-opportunities and challenges for drug development. *Front Pharmacol* 2012;**3**:16.
6. Sauve AA, Wolberger C, Schramm VL, Boeke JD. The biochemistry of sirtuins. *Annu Rev Biochem* 2006;**75**:435−65.
7. Gertz M, Steegborn C. Using mitochondrial sirtuins as drug targets: disease implications and available compounds. *Cell Mol Life Sci.* 2016;**73**:2871−96.
8. Moniot S, Schutkowski M, Steegborn C. Crystal structure analysis of human Sirt2 and its ADP-ribose complex. *J Struct Biol* 2013;**182**:136−43.
9. Cen Y. Sirtuins inhibitors: the approach to affinity and selectivity. *Biochim Biophys Acta* 2010;**1804**:1635−44.
10. Schuetz A, Min J, Antoshenko T, et al. Structural basis of inhibition of the human NAD + -dependent deacetylase SIRT5 by suramin. *Structure* 2007;**15**:377−89.
11. Trapp J, Meier R, Hongwiset D, Kassack MU, Sippl W, Jung M. Structure-activity studies on suramin analogues as inhibitors of NAD + -dependent histone deacetylases (sirtuins). *ChemMedChem* 2007;**2**:1419−31.
12. Kiviranta PH, Suuronen T, Wallen EA, et al. N(epsilon)-thioacetyl-lysine-containing tri-, tetra-, and pentapeptides as SIRT1 and SIRT2 inhibitors. *J Med Chem* 2009;**52**:2153−6.
13. Asaba T, Suzuki T, Ueda R, Tsumoto H, Nakagawa H, Miyata N. Inhibition of human sirtuins by in situ generation of an acetylated lysine-ADP-ribose conjugate. *J Am Chem Soc* 2009;**131**:6989−96.

14. Rauh D, Fischer F, Gertz M, et al. An acetylome peptide microarray reveals specificities and deacetylation substrates for all human sirtuin isoforms. *Nat Commun* 2013;**4**:2327.

15. Roessler C, Nowak T, Pannek M, et al. Chemical probing of the human sirtuin 5 active site reveals its substrate acyl specificity and peptide-based inhibitors. *Angew Chem Int Ed Engl* 2014;**53**:10728−32.

16. Yang T, Sauve AA. NAD metabolism and sirtuins: metabolic regulation of protein deacetylation in stress and toxicity. *AAPS J* 2006;**8**:E632−43.

17. Fischer F, Gertz M, Suenkel B, Lakshminarasimhan M, Schutkowski M, Steegborn C. Sirt5 deacylation activities show differential sensitivities to nicotinamide inhibition. *PLoS One* 2012;**7**:e45098.

18. Napper AD, Hixon J, McDonagh T, et al. Discovery of indoles as potent and selective inhibitors of the deacetylase SIRT1. *J Med Chem* 2005;**48**:8045−54.

19. Gertz M, Fischer F, Nguyen GT, et al. Ex-527 inhibits Sirtuins by exploiting their unique NAD + -dependent deacetylation mechanism. *Proc Natl Acad Sci USA* 2013;**110**:E2772−81.

20. Disch JS, Evindar G, Chiu CH, et al. Discovery of thieno[3,2-d]pyrimidine-6-carboxamides as potent inhibitors of SIRT1, SIRT2, and SIRT3. *J Med Chem* 2013;**56**:3666−79.

21. Nguyen GT, Gertz M, Steegborn C. Crystal structures of sirt3 complexes with 4'-bromo-resveratrol reveal binding sites and inhibition mechanism. *Chem Biol* 2013;**20**:1375−85.

22. Rumpf T, Schiedel M, Karaman B, et al. Selective Sirt2 inhibition by ligand-induced rearrangement of the active site. *Nat Commun* 2015;**6**:6263.

23. Moniot S, Forgione M, Lucidi A, et al. Development of 1,2,4-oxadiazoles as potent and selective inhibitors of the human deacetylase sirtuin 2: structure-activity relationship, X-ray crystal structure, and anticancer activity. *J Med Chem* 2017;**60**:2344−60.

24. Sundriyal S, Moniot S, Mahmud Z, et al. Thienopyrimidinone based sirtuin-2 (SIRT2)-selective inhibitors bind in the ligand induced selectivity pocket. *J Med Chem* 2017;**60**:1928−45.

25. Teng YB, Jing H, Aramsangtienchai P, et al. Efficient demyristoylase activity of SIRT2 revealed by kinetic and structural studies. *Sci Rep* 2015;**5**:8529.

26. Bonkowski MS, Sinclair DA. Slowing ageing by design: the rise of NAD + and sirtuin-activating compounds. *Nat Rev Mol Cell Biol* 2016;**17**(11):679−90.

27. Borra MT, Smith BC, Denu JM. Mechanism of human SIRT1 activation by resveratrol. *J Biol Chem* 2005;**280**:17187−95.

28. Pirola L, Frojdo S. Resveratrol: one molecule, many targets. *IUBMB Life* 2008;**60**:323−32.

29. Hubbard BP, Gomes AP, Dai H, et al. Evidence for a common mechanism of SIRT1 regulation by allosteric activators. *Science* 2013;**339**:1216−19.

30. Dai H, Case AW, Riera TV, et al. Crystallographic structure of a small molecule SIRT1 activator-enzyme complex. *Nat Commun* 2015;**6**:7645.

31. Lakshminarasimhan M, Rauth D, Schutkowski M, Steegborn C. Sirt1 activation by resveratrol is substrate sequence-selective. *Aging (Albany NY).* 2013;**5**:151−4.

32. Gertz M, Nguyen GT, Fischer F, et al. A molecular mechanism for direct sirtuin activation by resveratrol. *PLoS One* 2012;**7**:e49761.

33. Cao D, Wang M, Qiu X, et al. Structural basis for allosteric, substrate-dependent stimulation of SIRT1 activity by resveratrol. *Genes Dev* 2015;**29**:1316−25.

34. Mai A, Valente S, Meade S, et al. Study of 1,4-dihydropyridine structural scaffold: discovery of novel sirtuin activators and inhibitors. *J Med Chem* 2009;**52**:5496−504.

35. Feldman JL, Baeza J, Denu JM. Activation of the protein deacetylase SIRT6 by long-chain fatty acids and widespread deacylation by mammalian sirtuins. *J Biol Chem* 2013;**288**:31350−6.

36. You W, Rotili D, Li TM, et al. Structural basis of sirtuin 6 activation by synthetic small molecules. *Angew Chem Int Ed Engl* 2017;**56**:1007−11.

PHARMACOLOGICAL APPROACHES FOR MODULATING SIRTUINS

6

Alice E. Kane[1] and David A. Sinclair[1,2]

[1]*Harvard Medical School, Boston, MA, United States* [2]*The University of New South Wales, Sydney, NSW, Australia*

6.1 INTRODUCTION

The sirtuins were one of the first sets of genes identified that mediated the healthspan- and lifespan-extending effects of calorie restriction (CR). Since then, there has been considerable interest in identifying pharmaceutical compounds that can modulate these proteins.[1] This chapter outlines two main pharmacological approaches to the modulation of sirtuins: sirtuin-activating compounds (STACs) as allosteric activators of SIRT1 and compounds that modulate levels of nicotinamide adenine dinucleotide (NAD^+), collectively termed NAD^+-boosting molecules or "NBMs."

6.2 ALLOSTERIC ACTIVATORS OF SIRTUINS

The first generation of STACs were discovered in a high-throughput screen using a so-called "Fluor de Lys" peptide substrate.[2] Several classes of plant polyphenols were shown to activate recombinant SIRT1 and to extend the lifespan of *Saccharomyces cerevisiae*.[2] The most effective of these, activating SIRT1 by more than 10-fold, was resveratrol (3,5,4'-trihydroxystilbene),[2] a natural product found in grapes. A variety of stilbene derivatives with modifications at the 4' position of the B ring of resveratrol were synthesized that have lower toxicity and higher potency with respect to SIRT1 activation and lifespan extension in budding yeast, showing for the first time that it is possible to improve upon naturally occurring STACs.[3]

Drug design approaches and high-throughput screening have since identified thousands of both naturally occurring and synthesized STACs.[4] STACs of a variety of chemotypes, including oxazolo[4,5-b]pyridine, thiazolopyridine, and bridged urea have been developed with improved activation potency, physiochemical properties, and therapeutic potential.[5,6] Another unrelated class of 1,4-dihydropyridine-based compounds bearing a benzyl group at the N1 position are reported to activate SIRT1, SIRT2, and SIRT3.[7] Molecules such as SRT1720 and SRT2104 (from the third generation) have improved bioavailability and up to 1000 times the potency of resveratrol, and have been extensively tested in animals[8,9] (Fig. 6.1).

Introductory Review on Sirtuins in Biology, Aging, and Disease. DOI: https://doi.org/10.1016/B978-0-12-813499-3.00006-X

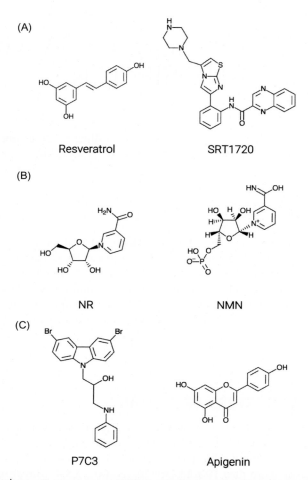

(A)

Resveratrol   SRT1720

(B)

NR   NMN

(C)

P7C3   Apigenin

FIGURE 6.1 Common NAD⁺-boosting molecules (NBMs) and sirtuin-activating compounds (STACs).

(A) STACs resveratrol and SRT1720; (B) NAD$^+$ precursors nicotinamide riboside (NR) and nicotinamide mononucleotide (NMN); (C) Potential nicotinamide phosphoribosyltransferase (NAMPT) activator P7C3 and CD38 inhibitor apigenin.

STACs act as allosteric modulators of SIRT1 (Fig. 6.2). Allosteric modulators bind to a site on a protein other than the substrate binding site and induce a conformational change that increases the binding affinity of the protein for the target substrate(s). In this way, STACs bind to an allosteric site of SIRT1 and increase the enzyme's affinity for target substrates.[6] Extensive research has identified the specific amino acid residues of the STAC binding domain and determined that they activate SIRT1 via a "bend-at-the-elbow" model in which the binding of an STAC exposes the substrate binding site of SIRT1.[6]

(A) Targets of NBMs

(B) Allosteric activation of SIRT1

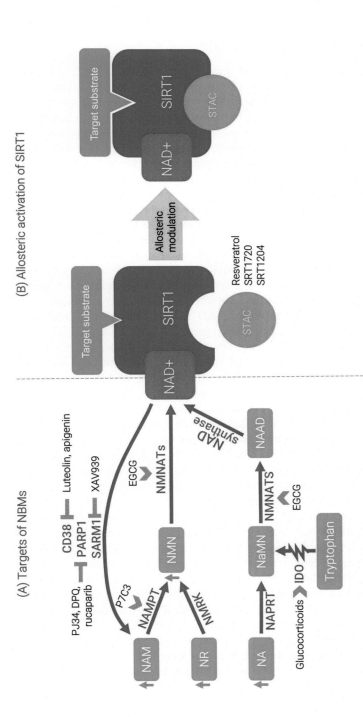

FIGURE 6.2 Pharmacological mechanisms of sirtuin modulation.

(A) Targets of NAD-boosting molecules or "NBMs." NAD$^+$ is synthesized de novo from tryptophan via a series of enzymatic reactions, including the initial conversion of tryptophan to kynurenine by the enzyme indoleamine 2,3-dioxygenase (IDO). NAD$^+$ is also salvaged from precursors by the conversion of nicotinamide (NAM) to nicotinamide mononucleotide (NMN) by nicotinamide phosphoribosyltransferase (NAMPT). NMN is then converted to NAD$^+$ by nicotinamide mononucleotide adenylyltransfereases (NMNATs). Nicotinamide riboside (NR) is converted to NMN by nicotinamide riboside kinase (NMRK), and nicotinic acid (NA) to nicotinic acid mononucleotide (NaMN) by nicotinic acid phosphoribosyltransferase (NAPRT). NaMN is converted to nicotinic acid adenine dinucleotide (NAAD) by NMNATs, and then NAAD to NAD$^+$ by NAD-synthase. NAD$^+$ is broken down by CD38, poly ADP-ribose polymerase 1 (PARP1), and sterile alpha and TIR motif containing protein 1 (SARM1). Known pharmacological agents that target these substrates and enzymes to increase NAD$^+$ levels are marked on the figure in black, or indicated with gray arrows. (B) Mechanism of allosteric sirtuin activation. Sirtuin-activating compounds (STACs) are allosteric modulators of SIRT1 that bind to a specific site on SIRT1 N-terminal to the catalytic domain, and induce a conformational change that increases the binding affinity of the N-terminus for the central domain, thereby increasing the affinity for the substrate by lowering K_m.

6.3 PHARMACOLOGICAL MODULATION OF NAD$^+$

Another approach to the pharmacological modulation of sirtuins is by increasing NAD$^+$, a cosubstrate necessary for sirtuin deacylase activity. The first evidence that NAD$^+$ boosting was beneficial came from studies in the yeast *S. cerevisiae*.[10] The yeast *PNC1* gene, which catalyzes the first step in the NAD$^+$ salvage pathway, is one of the most upregulated genes in response to environmental changes including CR. Given the known requirement of sirtuins in response to CR, it was reasoned that manipulating NAD$^+$ levels might mimic the benefits of this diet. Indeed, the constitutive overexpression of *PNC1* increased the stress resistance and lifespan of yeast cells, whereas deletion of *PNC1* rendered yeast cells unable to respond to caloric restriction.[10]

In mammals, NAD$^+$ is produced and recycled via pathways analogous to those in yeast. NAD$^+$ is either synthesized de novo from dietary tryptophan, or salvaged from the precursors nicotinamide (NAM), nicotinamide riboside (NR) and nicotinic acid (NA), as shown in Fig. 6.2.[11] NBMs boost NAD$^+$ levels by directly supplying NAD$^+$ precursors, or by activating enzymes that are involved in NAD$^+$ synthesis and salvage including nicotinamide mononucleotide adenylyltransfereases (NMNATs), nicotinamide phosphoribosyltransferase (NAMPT), and indoleamine 2,3-dioxygenase (IDO) (Fig. 6.2). Analogous to *PNC1*, Nampt is a stress-responsive gene whose expression dictates NAD$^+$ levels in mammalian cells and provides increased stress resistance by activating sirtuins.[12]

The supplementation of NAM, NA, and NR increases NAD$^+$ levels in preclinical studies and in humans.[13–15] NA and NAM are limited in their therapeutic potential by side effects, such as flushing. NR is promising as it is orally bioavailable and has few side effects, although it is relatively unstable.[13,15] There has been recent focus on NMN as a safe, bioavailable, and more stable therapeutic[14] (Fig. 6.1). The potential of the other intermediates of this pathway, such as NAAD or NaMN, to increase NAD$^+$ levels has not yet been explored.[1]

Glucocorticoids can reduce the symptoms of multiple sclerosis (MS) and the mechanism is attributed, at least in part, to the activation of IDO, the rate-limiting first step in the synthesis of NAD$^+$.[16] EGCG, an antioxidant found in green tea, is associated with neuroprotection and cancer prevention, and activates NMNATs.[17] The P7C3 class of aminopropyl carbazoles was identified by high-throughput screening of neuroprotective compounds.[18] Further studies suggest P7C3 acts as an NAMPT activator to increase intracellular levels of NAD$^+$ in neurons,[19] though this result has not yet been reproduced.

The final strategy to increase NAD$^+$ levels is to inhibit enzymes that contribute to its breakdown. There are several families of NADases that could be targeted for inhibition, and high-throughput screening has identified competitive antagonists for the NAD$^+$ binding sites of several proteins. CD38 is the primary NADase in mammals.[20] Aging is associated with an increase in CD38 in many tissues, and this results in a subsequent decrease in NAD$^+$ levels.[21] Pharmacological inhibition of CD38 with flavonoids including quercetin, apigenin, or luteolinidin, resulted in increased NAD$^+$ levels in mouse studies.[22,23] Thiazoloquin(az)olinones, in particular 78c, also inhibit CD38, and increase NAD$^+$ levels.[24] Poly ADP-ribose polymerases (PARPs), especially PARP1, are major users of NAD$^+$.[25] There are several small-molecule PARP inhibitors that increase NAD$^+$ levels including PJ34,[26] olaparib,[27] and several flavonoids.[28] Recent studies suggest that sterile alpha and TIR motif containing protein 1 (SARM1), a component of axon breakdown in response to nerve injury, has NAD$^+$ cleaving ability, and may be a target for inhibition.[29] XAV939 is a SARM1 inhibitor that increases NAD$^+$ in preclinical models.[29,30] Overall, NADase inhibitors are a promising area for therapeutics to increase NAD$^+$ levels.

6.4 BENEFICIAL HEALTH EFFECTS OF STACS

Activation of sirtuins with STACs has been associated with a wide range of health benefits. These have been extensively reviewed[8,9,31] and include protection against NAFLD and diabetes, increased exercise capacity, reduced muscle wasting, vascular and cardiac benefits, antiinflammatory effects, and protection against a range of neurodegenerative diseases.[8,31−33] Furthermore, resveratrol has been shown to extend lifespan in a range of species, and extend healthspan and delay age-related deterioration in mice (reviewed in Ref. 1). Resveratrol has also shown beneficial effects in non-human primates, including improved insulin sensitivity.[34] Other STACs, including SRT1720 and SRT2104, have been shown to increase lifespan in mice.[33,35,36]

STACs are now being tested in humans, with several clinical trials completed, and more underway. Resveratrol has been tested in at least six clinical trials which showed mostly positive effects on glucose metabolism and cardiovascular disease risk factors,[1,37,38] as well as a delay in cognitive decline in patients with Alzheimer's disease.[39] The different ways resveratrol was dosed in these human trials (e.g. with food or without, as a dry power or solubilized) likely resulted in observed differences in the bioavailability of resveratrol, an issue that should be addressed in future studies. SRT2104 has also been tested in a number of safety and now phase II trials,[40−42] where it improved triglyceride and cholesterol levels.[41,42] Several more clinical trials of resveratrol, SRT2104, and the fifth generation of STACs from GlaxoSmithKline are currently underway or will begin in 2018-2019.[1]

6.5 BENEFICIAL HEALTH EFFECTS OF NBMS

NAD^+ levels decline up to 50% by the time a rodent or human is past the halfway point in their lifespan.[21] Increasing NAD^+ above basal levels, especially in old age, with NAD^+-boosting molecules or "NBMs," has been shown to have considerable beneficial health effects in rodents (Table 6.1).

Pharmacologically increasing NAD^+ levels in preclinical animal models prevents detrimental age-related changes to the cardiovascular system, such as artery stiffness,[43] cardiac hypertrophy, and heart failure.[44] Additionally, NBMs protect against cardiovascular damage in animal models of ischemia/reperfusion injury[22] and cardiomyopathy.[45] NBMs, specifically NR and NAM, also protect against liver and kidney damage, including promoting liver regeneration in mice exposed to partial hepatectomy,[46] and improving function post renal ischemia.[47] NR and the PARP-inhibitor olaparib were shown to stop the progression of nonalcoholic fatty liver disease in mice fed a high-fat diet.[27,48] Furthermore, improved glucose homeostasis in diabetic mouse models occurs with treatment of PARP inhibitors,[26] CD38 inhibitors,[23] and NMN^5 (Table 6.1).

In preclinical models one of the most profound effects of treatment with NBMs is an improvement in muscle function (Table 6.1). For example, NMN improves muscle function and prevents age-related inflammation in old mice,[49] and NR improves exercise capacity in young[27] and old mice, ostensibly by boosting mitochondrial function.[50] Additionally, boosting NAD^+ levels is associated with increased DNA repair[51,52] and reduced tumor formation in some cancer models. Both NA and NAM increase NAD^+ levels, and subsequently inhibit tumor growth in SCID mice,[53] and PARP1 inhibitors have been

Table 6.1 Preclinical studies of NAD$^+$-boosting molecules (NBMs)

Pharmacological agent	Pharmacological mechanism	Health outcomes observed	References
NR	NAD$^+$ precursor	Prevents NAFLD, NASH	13,29,32,45,46,48,50,51,62,64,65
		↑ Regeneration post partial hepatectomy	
		↑ Glucose homeostasis in diabetic models	
		↑ Exercise capacity	
		↑ DNA repair	
		Antiinflammatory	
		Protects against autoimmune diseases	
		Neuroprotective	
		Delays neurodegenerative diseases	
		↑ Mitochondrial function	
		Lifespan extension (progeria models, *Caenorhabditis elegans*, mice)	
NA	NAD$^+$ precursor	↑ DNA repair	53
		↓ Tumor growth	
NAM	NAD$^+$ precursor	Protects against liver and kidney injury	47,53,56
		↓ Tumor growth	
		Immune protective	
		Antiinflammatory	
		Lifespan extension (*C. elegans*)	
NMN	NAD$^+$ precursor	↑ Glucose homeostasis in diabetic models	14,32,43,44,49,57,58,63
		↑ Mitochondrial function	
		↑ Muscle and physical function	
		↑ Vascular function	
		Cardioprotective	
		Protects against radiation-induced DNA damage	
		Antiinflammatory	
		Neuroprotective	
		Delays neurodegenerative diseases	
		↑ Mitochondrial function	
		Lifespan extension (progeria models)	

Table 6.1 Preclinical studies of NAD$^+$-boosting molecules (NBMs) *Continued*

Pharmacological agent	Pharmacological mechanism	Health outcomes observed	References
Glucocorticoids	↑ IDO activity	↓ Symptoms in acute MS attack	16
EGCG	NMNAT activator	Neuroprotective	17
		Cancer preventative	
P7C3	Potential NAMPT activator	Neuroprotective	19,59–61
		Delays neurodegenerative diseases	
Quercetin, apigenin, luteolinidin, thiazoloquin(az)olinones	CD38 inhibitors	↑ Glucose homeostasis in diabetic models	22–24
		Cardioprotective	
PJ34, olaparib, flavanoids	PARP inhibitors	Prevents/treats NAFLD, NASH	26–28
		↑ Glucose homeostasis in diabetic models	
		Cancer treatment	
		Lifespan extension (*C. elegans*)	
XAV939	SARM1 inhibitor	Neuroprotective	29,30

IDO, indoleamine 2,3-dioxygenase; NA, nicotinic acid; NAD$^+$, nicotinamide adenine dinucleotide; NAFLD, nonalcoholic fatty liver disease; NAM, nicotinamide; NAMPT, nicotinamide phosphoribosyltransferase; NASH, nonalcoholic steatohepatitis; NMN, nicotinamide mononucleotide; NMNAT, nicotinamide mononucleotide adenylyltransferease; NR, nicotinamide riboside; PARP, poly ADP-ribose polymerase; SARM1, sterile alpha and TIR motif containing protein 1.

shown to be effective treatments for many types of cancer.[54] However, there is also evidence that increasing NAD$^+$ and sirtuin activity may contribute to cancer development, and, in fact, targeting DNA repair is a potential cancer therapeutic approach.[55] Thus more evidence is needed to understand whether pharmacological modulation of NAD$^+$ and sirtuins will be beneficial in cancer or not, or in what circumstances.

NBMs also have beneficial immune and inflammation effects in preclinical models (Table 6.1). NAM protects against immune suppression post UV radiation.[56] NR reduces skeletal muscle inflammation in a mouse model of muscular dystrophy,[45] and protects, along with NMN, against autoimmune diseases in preclinical models.[51] NMN can also reduce age-related inflammation.[49] With regards to the nervous system, the effects of NBMs include protecting against neural and brain injury[57–59] and slowing the progression of neurodegenerative diseases. NMN and the potential NAMPT activator P7C3 are neuroprotective in animal models of Parkinson's disease[60] and ALS.[61] In Alzheimer's disease animal models, NR and NMN protect against declining cognition, reduce oxidative stress, and improve synaptic plasticity.[62,63] NBMs also improve motor function in mouse models of ataxia telangiectasia[51] and MS.[16]

Finally, there is evidence to suggest that NBMs also increase longevity. NR treatment improves mitochondrial function and DNA repair in the progeria model, Cockayne syndrome mice.[64] NR, along with NMN, also improves mitochondrial function and increases lifespan in worm and mouse progeria models.[32,51] Olaparib, NR, and NAM treatment in *Caenorhabditis elegans* all result in increased mitochondrial function and lifespan.[65] In studies of elderly mice, NR treatment improves stem cell function and modestly extends lifespan[50] and NMN treatment improves insulin sensitivity and physical activity.[14] This evidence indicates that NBMs may be possible antiaging interventions.

Some NBMs, such as NR and NMN, have been investigated in clinical trials,[13,66] and there are more currently underway. Over the next few years there will be a large amount of data coming out from these human trials addressing the safety and efficacy of NBMs in older people and people with specific diseases related to aging.

6.6 CONCLUSIONS

The pharmacological modulation of sirtuins, via either allosteric activation with STACs or increasing available NAD^+ levels via NBMs, has shown benefits across a wide range of diseases in preclinical studies. Evidence from clinical studies is now beginning to support these findings. Future research will contribute to further understanding of these therapeutics and their clinical potential in treating age-related diseases, and perhaps aging itself.

REFERENCES

1. Bonkowski MS, Sinclair DA. Slowing ageing by design: the rise of NAD + and sirtuin-activating compounds. *Nat Rev Mol Cell Biol* 2016;**230**:2−3.
2. Howitz KT. Small molecule activators of sirtuins extend *Saccharomyces cerevisiae* lifespan. *Nature* 2003;**425**:191−6.
3. Yang H, Baur JA, Chen A, Miller C, Sinclair DA. Design and synthesis of compounds that extend yeast replicative lifespan. *Aging Cell* 2007;**6**:35−43.
4. Nayagam VM. SIRT1 modulating compounds from high-throughput screening as anti-inflammatory and insulin-sensitizing agents. *J Biomol Screen* 2006;**11**:959−67.
5. Hubbard BP, Gomes AP, Dai H, Li J, Case AW, Considine T, et al. Evidence for a common mechanism of SIRT1 regulation by allosteric activators. *Science* 2013;**339**:1216−19.
6. Dai H. Crystallographic structure of a small molecule SIRT1 activator-enzyme complex. *Nat Commun* 2015;**6**:7645.
7. Valente S, Mellini P, Spallotta F, Carafa V, Nebbioso A, Polletta L, et al. 1,4-Dihydropyridines active on the SIRT1/AMPK pathway ameliorate skin repair and mitochondrial function and exhibit inhibition of proliferation in cancer cells. *J Med Chem* 2016;**59**:1471−91.
8. Milne JC. Small molecule activators of SIRT1 as therapeutics for the treatment of type 2 diabetes. *Nature* 2007;**450**:712−16.
9. Hubbard BP, Sinclair DA. Small molecule SIRT1 activators for the treatment of aging and age-related diseases. *Trends Pharmacol Sci* 2014;**35**:146−54.
10. Anderson RM, Bitterman KJ, Wood JG, Medvedik O, Sinclair DA. Nicotinamide and PNC1 govern lifespan extension by calorie restriction in *Saccharomyces cerevisiae*. *Nature* 2003;**423**:181−5.
11. Revollo JR, Grimm AA, Imai S. The regulation of nicotinamide adenine dinucleotide biosynthesis by Nampt/PBEF/visfatin in mammals. *Curr Opin Gastroenterol* 2007;**23**:164−70.

12. Yang H, Yang T, Baur J, Perez E, Matsui T, Carmona JJ, et al. Nutrient-sensitive mitochondrial NAD + levels dictate cell survival. *Cell* 2007;**130**:1095–107.

13. Trammell SAJ, Schmidt MS, Weidemann BJ, Redpath P, Jaksch F, Dellinger RW, et al. Nicotinamide riboside is uniquely and orally bioavailable in mice and humans. *Nat Commun* 2016;**7**:12948.

14. Mills KF, Yoshida S, Stein LR, Uchida K, Mills KF, Yoshida S, et al. Long-term administration of nicotinamide mononucleotide mitigates age-associated physiological decline in mice article long-term administration of nicotinamide mononucleotide mitigates age-associated physiological decline in mice. *Cell Metab* 2016;**24**:795–806.

15. Conze D, Crespo-Barreto J, Kruger C. Safety assessment of nicotinamide riboside, a form of vitamin B 3. *Hum Exp Toxicol* 2016;**35**:1149–60.

16. Penberthy WT. Nicotinic acid-mediated activation of both membrane and nuclear receptors towards therapeutic glucocorticoid mimetics for treating multiple sclerosis. *PPAR Res* 2009;853707.

17. Berger F, Lau C, Dahlmann M, Ziegler M. Subcellular compartmentation and differential catalytic properties of the three human nicotinamide mononucleotide adenylyltransferase isoforms. *J Biol Chem* 2005;**280**:36334–41.

18. Pieper AA, McKnight SL, Ready JM. P7C3 and an unbiased approach to drug discovery for neurodegenerative diseases. *Chem Soc Rev* 2014;**43**:6716–26.

19. Wang G. P7C3 neuroprotective chemicals function by activating the rate-limiting enzyme in NAD salvage. *Cell* 2014;**158**:1324–34.

20. Aksoy P, White TA, Thompson M, Chini EN. Regulation of intracellular levels of NAD: a novel role for CD38. *Biochem Biophys Res Commun* 2006;**345**:1386–92.

21. Camacho-Pereira J. CD38 dictates age-related NAD decline and mitochondrial dysfunction through an SIRT3-dependent mechanism. *Cell Metab* 2016;**23**:1127–39.

22. Boslett J, Hemann C, Zhao Y, Lee H, Zweier J. Luteolinidin protects the postischemic heart through CD38 inhibition with preservation of NAD(P)(H). *J Pharmacol Exp Ther* 2017;**361**:99–108.

23. Escande C. Flavonoid apigenin is an inhibitor of the NAD + ase CD38: implications for cellular NAD + metabolism, protein acetylation, and treatment of metabolic syndrome. *Diabetes* 2013;**62**:1084–93.

24. Haffner CD. Discovery, synthesis, and biological evaluation of thiazoloquin(az)olin(on)es as potent CD38 inhibitors. *J Med Chem* 2015;**58**:3548–71.

25. Bai P, Canto C. The role of PARP-1 and PARP-2 enzymes in metabolic regulation and disease minireview. *Cell Metab* 2012;**16**:290–5.

26. Bai P, Canto C, Oudart H, Brunyanszki A, Cen Y, Thomas C, et al. PARP-1 inhibition increases mitochondrial metabolism through SIRT1 activation. *Cell Metab* 2011;**13**:461–8.

27. Gariani K, Ryu D, Menzies KJ, Yi H, Stein S, Zhang H, et al. Inhibiting poly ADP-ribosylation increases fatty acid oxidation and protects against fatty liver disease. *J Hepatol* 2017;**66**:132–41.

28. Geraets L, Moonen HJJ, Brauers K, Wouters EFM, Bast A, Hageman GJ. Dietary flavones and flavonols are inhibitors of poly(ADP-ribose) polymerase-1 in pulmonary. *J Nutr* 2007;**137**:2190–5.

29. Gerdts J, Brace EJ, Sasaki Y, DiAntonio A, Milbrandt J. Neurobiology. SARM1 activation triggers axon degeneration locally via NAD + destruction. *Science* 2015;**348**:453–7.

30. Huang SM, Mishina YM, Liu S, Cheung A, Stegmeier F, Michaud GA, et al. Tankyrase inhibition stabilizes axin and antagonizes Wnt signalling. *Nature* 2009;**461**:614–20.

31. Dai H. SIRT1 activation by small molecules: kinetic and biophysical evidence for direct interaction of enzyme and activator. *J Biol Chem* 2010;**285**:32695–703.

32. Fang EF, Scheibye-knudsen M, Brace LE, Kassahun H, Sengupta T, Nilsen H, et al. Defective mitophagy in XPA via PARP-1 hyperactivation and NAD + / SIRT1 reduction. *Cell* 2014;**157**:882–96.

33. Mercken EM. SRT2104 extends survival of male mice on a standard diet and preserves bone and muscle mass. *Aging Cell* 2014;**13**:787–96.

34. Jimenez-Gomez Y. Resveratrol improves adipose insulin signaling and reduces the inflammatory response in adipose tissue of rhesus monkeys on high-fat, high-sugar diet. *Cell Metab* 2013;**18**:533−45.

35. Minor RK, Baur JA, Gomes AP, Ward TM, Csiszar A, Mercken EM, et al. SRT1720 improves survival and healthspan of obese mice. *Sci Rep* 2011;70.

36. Mitchell SJ. The SIRT1 activator SRT1720 extends lifespan and improves health of mice fed a standard diet. *Cell Rep* 2014;**6**:836−43.

37. Wong RH. Acute resveratrol supplementation improves flow-mediated dilatation in overweight/obese individuals with mildly elevated blood pressure. *Nutr Metab Cardiovasc Dis* 2011;**21**:851−6.

38. Poulsen MM. High-dose resveratrol supplementation in obese men: an investigator-initiated, randomized, placebo-controlled clinical trial of substrate metabolism, insulin sensitivity, and body composition. *Diabetes* 2013;**62**:1186−95.

39. Turner RS. A randomized, double-blind, placebo-controlled trial of resveratrol for Alzheimer disease. *Neurology* 2015;**85**:1383−91.

40. Hoffmann E. Pharmacokinetics and tolerability of SRT2104, a first-in-class small molecule activator of SIRT1, after single and repeated oral administration in man. *Br J Clin Pharmacol* 2013;**75**:186−96.

41. Libri V. A pilot randomized, placebo controlled, double blind phase I trial of the novel SIRT1 activator SRT2104 in elderly volunteers. *PLoS One* 2012;**7**:e51395.

42. Venkatasubramanian S. Cardiovascular effects of a novel SIRT1 activator, SRT2104, in otherwise healthy cigarette smokers. *J Am Hear Assoc* 2013;**2**:e000042.

43. de Picciotto NE, Gano LB, Johnson LC, Martens CR, Sindler AL, Mills KF, et al. Nicotinamide mononucleotide supplementation reverses vascular dysfunction and oxidative stress with aging in mice. *Aging Cell* 2016;**15**:522−30.

44. Horton JL, Martin OJ, Lai L, Riley NM, Richards AL, Vega RB, et al. Mitochondrial protein hyperacetylation in the failing heart. *JCI Insight* 2016;**1**:1−14.

45. Ryu D, Zhang H, Ropelle ER, Sorrentino V, Mázala DAG, Mouchiroud L, et al. NAD + repletion improves muscle function in muscular dystrophy and counters global PARylation. *Sci Transl Med* 2016;**8** 361ra139.

46. Mukherjee S, Chellappa K, Moffitt A, Ndungu J, Dellinger RW, Davis JG, et al. Nicotinamide adenine dinucleotide biosynthesis promotes liver regeneration. *Hepatology* 2017;**65**:616−30.

47. Tran MT, Zsengeller ZK, Berg AH, Khankin EV, Bhasin MK, Kim W, et al. PGC1α-dependent NAD biosynthesis links oxidative metabolism to renal protection. *Nature* 2016;**531**:528−32.

48. Zhou C, Yang X, HUa X, Liu J, Fan M, Li G-Q, et al. Hepatic NAD + deficiency as a therapeutic target for non-alcoholic fatty liver disease in ageing Tables of Links. *Br J Pharmacol* 2016;**173**:2352−68.

49. Gomes AP. Declining NAD + induces a pseudohypoxic state disrupting nuclear-mitochondrial communication during aging. *Cell* 2013;**155**:1624−38.

50. Zhang H. NAD + repletion improves mitochondrial and stem cell function and enhances life span in mice. *Science* 2016;**352**:1436−43.

51. Fang EF, Kassahun H, Croteau DL, Mattson MP, Nilsen H, Fang EF, et al. NAD + replenishment improves lifespan and healthspan in ataxia telangiectasia models via article NAD + replenishment improves lifespan and healthspan in ataxia telangiectasia models via mitophagy and DNA repair. *Cell Metab* 2016;**24**:566−81.

52. Li J, Bonkowski MS, Moniot S, Zhang D, Hubbard BP, Ling AJY, et al. A conserved NAD + binding pocket that regulates protein-protein interactions during aging. *Science* 2017;**355**:1312−17.

53. Santidrian AF, Matsuno-yagi A, Ritland M, Seo BB, Leboeuf SE, Gay LJ, et al. Mitochondrial complex I activity and NAD+/NADH balance regulate breast cancer progression. *J Clin Invest* 2013;**123**:1068−81.

54. Rajawat J, Shukla N, Mishra D. Therapeutic targeting of poly(ADP-ribose) polymerase-1 (PARP1) in cancer: current developments, therapeutic strategies, and future opportunities. *Med Res Rev* 2017;**37**:1461−91.

55. Gavande NS, Vandervere-carozza PS, Hinshaw HD, Jalal SI, Sears CR, Pawelczak KS, et al. Pharmacology & therapeutics DNA repair targeted therapy: the past or future of cancer treatment? *Pharmacol Ther* 2016;**160**:65−83.

56. Yiasemides E, Sivapirabu G, Halliday GM, Park J, Damian DL. Oral nicotinamide protects against ultraviolet radiation-induced immunosuppression in humans. *Carcinogenesis* 2009;**30**:101−5.

57. Park JH, Long A, Owens K, Kristian T. Neurobiology of disease nicotinamide mononucleotide inhibits post-ischemic NAD + degradation and dramatically ameliorates brain damage following global cerebral ischemia. *Neurobiol Dis* 2016;**95**:102−10.

58. Wei CC, Kong YY, Hua X, Li GQ, Zheng SL, Cheng MH, et al. NAD replenishment with nicotinamide mononucleotide protects blood-brain barrier integrity and attenuates delayed tissue plasminogen activator-induced haemorrhagic transformation after cerebral ischaemia. *Br J Pharmacol* 2017;**174**:3823−36.

59. Yin TC, Britt JK, De Jesús-Cortés H, Lu Y, Genova RM, Khan MZ, et al. P7C3 neuroprotective chemicals block axonal degeneration and preserve function after traumatic brain injury. *Cell Rep* 2014;**8**:1731−40.

60. De Jesús-Cortés H, Xu P, Drawbridge J, Estill SJ, Tran S, Britt J, et al. Neuroprotective efficacy of aminopropyl carbazoles in a mouse model of Parkinson disease. *PNAS* 2012;**109**:17010−15.

61. Tesla R, Wolf HP, Xu P, Drawbridge J, Jo S, Huntington P, et al. Neuroprotective efficacy of aminopropyl carbazoles in a mouse model of amyotrophic lateral sclerosis. *PNAS* 2012;**109**:17016−21.

62. Gong B. Nicotinamide riboside restores cognition through an upregulation of proliferator-activated receptor-gamma coactivator 1[alpha] regulated beta-secretase 1 degradation and mitochondrial gene expression in Alzheimer's mouse models. *Neurobiol Aging* 2013;**34**:1581−8.

63. Long AN, Owens K, Schlappal AE, Kristian T, Fishman PS, Schuh RA. Effect of nicotinamide mononucleotide on brain mitochondrial respiratory deficits in an Alzheimer's disease-relevant murine model. *BMC Neurol* 2015;**15**:19.

64. Scheibye-Knudsen M. A high-fat diet and NAD + activate Sirt1 to rescue premature aging in cockayne syndrome. *Cell Metab* 2014;**20**:840−55.

65. Mouchiroud L. The NAD + /sirtuin pathway modulates longevity through activation of mitochondrial UPR and FOXO signaling. *Cell* 2013;**154**:430−41.

66. Menezes R, Rodriguez-Mateos A, Kaltsatou A, González-Sarrías A, Greyling A, Giannaki C, et al. Impact of flavonols on cardiometabolic biomarkers: a meta-analysis of randomized controlled human trials to explore the role of inter-individual variability. *Nutrients* 2017;**9**.

REACTIVE ACYL-COA SPECIES AND DEACYLATION BY THE MITOCHONDRIAL SIRTUINS

Kathleen A. Hershberger and Matthew D. Hirschey

Duke University Medical Center, Durham, NC, United States

7.1 INTRODUCTION

The sirtuins are a family of nicotinamide adenine dinucleotide (NAD^+)-dependent enzymes that regulate gene silencing, chromosome stability, and metabolism by removing acyl-lysine modifications from a variety of proteins.[1] The *SIR* (silent information regulator) family of genes was first discovered in *Saccharomyces cerevisiae* and was identified as necessary for repressing translation of the silent mating type loci; indeed, inactivation of any of these genes results in sterility.[2,3] *Sir2*, the only *SIR* gene conserved in higher eukaryotes, was shown to promote longevity in *S. cerevisiae* and spurred great interested in the field[4] of Sir2-like proteins (sirtuins). The identification of Sir2 as an NAD^+-dependent deacetylase initiated the idea that this class of proteins may serve to sense the energy or nutrient status of cells.[5]

The sirtuins are highly conserved from prokaryotes to mammals. In mammals, seven sirtuins (SIRT1−7) occupy different cellular compartments. SIRT1, 6, and 7 reside in the nucleus and nucleolus; SIRT2 resides primarily in the cytoplasm. Three sirtuins (SIRT3, SIRT4, and SIRT5) are localized in mitochondria and are of particular interest due to their role in regulating metabolism by removal of acyl-lysine posttranslational modifications. The sirtuins are divided into classes based on the conservation of a core domain that is approximately 250 amino acids.[6] Of the seven mammalian sirtuins, SIRT1−3 are class I, SIRT4 is class II, SIRT5 is class III, and SIRT6−7 are class IV. An additional class (class U) contains no mammalian sirtuins and is thought to be a precursor of class I and class IV sirtuins.[7] The seven mammalian sirtuins are thought to have evolved from the engulfment of a Proteobacterium (from which mitochondria originated) containing a class II and class U sirtuin by an Archean containing a class III sirtuin.[7]

7.2 ENZYMATIC ACTIVITY OF THE MITOCHONDRIAL SIRTUINS

The sirtuins were initially described as deacetylases that target only acetyl-lysine for removal, but the enzymatic activities of this family of proteins is broader than originally realized. Sirtuins are now

Introductory Review on Sirtuins in Biology, Aging, and Disease. DOI: https://doi.org/10.1016/B978-0-12-813499-3.00007-1

appreciated to target several acyl-lysine modifications for removal, including malonyl-,[8,9] succinyl-,[9] and glutaryl-lysine[10] modifications, as well as long-chain acyl-modifications[11,12] (Fig. 7.1). Acylation, referring to a wide range of modifications, of mitochondrial proteins is generally repressive,[13] though lysine acylation has been described to activate enzyme activity in a few cases.[14]

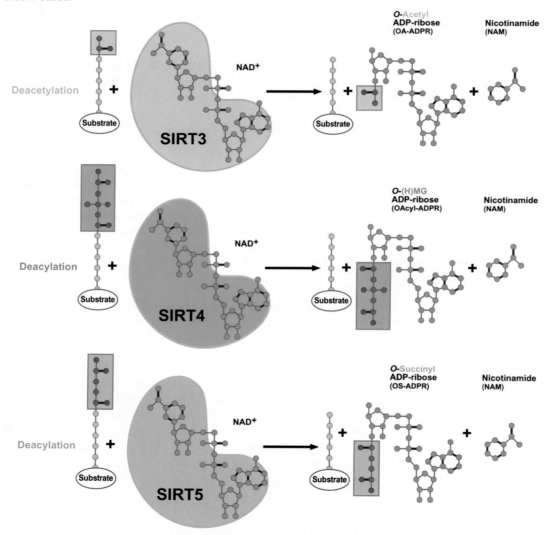

FIGURE 7.1 Mitochondrial sirtuins deacylate distinct protein modifications.

Each sirtuin deacylation reaction targets distinct protein acyl-modifications for removal and generates the corresponding *O*-acyl-ADP-ribose moiety; (top) SIRT3 is a potent deacetylase; (middle) SIRT4 is a dehydroxymethylglutarylase and demethylglutarylase; (bottom) SIRT5 is a potent desuccinylase.

SIRT3 is a lysine deacetylase that controls several pathways, including lipid metabolism and oxidative stress.[15] The enzymatic activity of SIRT4 was previously described as an ADP-ribosyltransferase,[16] a deacetylase,[17] and a lipoamidase;[18] more recently, we found robust deacylase activity, where SIRT4 removes methylglutaryl, hydroxymethylglutaryl, and 3-methylglutaconyl groups from lysine residues.[19] SIRT4 has been shown to influence the metabolism of amino acids,[19] metabolism of lipids, and the TCA cycle.[16,20] SIRT5 is a lysine demalonylase, desuccinylase, and deglutarylase which controls several metabolic pathways, including the urea cycle.[10]

7.3 PROTEOMIC CHARACTERIZATION OF ACYL-LYSINE MODIFICATIONS

While the enzymatic activities of mitochondria are becoming better characterized, the breadth and depth of acylation remains a nascent field of investigation. In order to begin to answer this question, a number of proteomic surveys have been conducted to determine the landscape of acetylation, succinylation, malonylation, and glutarylation. In one early study, acetylation accumulated on mitochondrial proteins in the absence of SIRT3.[21] Using an unbiased approach, glutamate dehydrogenase was identified as an acetylated protein regulated by SIRT3. Since these initial discoveries of SIRT3 substrates, more sensitive and high-throughput proteomic approaches to identify acylated proteins have been developed. In 2009, 3600 lysine acetylation sites on 1750 proteins were identified in cellular processes including DNA damage and repair, chromatin remodeling, and cell cycle.[22] This study showed the breadth of lysine acetylation. More recently, acyl proteomics performed on SIRT3- and SIRT5-deficient tissues compared to wildtype tissues have provided insight as to what proteins and pathways are regulated by sirtuin-mediated deacylation.

In a study of SIRT3KO and WT mouse livers, 2187 sites of lysine acetylation were identified. Interestingly, about 14% and 17.5% of these sites were unique to WT and SIRT3KO livers, respectively, suggesting that the landscape of acetylation is dynamic and partially regulated by SIRT3. By analyzing proteins with twofold or higher changes in acetylation in the absence of SIRT3, the authors of this study identified a number of metabolic pathways that are likely regulated by SIRT3. The top pathways include fatty acid oxidation, the TCA cycle, branched chain amino acid catabolism, ketone body metabolism, and the electron transport chain.[23] Similar approaches have identified SIRT3 as a potential regulator of metabolic fuel switching and have suggested that regulation may be tissue-specific.[24]

In contrast to SIRT3-mediated deacetylation, SIRT5 removes succinyl, malonyl, and glutaryl groups. Addition of succinyl, malonyl, and glutaryl groups changes the charge of the modified lysine from a positive to negative change. The unique acyl-binding pocket of SIRT5 coordinates the interaction of carboxylates with this enzyme in order for the deacylation reaction to occur.[9] The role of SIRT5-mediated desuccinylation is currently the most well explored of these three posttranslational modifications. A comprehensive survey of the succinylome in mouse liver tissue revealed 2565 succinylation sites on 779 proteins, indicating that the number of succinylated proteins is comparable to the number of acetylated proteins. Similar to acetylation, succinylation in SIRT5KO tissue occurs on a statistically significant number of proteins in metabolic pathways including branched chain amino acid degradation, the TCA cycle, and fatty acid oxidation.[14] Complementing this succinylome survey, another succinylome survey in SIRT5KO and WT mouse livers identified

1576 sites of lysine succinylation on 392 proteins. Pathway analysis of SIRT5 targets (twofold higher succinylation in the SIRT5KO liver) indicated that SIRT5 likely regulates multiple metabolic pathways including fatty acid oxidation, branched chain amino acid catabolism, the TCA cycle, ATP synthesis, and ketone body synthesis.[25] Interestingly, investigation of the malonylome identified 1137 sites of lysine malonylation on 430 proteins with many sites occurring on cytoplasmic proteins.[26] In this study, it was determined that glycolysis was the main pathway regulated by SIRT5-mediated demalonylation.[26] Characterization of the glutarylome reported fewer sites of lysine glutarylation (683 lysines in 191 proteins), suggesting that this modification is not abundant or development of a more sensitive *pan*-glutaryl-lysine antibody is necessary to fully understand the glutarylation landscape.[10]

7.4 THE ROLES OF MITOCHONDRIAL SIRTUINS UNDER BASAL CONDITIONS

The proteomic studies described above provide a roadmap of the possible physiological targets of the mitochondrial sirtuins. Of the mitochondrial sirtuins, SIRT3 has been studied most extensively (for a recent review see Ref. [27]), with extensive integration of SIRT3 targets with overall physiological phenotypes. Until the recent discovery of SIRT4 enzymatic activity, a complete understanding of SIRT4 and integration with physiologic roles has been difficult. Despite a thorough investigation of the acylation landscape and understanding of SIRT5 enzymatic activity, comparatively little is known about the physiological role of SIRT5. Part of this challenge stems from the fact that genetic mouse models of SIRT3 ablation, SIRT4 ablation, or SIRT5 ablation show few phenotypes at baseline. Indeed, no overt phenotypes are known in any of the mitochondrial sirtuin mouse knockout models under basal conditions. This is somewhat surprising given the dramatic increase in acyl-lysine modifications regulated by the sirtuins—hyperacetylation in SIRT3KO mice;[15] and hypermalonylation,[26] hypersuccinylation,[25,26,28] and hyperglutarylation[10] in SIRT5KO mice. Instead, it appears that hyperacylation may contribute to the susceptibility of disease states and contribute to diminished healthspan. For example, loss of SIRT3 has been shown to accelerate the development of metabolic syndrome in a mouse model of high-fat diet feeding,[29] and is associated with accelerated progression of several diseases of aging.[30] Additionally, SIRT5 ablation may lead to reduced flux through fatty acid oxidation[25,31] and ketogenesis,[25] and accelerated progression of some diseases of aging, including cardiac hypertrophy.[31]

7.5 PHYSIOLOGICAL ROLES OF MITOCHONDRIAL SIRTUINS IN THE HEART

The heart has been commonly used to study control of energy metabolism by the mitochondrial sirtuins. The adult heart is considered an "omnivore" and metabolically flexible, in that it can use many energy substrates including fatty acids, glucose, lactate, pyruvate, ketones, and amino acids to generate ATP to meet energy demands.[32] The heart requires an enormous amount of ATP to maintain constant pumping to deliver oxygenated blood throughout the body. Although metabolically flexible, the preferred substrates of the heart are fatty acids and glucose. Under normal conditions, 60%−90% of ATP generated comes from oxidation of fatty acids, and 10%−40% of ATP is generated from glucose oxidation.[33] Reserve energy is stored in the heart as phosphocreatine, and phosphoryl transfer to ADP via creatine kinase generates ATP approximately 10 times faster than ATP synthesis in the mitochondria and is thus important to meet the demands of the heart during increased workloads.[34]

Compared to the known cardioprotective role of SIRT3 in the heart,[35] the roles of SIRT4 and SIRT5 in maintaining normal cardiac function have only recently been studied. Preliminary work on the role of SIRT4 in the heart suggests that it might be detrimental to cardiac function. In one study, Ang II-induced cardiac hypertrophy and fibrosis were reduced in a mouse model of SIRT4 deficiency.[36] Additionally, early studies with a mouse model of SIRT4 overexpressed in cardiomyocyte show more pronounced cardiac dysfunction and dilation after 12 weeks of pressure-overload-induced cardiac hypertrophy.[37] Finally, inhibition of SIRT4 by microRNA-497 inhibits cardiac hypertrophy.[38] While understanding the specific mechanisms by which SIRT4 functions in the heart requires further investigation, these early studies suggest SIRT4 controls metabolic pathways that are detrimental to some models of cardiac stress.

In contrast to SIRT4, early studies on SIRT5 in the heart are showing a cardioprotective role. The level of SIRT5 protein is high in the heart and succinylation increases dramatically in SIRT5KO compared to WT hearts, suggesting that SIRT5-mediated desuccinylation may be important in cardiac function and metabolism. In an early study describing the metabolic characterization of *Sirt5*$^{-/-}$ (SIRT5KO) mice, no overt cardiac differences were found, suggesting that SIRT5 is metabolically unremarkable.[28] Specifically, heart weight, heart rate, and systolic blood pressure of SIRT5KO animals undergoing the stress of a chronic high-fat diet remained unchanged compared to WT controls. However, phenotypes in sirtuin knockout mouse models often require a stress to elicit a phenotype; e.g., a recent study showed a cardiac stressor is required to study SIRT5 in heart.[39] Specifically, in a model of ischemia-reperfusion, SIRT5KO mouse hearts had a greater infarct area postreperfusion compared to WT controls, as well as slightly elevated ROS.[39] In this study, succinylation activated succinate dehydrogenase (SDH) in the SIRT5KO heart, and inhibition of SDH normalized infarct size to wildtype levels.[39] Together, these data suggest that SIRT5 influences the cardiac stress response by decreasing SDH activity via desuccinylation. Beyond ischemia-reperfusion, aging was identified as another stress that elicits a phenotype in SIRT5KO mouse hearts.[31] At 39 weeks of age, the hearts of SIRT5KO mice showed increased hypertrophy and fibrosis compared to wildtype controls. Additionally, echocardiography showed that shortening fraction and ejection fraction in SIRT5KO hearts were decreased at 8 and 39 weeks of age, compared to wildtype controls.[31] In this study, inhibitory succinylation of the mitochondrial trifunctional protein alpha subunit was identified as one mechanism contributing to the observed cardiomyopathy in the SIRT5KO mouse heart.

We recently extended these findings by determining the role of SIRT5 in cardiac stress responses in an established model of pressure overload-induced heart muscle hypertrophy caused by transverse aortic constriction (TAC). Remarkably, we found that SIRT5KO mice had reduced survival upon TAC compared with wildtype mice, but exhibited no mortality when undergoing a sham control operation.[40] The increased mortality with TAC was associated with some signs of pathological hypertrophy and abnormalities in both cardiac performance and ventricular compliance. By combining high-resolution MS-based metabolomic and proteomic analyses of cardiac tissues from wildtype and SIRT5KO mice, we found evidence of biochemical abnormalities exacerbated in the SIRT5KO mice, including apparent decreases in fatty acid oxidation and glucose oxidation, as well as an overall decrease in mitochondrial NAD^+/NADH. Together, these findings support the model that SIRT5 deacylates protein substrates involved in cellular oxidative metabolism to maintain mitochondrial energy production. Overall, the functional and metabolic data might explain the finding that SIRT5KO mice develop cardiac dysfunction in response to TAC at an accelerated rate, explaining increased mortality upon cardiac stress. Thus, SIRT5 has emerged as a protein positioned to control cardiac oxidative metabolism in response to stress.

7.6 REACTIVE ACYL-COA SPECIES AND CARBON STRESS

While strong progress has been made towards understanding the role of sirtuin deacylation in metabolic control, less is known about the mechanisms leading to mitochondrial protein acylation. Emerging evidence supports the idea that lysine acylation can occur nonenzymatically in mitochondria.[41–43] We recently discovered that some metabolites are intrinsically more reactive than others,[42,44] and this intrinsic reactivity can explain some protein acylation. Thus, we propose a model where *reactive acyl-CoA species* (RACS) represent a rich source of diverse protein modifications. We predict increases in acyl-CoA levels would occur during the oxidation of fatty acids, glucose, and amino acids, and if not handled appropriately, would lead to increases in protein acylation and subsequent inhibition of protein function.[42] This is especially apparent in several genetic models of inborn errors in metabolism, which lack acyl-CoA-consuming enzymes; in several cases, increases in protein modifications corresponding to the increases in reactive acyl-CoA species are seen.[10,44,45] Consequently, enzyme function could be inhibited by acylation due to a change in enzyme structure or by affecting substrate or cofactor binding.[13]

If RACS are a common mechanism leading to protein modifications, then the mitochondrial sirtuins might serve to remove acyl-lysine modifications as a way to deal with carbon stress that comes from reactive species.[42] Indeed, metabolic stress is often associated with increased carbon metabolism. Elevated acyl-CoA species generated from nutrient oxidation may exceed acyl-CoA use, resulting in hyperacylation and decreased enzyme function. In this setting, sirtuins might act in response to these stressors to remove acyl-lysine modifications in order to restore enzymatic activity and maintain metabolic homeostasis (Fig. 7.2). While this is only a working model, the known role of mitochondrial sirtuins in stress response, coupled with the consistent links between acyl-CoA generation and protein modification, support this idea; future studies aimed directly at testing this model directly will determine how the sirtuins control metabolic homeostasis.

7.7 RACS, METABOLISM, AND HEART FAILURE

In the heart, many stimuli, including pressure overload, ischemia, and diabetes, can lead to altered metabolic fluxes.[46] With pressure overload, these shifts in metabolism lead to alterations in fatty acid and glucose oxidation, and concomitant, inverse shifts in glycolysis.[47] Shifts in cardiac metabolism with pathological hypertrophy appear to be driven by transcriptional changes, with downregulation of enzymes in fatty acid oxidation, an increase in glucose uptake by upregulation of GLUT1,

Reactive species	Modification	Structural motif	Pathway	Sirtuin
Acetyl-CoA	Acetyl		Pyruvate, lipid	SIRT3
(H)MG-CoA	(H)MG	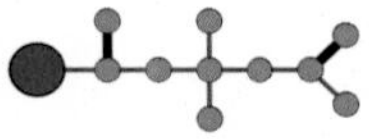	Leucine, lysine	SIRT4
Succinyl-CoA	Succinyl	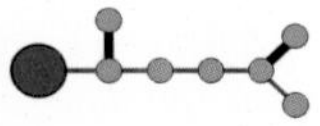	TCA cycle, BCAA	SIRT5

FIGURE 7.2 Reactive acyl-CoA species and carbon stress.

Reactive acyl-CoA species generated in discrete metabolic pathways lead to a unique protein modification that can be targeted by the mitochondrial sirtuins for removal.

and an increase in inhibition of PDH by PDK-mediated phosphorylation.[33] Importantly, an uncoupling of glycolysis from glucose oxidation is detrimental to cardiac efficiency[32] and defects in oxidative metabolism often precede the onset of cardiac dysfunction.[33] Furthermore, NAD^+ levels decline with cardiac hypertrophy,[48] which could contribute to a reduction in mitochondrial function[49] and disease pathogenesis.[50,51] Thus, the combined increases in RACS and reductions in NAD^+, and therefore sirtuin activity, are emerging as key contributors to heart failure (Fig. 7.3). Active studies on NAD^+ supplementation in heart failure will determine the therapeutic potential of manipulating this pathway.[35]

7.8 CONCLUSIONS

One major question in the field is whether acylation regulates metabolism, or is a consequence of normal and/or pathological shifts in metabolism. In the study of cardiac metabolism, one recent study proposed that alterations in acylation in the newborn heart contribute to shifts in metabolism, including a decrease in glycolysis and an increase in fatty acid oxidation.[52] The authors show that acetylation and succinylation increased in newborn rabbit hearts from postnatal day 1 to postnatal day 21. The data presented from this study are largely correlative, and changes in transcriptional regulation, such as an increase in PPARα protein expression could also explain the regulation of shifts in metabolism. The authors observed an increase in acetyl-CoA generated from fatty acid oxidation over this time period, which could explain the concomitant increase in protein acetylation. While succinyl-CoA is not directly measured, it is also plausible that shifts in metabolism (from glycolytic to more oxidative in the newborn heart) contribute to increased succinyl-CoA that leads to protein succinylation. Thus, more work is required to determine if, or how, acylation influences mitochondrial metabolism.

Further highlighting this major question are several succinyl proteomic studies in whole-body SIRT5KO mouse hearts, which revealed pathways in oxidative metabolism contain many proteins

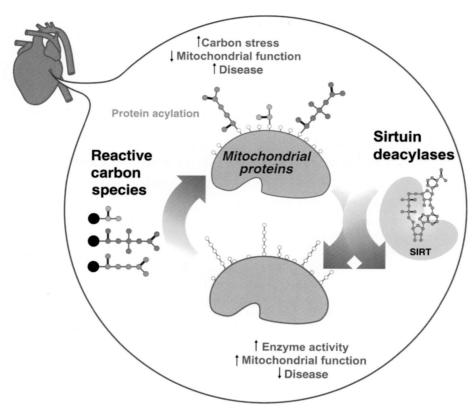

FIGURE 7.3 Overall working model of RACS, carbon stress, and cardiac metabolism.

In the setting of cardiac stress, reactive acyl-CoA species increase from elevated carbon metabolism and/or reduced consumption, which leads to increases in protein posttranslational modifications. The mitochondrial sirtuins are then positioned to remove protein modifications, restore enzymatic activity, and maintain mitochondrial function.

with multiple sites of succinylation. Major pathways that are succinylated in the absence of SIRT5 include fatty acid oxidation and the TCA cycle, suggesting that a main function of SIRT5 could be to maintain oxidative metabolism by its desuccinylation activity. However, one caveat of using pathway analyses is that only one site of succinylation per protein is analyzed. When examining individual pathways, such as fatty acid oxidation, we find that a majority of identified sites of lysine succinylation have small changes in succinylation when SIRT5 is ablated.[40] In contrast, only a few sites of succinylation increase over 100-fold in SIRT5KO mouse hearts. How succinylation at these sites influences enzyme function remains unknown. One previous study found a key role for succinylation in controlling the fatty acid oxidation enzyme HADHA; they further found that the mutation of one lysine to arginine (K351) led to a significant loss of HADHA function.[31] These types of data reveal a major question that remains in the field: whether the effect of hypersuccinylation is due to specific SIRT5 targets, the sum of all succinylation events, or if some sites of acylation are simply phenomenological.

In conclusion, the relationship between reactive acyl-CoA species and protein acylation is becoming well-established. Several acyl-CoA species are known to be or have the chemical properties to be reactive. These reactive carbon species (RACS) are well-positioned to modify proteins and induce a wide range of posttranslational modifications. Further studies are required to better understand if specific sites of lysine acylation contribute to decreased enzyme function and how the global acyl-lysine landscape and sirtuin loss leads to increased susceptibility to disease. Indeed, several exciting, unanswered questions in this field await discovery.

ACKNOWLEDGMENTS

We would like to thank Dr. Frank K. Huynh for thoughtful feedback. Funding on cardiac metabolism and RACS in the Hirschey lab comes from the American Heart Association grants 12SDG8840004 and 12IRG9010008, the National Institutes of Health and the NIA grant R01AG045351; KAH was supported by an NIH/NIGMS training grant to Duke University Pharmacological Sciences Training Program (5T32GM007105-40) and is supported by 1F31HL127959-01. The authors apologize to colleagues whose work was not cited due to space limitations or oversight. Please bring errors and egregious omissions to our attention.

REFERENCES

1. Anderson KA, Green MF, Huynh FK, Wagner GR, Hirschey MD. SnapShot: mammalian sirtuins. *Cell* 2014;**159**(4):956−956.e1.
2. Ivy JM, Klar AJ, Hicks JB. Cloning and characterization of four SIR genes of *Saccharomyces cerevisiae*. *Mol Cell Biol ASM* 1986;**6**(2):688−702.
3. Rine J, Herskowitz I. Four genes responsible for a position effect on expression from HML and HMR in *Saccharomyces cerevisiae*. *Genetics* 1987;**116**(1):9−22.
4. Kaeberlein M, McVey M, Guarente L. The SIR2/3/4 complex and SIR2 alone promote longevity in *Saccharomyces cerevisiae* by two different mechanisms. *Genes Dev* 1999;**13**(19):2570−80.
5. Imai S, Armstrong CM, Kaeberlein M, Guarente L. Transcriptional silencing and longevity protein Sir2 is an NAD-dependent histone deacetylase. *Nature* 2000;**403**(6771):795−800.
6. Frye RA. Phylogenetic classification of prokaryotic and eukaryotic Sir2-like proteins. *Mol Cell* 2000;**273**(2):793−8.
7. Frye RA. *Evolution of Sirtuins From Archaea to Vertebrates*. In: Verdin E. (eds) Histone Deacetylases. Cancer Drug Discovery and Development. Springer: Humana Press; 2006. p. 183−202.
8. Peng C, Lu Z, Xie Z, Cheng Z, Chen Y, Tan M, et al. The first identification of lysine malonylation substrates and its regulatory enzyme. *Mol Cell Proteomics* 2011 Dec 6;**10**(12). M111.012658−8.
9. Du J, Zhou Y, Su X, Yu JJ, Khan S, Jiang H, et al. Sirt5 is a NAD-dependent protein lysine demalonylase and desuccinylase. *Science* 2011;**334**(6057):806−9.
10. Tan M, Peng C, Anderson KA, Chhoy P, Xie Z, Dai L, et al. Lysine glutarylation is a protein posttranslational modification regulated by SIRT5. *Cell Metabol* 2014;**19**(4):605−17.
11. Feldman JL, Baeza J, Denu JM. Activation of the protein deacetylase SIRT6 by long-chain fatty acids and widespread deacylation by mammalian sirtuins. *J Biol Chem* 2013;**288**(43):31350−6.

12. Madsen AS, Andersen C, Daoud M, Anderson KA, Laursen JS, Chakladar S, et al. Investigating the sensitivity of NAD + -dependent sirtuin deacylation activities to NADH. *J Biol Chem* 2016;**291**(13):7128−41.
13. Chhoy P, Anderson KA, Hershberger KA, Huynh FK, Martin AS, McDonnell E, et al. Deacetylation by SIRT3 relieves inhibition of mitochondrial protein function. In: Sirtuins. Sirtuins, 2016, pp. 105−38.
14. Park J, Chen Y, Tishkoff DX, Peng C, Tan M, Dai L, et al. SIRT5-mediated lysine desuccinylation impacts diverse metabolic pathways. *Mol Cell* 2013;**50**(6):919−30.
15. Hirschey MD, Shimazu T, Goetzman E, Jing E, Schwer B, Lombard DB, et al. SIRT3 regulates mitochondrial fatty-acid oxidation by reversible enzyme deacetylation. *Nature* 2010;**464**(7285):121−5.
16. Haigis MC, Mostoslavsky R, Haigis KM, Fahie K, Christodoulou DC, Murphy AJ, et al. SIRT4 inhibits glutamate dehydrogenase and opposes the effects of calorie restriction in pancreatic beta cells. *Cell* 2006;**126**(5):941−54.
17. Michishita E, Park JY, Burneskis JM. Evolutionarily conserved and nonconserved cellular localizations and functions of human SIRT proteins. *Mol Biol Cell* 2005;**16**(10):4623−35.
18. Mathias RA, Greco TM, Oberstein A, Budayeva HG, Chakrabarti R, Rowland EA, et al. Sirtuin 4 is a lipoamidase regulating pyruvate dehydrogenase complex activity. *Cell* 2014;**159**(7):1615−25.
19. Anderson KA, Huynh FK, Fisher-Wellman K, Stuart JD, Peterson BS, Douros JD, et al. SIRT4 is a lysine deacylase that controls leucine metabolism and insulin secretion. *Cell Metabol* 2017;**25**(4). 838−855.e15.
20. Jeong SM, Xiao C, Finley LWS, Lahusen T, Souza AL, Pierce K, et al. SIRT4 has tumor-suppressive activity and regulates the cellular metabolic response to DNA damage by inhibiting mitochondrial glutamine metabolism. *Cancer Cell* 2013;**23**(4):450−63.
21. Lombard DB, Alt FW, Cheng H-L, Bunkenborg J, Streeper RS, Mostoslavsky R, et al. Mammalian Sir2 homolog SIRT3 regulates global mitochondrial lysine acetylation. *Mol Cell Biol* 2007;**27**(24):8807−14.
22. Choudhary C, Kumar C, Gnad F, Nielsen ML, Rehman M, Walther TC, et al. Lysine acetylation targets protein complexes and co-regulates major cellular functions. *Science* 2009;**325**(5942):834−40.
23. Rardin MJ, Newman JC, Held JM, Cusack MP, Sorensen DJ, Li B, et al. Label-free quantitative proteomics of the lysine acetylome in mitochondria identifies substrates of SIRT3 in metabolic pathways. *Proc Natl Acad Sci U S A* 2013 Apr 1;**110**(16):6601−6.
24. Dittenhafer-Reed KE, Richards AL, Fan J, Smallegan MJ, Fotuhi Siahpirani A, Kemmerer ZA, et al. SIRT3 mediates multi-tissue coupling for metabolic fuel switching. *Cell Metabol* 2015;**21**(4):637−46.
25. Rardin MJ, He W, Nishida Y, Newman JC, Carrico C, Danielson SR, et al. SIRT5 regulates the mitochondrial lysine succinylome and metabolic networks. *Cell Metabol* 2013;**18**(6):920−33.
26. Nishida Y, Rardin MJ, Carrico C, He W, Sahu AK, Gut P, et al. SIRT5 regulates both cytosolic and mitochondrial protein malonylation with glycolysis as a major target. *Mol Cell.* 2015 Jun 9;**59**(2):321−32.
27. van de Ven RAH, Santos D, Haigis MC. Mitochondrial sirtuins and molecular mechanisms of aging. *Trends Mol Med* 2017;**23**(4):320−31.
28. Yu J, Sadhukhan S, Noriega LG, Moullan N, He B, Weiss RS, et al. Metabolic characterization of a Sirt5 deficient mouse model. *Sci Rep* 2013;**3**:2806.
29. Hirschey MD, Shimazu T, Jing E, Grueter CA, Collins AM, Aouizerat B, et al. SIRT3 deficiency and mitochondrial protein hyperacetylation accelerate the development of the metabolic syndrome. *Mol Cell* 2011;**44**(2):177−90.
30. McDonnell E, Peterson BS, Bomze HM, Hirschey MD. SIRT3 regulates progression and development of diseases of aging. *Trends Endocrinol Metab* 2015 Jun 29;**26**(9):486−92.
31. Sadhukhan S, Liu X, Ryu D, Nelson OD, Stupinski JA, Li Z, et al. Metabolomics-assisted proteomics identifies succinylation and SIRT5 as important regulators of cardiac function. *Proc Natl Acad Sci U S A* 2016 Apr 5;**113**(16):4320−5.
32. Masoud WG, Clanachan AS, Lopaschuk GD. *The Failing Heart: Is It an Inefficient Engine or an Engine Out of Fuel?* New York: Springer; 2013. p. 65−84.

33. Sankaralingam S, Lopaschuk GD. Cardiac energy metabolic alterations in pressure overload-induced left and right heart failure (2013 Grover Conference Series). *Pulm Circ* 2015;**5**(1):15—28.

34. Ingwall JS. Energy metabolism in heart failure and remodelling. *Cardiovas Res* 2009;**81**(3):412—19.

35. Hershberger KA, Martin AS, Hirschey MD. Role of NAD(+) and mitochondrial sirtuins in cardiac and renal diseases. *Nat Publishing Group* 2017;**13**(4):213—25.

36. Luo Y-X, Tang X, An X-Z, Xie X-M, Chen X-F, Zhao X, et al. Sirt4 accelerates Ang II-induced pathological cardiac hypertrophy by inhibiting manganese superoxide dismutase activity. *Eur Heart J* 2016 Apr 20;**38**(18):1389—98.

37. Koentges C, Doerfer E, Pfeil K, Birkle S, Hoelscher M, Hoffmann MM, et al. P453: overexpression of SIRT4 accelerates the development of heart failure. Eur Heart J [Internet]. 2017;38(1):1—1. Available from: https://doi.org/10.1093/eurheartj/ehx501.P453.

38. Xiao Y, Zhang X, Fan S, Cui G, Shen Z. MicroRNA-497 inhibits cardiac hypertrophy by targeting sirt4. *PLoS One* 2016;**11**(12):e0168078.

39. Boylston JA, Sun J, Chen Y, Gucek M, Sack MN, Murphy E. Characterization of the cardiac succinylome and its role in ischemia-reperfusion injury. *J Mol Cell Cardiol* 2015;**88**:73—81.

40. Hershberger KA, Abraham DM, Martin AS, Mao L, Liu J, Gu H, et al. Sirtuin 5 is required for mouse survival in response to cardiac pressure overload. *J Biol Chem* 2017 Oct 2. Available from: http://dx.doi.org/10.1074/jbc.M117.809897.

41. Wagner GR, Payne RM. Mitochondrial acetylation and diseases of aging. *J. Aging Res* 2011;**2011**(5):1—13 Hindawi Publishing Corporation.

42. Wagner GR, Hirschey MD. Nonenzymatic protein acylation as a carbon stress regulated by sirtuin deacylases. *Mol Cell* 2014;**54**(1):5—16 Elsevier Inc.

43. Ghanta S, Grossmann RE, Brenner C. Mitochondrial protein acetylation as a cell-intrinsic, evolutionary driver of fat storage: biocehmical and metabolic logic of acetyl-lysine modifications. Crit Rev Biochem Mol Biol. 2013 Nov-Dec;48(6):561—74.

44. Wagner GR, Bhatt DP, O'Connell TM, Thompson JW, Dubois LG, Backos DS, et al. A class of reactive acyl-coa species reveals the non-enzymatic origins of protein acylation. *Cell Metabol* 2017;**25**(4):823—8.

45. Pougovkina O, Brinke te H, Wanders RJA, Houten SM, de Boer VCJ. Aberrant protein acylation is a common observation in inborn errors of acyl-CoA metabolism. *J Inherit Metab Dis* 2014;**37**(5):709—14.

46. Taegtmeyer H, Sen S, Vela D. Return to the fetal gene program: a suggested metabolic link to gene expression in the heart. *Ann NY Acad Sci* 2010;**1188**:191—8.

47. Lopaschuk GD. Metabolic modulators in heart disease: past, present, and future. *Can J Cardiol* 2017;**33**(7):838—49.

48. Katsyuba E, Auwerx J. Modulating NAD(+) metabolism, from bench to bedside. *EMBO J* 2017 Aug 7;**36**(18):2670—83.

49. Braidy N, Guillemin GJ, Mansour H, Chan-Ling T, Poljak A, Grant R. Age related changes in NAD + metabolism oxidative stress and Sirt1 activity in wistar rats. *PLoS One* 2011;**6**(4):e19194.

50. Mericskay M. Nicotinamide adenine dinucleotide homeostasis and signalling in heart disease: pathophysiological implications and therapeutic potential. *Arch Cardiovas Dis* 2015 Dec 18;**109**(3):207—15.

51. Forbes JM. Mitochondria-power players in kidney function? *Trends Endocrinol Metab* 2016;**27**(7):441—2.

52. Fukushima A, Alrob OA, Zhang L, Wagg CS, Altamimi T, Rawat S, et al. Acetylation and succinylation contribute to maturational alterations in energy metabolism in the newborn heart. *AJP: Heart Circul Physiol* 2016;**311**(2):H347—63.

MITOCHONDRIAL SIRTUINS: COORDINATING STRESS RESPONSES THROUGH REGULATION OF MITOCHONDRIAL ENZYME NETWORKS

8

Wen Yang[1,2], Robert A.H. van de Ven[1] and Marcia C. Haigis[1]

[1]Harvard Medical School, Boston, MA, United States [2]Shanghai Jiao Tong University, School of Medicine, Shanghai, China

8.1 INTRODUCTION

Mitochondria are the hub for numerous metabolic pathways such as fatty acid oxidation (FAO), the tricarboxylic acid (TCA) cycle, electron transport (ETC), and ATP synthesis. Many metabolic intermediates generated in the mitochondria serve as precursors in anabolic and catabolic processes. Other metabolic intermediates such as acyl-CoA, a group of coenzymes including succinyl-CoA and acetyl-CoA, can be used for the posttranslational modification of numerous mitochondrial enzymes. These modifications occur on lysine residues and have the potential to positively or negatively influence the activity of the target protein. Interestingly, one-third of the mitochondrial proteins are acetylated and/or succinylated and the majority of the proteins in the acetylome and succinylome are directly involved in energy metabolism.[1−3]

Given that the majority of mitochondrial proteins are synthesized in the cytosol and have to be imported into mitochondria to perform their function, one could argue that mitochondrial proteins are acylated before entering the mitochondria. However, the finding that proteins encoded in the mtDNA, and therefore are translated in the mitochondria, are acetylated, such as the ATP synthase subunit 8 (MT-ATP8), demonstrates that mitochondria harbor functional acetylation machinery.[4] While the lysine acetyltransferases (KAT) family mediates protein lysine acetylation in the nucleus and the cytosol, mitochondrial protein acetylation may not be mediated through the enzymatic activity of acetyltransferases but instead may occur nonenzymatically.[5] While the exact machinery behind mitochondrial protein acylation remains to be determined, the biologies of deacetylation and deacylation in the mitochondria are better understood. Research over the past decade has identified the mitochondrial sirtuins as the major deacylases in the mitochondria.

Introductory Review on Sirtuins in Biology, Aging, and Disease. DOI: https://doi.org/10.1016/B978-0-12-813499-3.00008-3

Sir2 was originally characterized as one of the four genes that regulate the yeast mating type system in 1970−80.[6] Later studies established Sir2 as an NAD$^+$-dependent protein deacetylase.[7] In humans, there are seven Sir2 homologs (SIRT1−7); all of them share a catalytic domain similar to yeast Sir2. Of these, SIRT3 was first identified as a mitochondrial sirtuin, based on its N-terminal mitochondrial targeting sequence and mitochondrial matrix localization.[8] Subsequent studies identified SIRT4 and SIRT5 as mitochondrially localized sirtuins.[9] Later studies showed that both SIRT3 and SIRT5 display additional nonmitochondrial localization and functions.[10,11] In particular for SIRT5, growing evidence indicates this enzyme functions both in mitochondria and in cytosol.[12]

While SIRT3−5 all reside in the mitochondria, they possess nonredundant and distinct enzymatic activities. While knockout of SIRT4 or SIRT5 did not change the global acetylation status of mitochondrial proteins, cells deficient for SIRT3 demonstrated elevated acetylation, indicating SIRT3 to be a major mitochondrial deacetylase.[13] Analysis of the enzymatic activity of the different mitochondrial sirtuins revealed that SIRT5 possesses different deacylase activities, of which the desuccinylase activity is the most notable.[14] In contrast to SIRT3 and SIRT5, SIRT4 harbors low enzymatic activity in vitro, but has been shown to have ADP-ribosyl transferase and deacylase activity.[14−17]

In the following sections, the enzymatic activity and regulation of the different mitochondrial sirtuins will be discussed. Next, we focus on the mechanisms by which SIRT3−5 modulate mitochondrial homeostasis. Finally, we highlight the involvement of mitochondrial sirtuins in human diseases. For an overview of the mitochondrial programs and targets that are under control of the mitochondrial sirtuins, see Fig. 8.1.

8.2 SIRT3

SIRT3 is a major mitochondrial deacetylase and has been the best studied mitochondrial sirtuin since the first paper reporting on SIRT3 activity.[8] About 30 mitochondrial proteins have been individually studied as SIRT3 substrates (Fig. 8.1) and many more proteins have been identified by acetylomic and proteomic studies.[1,2,18] The amino acid sequence of SIRT3 shares a high degree of homology to the cytosolic sirtuin SIRT2 and while the major enzymatic function of SIRT3 is deacetylation, additional enzymatic activities toward different acyl-chain lysine modification, such as decanoylase, have been identified in vitro.[14] Through its enzymatic activity, SIRT3 regulates many aspects of mitochondrial biology including FAO, ETC, and ATP production. Not surprisingly, loss of SIRT3 has been implicated in many diseases including cardiac diseases, neurodegenerative diseases, and cancer, which further underlines the central function of SIRT3 in mitochondrial homeostasis.

8.2.1 SIRT3 SUBSTRATES AND PATHWAYS

8.2.1.1 SIRT3 regulates mitochondrial fuel utilization

Over the last few decades, numerous binding partners and substrates have been identified for SIRT3, including enzymes involved in FAO, amino acid metabolism, TCA cycle, ETC, and reactive oxygen species (ROS) defense. The first SIRT3 substrate identified was acetyl-CoA

SIRT3	
Program	*Target*
TCA cycle	IDH2, ACO2, OGDH, SDH, PDH complex
ETC/OXPHOS	Complex I, II, III, IV, ATP synthase
Amino acid metabolism	GDH, GOT2, OTC
Lipid metabolism	AceCS2, HMGCS2, LCAD, VLCAD
Protein homeostasis	MRPL10, LKB1, LONP1
Mitochondrial integrity	Ku70, Cyclophilin D
ROS balance	IDH2, FOXO3A, SOD2

SIRT4	
Program	*Target*
Amino acid metabolism	GDH, MCCC complex
Lipid metabolism	MCD
TCA cycle	PDH complex

SIRT5	
Program	*Target*
Lipid metabolism	HMGCS2, HADHA
Glycolysis	GAPDH, PKM2
ROS balance	IDH2, G6PD, SOD1
Urea cycle	CPS1

FIGURE 8.1

Numerous mitochondrial sirtuin substrates and interaction partners have been identified over recent decades shown are the different mitochondrial programs that are under the control of SIRT3, SIRT4, and SIRT5. Please note the overlap between programs and substrates. The proteins shown are either experimentally validated as bona fide substrates or as high-confidence interaction partners (see text for details and additional sirtuin targets). *ETC*, electron transport chain; *OXPHOS*, oxidative phosphorylation; *ROS*, reactive oxygen species; *TCA cycle*, tricarboxylic acid cycle.

synthetase (AceCS2), an enzyme involved in fatty acid metabolism.[19] Later studies identified long-chain acyl coenzyme A dehydrogenase (LCAD), which breaks down long-chain fatty acids, as a SIRT3 target.[20] In addition, SIRT3 activity was found to be necessary to localize very long-chain acyl-CoA dehydrogenase (VLCAD) to the mitochondrial membrane by increasing its binding to cardiolipin and thereby increasing its activity.[21] SIRT3 was also identified as a regulator of ketogenesis through deacetylation and activation of 3-hydroxy-3-methylglutaryl CoA synthase 2 (HMGCS2).[22] Together, these studies establish SIRT3 as a positive regulator of FA metabolism.

Around the same time that the SIRT3 substrates involved in fatty acid metabolism were identified, the first SIRT3 substrate involved in amino acid metabolism, glutamate dehydrogenase (GDH or GLUD1), was identified.[23] GDH connects amino acid metabolism to the TCA cycle by deaminating glutamate to form α-ketoglutarate. SIRT3 deacetylates and activates GDH and thereby functions as a positive regulator of amino acid metabolism.[23] In addition, SIRT3 deacetylases and activates ornithine transcarbamoylase to promote flux through the urea cycle.[19]

Another major metabolism pathway involved in energy metabolism regulated by SIRT3 is glucose metabolism. The pyruvate dehydrogenase (PDH) complex has been reported to be deacetylated and activated by SIRT3.[24] Moreover, mitochondrial pyruvate carrier which mediates the translocation of pyruvate to mitochondria[25] and pyruvate phosphatase (PDP1), which dephosphorylates and activates PDH, are also activated by SIRT3.[26] While the above studies support a model in which SIRT3 is a positive regulator of oxidative glucose metabolism, conflicting data have been reported. For instance, the binding between malate dehydrogenase (MDH2) and glutamic-oxaloacetic transaminase (GOT2) was observed to be decreased with SIRT3-mediated deacetylation of GOT2. In turn, the malate-asparate shuttle is inhibited and the mitochondrial utilization of NADH generated from glycolysis is decreased.[27] However, further studies are needed to understand the physiological relevance of this interaction.

8.2.1.2 SIRT3 promotes the TCA cycle and OXPHOS

Many substrates of SIRT3 operate in the TCA cycle including aconitase (ACO2),[28] isocitrate dehydrogenase (IDH2),[29] oxoglutarate dehydrogenase (OGDH),[18] and succinate dehydrogenase (SDH).[30] All these substrates, except for ACO2, are activated upon deacetylation by SIRT3. In addition, citrate synthase (CS) is a SIRT3 interaction partner[18] and the acetylation level of CS was significantly increased in SIRT3 KO animals, suggesting that CS is a SIRT3 substrate.[1,31] In addition, the concentration of citrate, aconitate, isocitrate, and succinate was decreased in SIRT3 KO mouse embryonic fibroblasts (MEFs).[30] Together, these studies show that SIRT3 functions as an important activator of the TCA cycle through deacetylation and activation of many key enzymes.

The NADH generated by the TCA cycle is used by the ETC to build-up a proton gradient and eventually produce ATP through the action of ATP synthase. Not surprisingly, SIRT3 also promotes ETC components and ATP synthase activity. Downregulation of SIRT3 levels decreased mitochondrial membrane potential, prolonged membrane potential recovery, and decreased ATP production, indicating a central role for SIRT3 in coordination of ETC complex function.[18,32,33] SIRT3 KO animals display a twofold decrease in basal ATP levels and hyperacetylation of multiple complex I subunits, including NDUFA9, which further highlights SIRT3 as a regulator of OXPHOS.[34] Consistent with these findings, complex II, complex III, and complex V are all regulated by SIRT3.[24,30,35,36]

While deacetylation by SIRT3 activates most characterized substrates there are exceptions. Next to ACO2 and GOT2, acetylation activates argininosuccinate lyase, hydratase/3-hydroxyacyl−coenzyme A, and MDH.[37] In these scenarios, SIRT3-mediated deacetylation is expected to lower the activity of these enzymes, which can be explained by the notion that acetylation of specific lysine residues may cause different effects on target protein function. In addition, the observed changes in acetylation levels observed by Zhao et al. may not be all dependent on mitochondrial sirtuin activity.

Taken together, a bigger picture of SIRT3 function in energy metabolism emerges. The majority of enzymes contributing to energy metabolism are deacetylated and activated by SIRT3. Through regulation of these key enzymes, SIRT3 coordinates energy production through boosting the TCA cycle, thereby increasing ETC efficiency and ATP synthesis.

8.2.1.3 SIRT3 promotes mitochondrial integrity through balancing of ROS levels

By boosting mitochondrial metabolism, SIRT3 activity also increases a potentially dangerous side product of ETC function, namely ROS. In order to combat elevated ROS levels, SIRT3 directly and indirectly upregulates mechanisms involved in antioxidant responses.[38-40] SIRT3 directly activates superoxide dismutatse 2 (SOD2) to accelerate the clearance of ROS.[41] In addition, SIRT3 can indirectly contribute to glutathione production by activating IDH2 and increase NADPH levels.[29] SIRT3 also can prevent ROS damage to mitochondrial DNA (mtDNA) by deacetylation and stabilization of 8-oxoguanine-DNA glycosylase 1 (OGG1).[42] Through this regulation, SIRT3 improves mtDNA damage repair and protects the mitochondrial integrity under oxidative stress conditions. Increased levels of mitochondrial ROS will eventually lead to cell death. Through its interaction with cyclophilin D, Ku70, and FOXO3A, SIRT3 maintains mitochondrial integrity and promotes cell survival through mitochondrial and retrograde signaling. SIRT3 deacetylates cyclophilin D and prevents opening of the mitochondrial permeability transition pore (PTP), thereby preventing rapid loss of membrane potential, mitochondrial swelling, and cell death.[43] SIRT3 can also activate transcription of nuclear-encoded antioxidant genes through the activity of the transcription factor FOXO3A.[44] Finally, SIRT3 binds and deacetylates Ku70 to promote binding to BAX and suppresses apoptosis.[45] However, recent data have shown that SIRT3 deacetylates and activates glycogen synthase kinase-3β (GSK-3β) to induce expression and mitochondrial translocation of BAX to promote apoptosis in hepatocellular carcinoma.[46] Nonetheless, in general, SIRT3 functions as a promoter of mitochondrial integrity by balancing ROS levels and preventing mitochondrial membrane potential loss.

8.2.1.4 SIRT3 and protein homeostasis

SIRT3 further contributes to mitochondrial function by controlling mitochondrial protein homeostasis. SIRT3 has been demonstrated to associate with and to deacetylate the mitochondrial ribosome subunit MRPL10.[47] Here, deacetylation of MRPL10 by SIRT3 inhibits mitochondrial protein synthesis instead of increasing it. A similar effect of SIRT3 function on protein synthesis has been observed. Here, SIRT3 deacetylates liver kinase B1 (LKB1), thereby reducing the activity of the LKB1−AMPK pathway, and preventing subsequent induction of mTOR-mediated protein synthesis.[48] In addition, the protease LONP1, an enzyme governing the mitochondrial unfolded protein response (mitoUPR), has been suggested as an SIRT3 target.[49] Collectively, these studies highlight the role of SIRT3 in promoting mitochondrial energy metabolism, combating increased ROS levels, preventing mtDNA damage, preservation of mitochondrial integrity, and balancing mitochondrial protein synthesis. However, these findings may only reflect part of the wide range of SIRT3 functions. Acetylome studies in SIRT3 KO cells and interactome studies suggest that SIRT3 is also touching other aspects of mitochondrial biology including mitochondrial transcription machinery, protein folding and degradation, and protein import machinery.[1,2,18]

8.2.2 REGULATION OF SIRT3

Given the large number of SIRT3 substrates operating in the same metabolic pathway, it is likely that SIRT3 does not deacetylate individual substrates but instead simultaneously regulates clusters of enzymes to upregulate specific nodes of mitochondrial metabolism. This raises the important question of how SIRT3 is regulated under stress conditions. Similar to all the other members of the sirtuin family, SIRT3 is activated by NAD^+ and inhibited by nicotinamide.[50] During recent years, a number of additional regulatory mechanisms have been identified. Exercise, short-term food withdrawal, and caloric restriction (CR) have been reported to induce SIRT3 transcript levels in mice.[42,51,52] In the liver, SIRT3 levels increased upon fasting,[20] while a high-fat diet lowered SIRT3 levels.[53] In addition, increased ROS levels induced by antimycin A also increased the SIRT3 expression level, whereas treatment with the antioxidant N-acetylcysteine (NAC) reduced SIRT3 levels in neurons.[54] Loss of SIRT3 in MEFs induced increased ROS levels leading to genomic instability.[55] In addition, increased ROS due to SIRT3 loss can stabilize the transcription factor hypoxia inducible factor 1 alpha (HIF1α), thereby contributing to the Warburg effect.[56,57] A number of factors that regulate mitochondrial biogenesis control SIRT3 expression. In collaboration with NRF-2 and ERRα, PGC-1α increases SIRT3 transcription.[58] Moreover, SIRT1 can regulate SIRT3 expression through PGC1α or RELB in human blood monocytes upon oleic acid treatment and mouse splenocytes.[59,60]

At the posttranscriptional level it is currently unclear which mechanisms are regulating SIRT3 levels. Interestingly, loss of FABP4/aP2 in macrophages elevated the SIRT3 protein level without an apparent increase in SIRT3 mRNA.[61] Although the exact mechanisms are unclear, this study indicates mechanisms exist that regulate SIRT3 protein stability. An alternate mode of regulation stems from the observation that SIRT3 binds to the ATP synthase subunit ATP5O in a manner dependent on the integrity of the mitochondrial membrane potential.[18] Upon loss of membrane potential, and the subsequent pH decrease in the matrix, SIRT3 dissociates from ATP5O to upregulate mitochondrial metabolism and restore the proton gradient. Collectively, it is interesting to notice that the mechanisms regulating SIRT3 levels and function are related to mitochondrial biogenesis or energy metabolism, further highlighting SIRT3 as a central node in the regulation of mitochondrial homeostasis and critical to cellular stress responses.

8.2.3 SIRT3-RELATED DISEASES

SIRT3 is highly expressed in organs with a high metabolic turnover such as heart, liver, brain, brown adipose tissue, and kidney.[13,34] Not surprisingly, diseases associated with altered SIRT3 activity include heart diseases, neurodegenerative diseases, and cancer.

8.2.3.1 Heart disease

A growing body of work implicates SIRT3 in cardiac function. The expression level of SIRT3 is highly correlated with metabolic states and energy demand, such as the heart.[13] For example, SIRT3 loss correlated with decreased ATP production and decreased respiration in cardiomyocytes.[62] Interestingly, SIRT3 levels are downregulated in failing human hearts, although the specific mechanism behind this observation remains obscure.[63] However, given that SIRT3 and other mitochondrial proteins are decreased in mouse heart, the possibility remains that reduced levels of SIRT3 in this system are the consequence of mitochondrial degradation after heart failure.[64] Interestingly, SIRT3 levels are increased during induction of cardiac hypertrophy such as chronic infusion of the hypertrophic agonists phenylephrine (PE) or Ang II, and forced exercise,[45] and switching of SIRT3 isoforms was observed during hypertrophy.[65] While the shorter and longer SIRT3 isoforms are increased during mild hypertrophy, the shorter SIRT3 isoform is specifically downregulated during severe hypertrophy. Together, these observations indicate a potential protective role of SIRT3 during cardiomyocyte stress. However, SIRT3 deficiency by itself does not cause heart failure, as young mice display normal cardiac function with little sign of hypertrophy and interstitial fibrosis.[44,62] However, lack of SIRT3 renders the heart more vulnerable to aging and stress conditions.[65] In addition, application of hypertrophic stimuli caused a much more severe cardiac hypertrophic response in SIRT3 KO mice compared to wildtype mice.[44] Transverse aortic constriction caused cardiac hypertrophy, fibrosis, and overall reduced heart function in SIRT3 KO mice compated to wildtype mice, indicating reduced myocardial energetics.[62] Moreover, SIRT3 heterozygotes were more vulnerable to ischemia/reperfusion injury.[66] Additionally, aging is a major factor that distinguishes SIRT3 KO hearts: at 24 weeks, SIRT3 KO mice developed a decrease of ejection fraction and an increase in end diastolic volume.[62] On the contrary, restoration or overexpression of SIRT3 levels protected mice from heart injury, attenuated lipid accumulation, and protected mice from hypertrophic stimuli.[44,65] Interestingly, a number of treatments that improve the heart condition appear to function through SIRT3. NAD^+ restoration treatment blocked agonist-induced cardiac hypertrophy through SIRT3 activation.[48] Additionally, inhibition of cardiac hypertrophy by the small molecule honokiol requires SIRT3 activity.[67] Moreover, ANG-(1−7) can attenuate ANG II-induced mitochondrial ROS, and this beneficial effect was abolished by SIRT3 loss.[68] Despite the obvious beneficial effect of SIRT3 on cardiomyocyte and heart function, the exact mechanism remains largely unknown. Given that heart function relies on continuous high-energy delivery from OXPHOS, a possible explanation is that reduced SIRT3 activity hampers energy production. Indeed, SIRT3 loss decreased respiration as well as ATP production in cardiomyocytes accompanied by increased acetylation of mitochondrial proteins in cardiomyocytes.[62] However, this explanation alone is not sufficient to explain the observation that SIRT3 KO animals display no obvious defects in young mice but develop severe heart problems during aging and other stress conditions. One alternative explanation is that SIRT3 functions as a safeguard of mitochondrial energy homeostasis in tissues that consume a lot of energy and experience sporadical peaks of higher energy demand. In this situation, decreased SIRT3 levels would render tissues more vulnerable to environmental stress and the development of heart disease. In line with this hypothesis are the findings that SIRT3 is activated both transcriptionally and posttranslationally under stress conditions (see Section 8.2.2 for details). NAD^+ concentration and mitochondrial membrane potential can serve as regulatory factors to regulate SIRT3 function to reinstate energy homeostasis under stress conditions.[18] Next to global regulation of mitochondrial protein acetylation levels and maintenance of energy homeostasis, other SIRT3 substrates have drawn attention in the context of cardiomyocyte function. Through deacetylation of CypD, SIRT3 inhibits the opening of the PTP and prevents loss of mitochondrial integrity in aged cardiac muscle.[43] SIRT3 can also deacetylate and activate LKB1, thereby regulating the activity of the LKB1−AMPK pathway.[48] In addition, a dominant-negative mutant of Foxo3a eliminated the antihypertrophic effects of Sirt3 under PE treatment, indicating Sirt3 blocked cardiac hypertrophy by inducing antioxidant defence in a Foxo3a-dependent manner.[44] In summary, while SIRT3 activity has an overall protective effect on cardiomyocytes during stress, the precise mechanisms by which SIRT3 acts under hypertrophic and ischemic conditions remain largely unknown.

8.2.3.2 Neurodegenerative diseases

The central nervous system consumes a very high percentage of energy and neurons are particularly susceptible to oxidative damage and ROS has been implicated in many different neurodegenerative diseases.[69] Given the regulatory function of SIRT3 in the antioxidant response and energy metabolism, deregulation of SIRT3 activity has been associated with a number of neurological diseases including neurodegenerative diseases, hearing loss, and stroke-induced neuron damage. In Alzheimer's disease (AD), SIRT3 levels are downregulated in the frontal cortex, and are further decreased in Apolipoprotein E4 (APOE4) alleles patients. Mouse models for AD (3 × Tg AD and APOE4) phenocopied the downregulation of SIRT3.[70] A similar decrease of SIRT3 has been observed in the cortex of APP/PS1 double transgenic mice when compared to wildtype littermates,[71] but not in the cortex of the PDAPP mouse model.[54] Pituitary adenylate cyclase activating polypeptide (PACAP), a potent neurotrophic and neuroprotective, is downregulated in human AD and triple transgenic mouse (3 × TG, Psen1/APPSwe/TauP301L) brains. In addition, PACAP treatment protects cultured mouse cortical neurons against Aβ42 toxicity in an SIRT3-dependent manner.[72]

SIRT3 has also been linked to prevention of Parkinson's disease (PD). Age-dependent increases in mitochondrial oxidative stress are a major cause of dopaminergic neuron loss in PD.[73] Through deacetylation and activation of SOD2, SIRT3 has been postulated to play a beneficial role in PD.[74] Deletion of SIRT3 increased oxidative stress and decreased the membrane potential in the dopaminergic neuron, while overexpressing of K68R MnSOD, mimicking nonacetylated MnSOD, rescued these phenotypes.[74] Treatment with 1-methyl-4-phenyl-1,2,3,6-tetrahydropyridine (MPTP) increased mitochondrial oxidative stress and induced the PD phenotype in mice.[31] Similarly, SIRT3 deficiency dramatically exacerbated the degeneration of nigrostriatal dopaminergic neurons in MPTP-treated mice.[75,76] MPTP treatment increased protein acetylation of CS and IDH2 and reduced their enzymatic activities while overexpressed SIRT3 partially reversed the decline of CS activity.[31] There are fewer reports linking SIRT3 with Huntington disease. In cell model, overexpression of mutant huntingtin (HTT) reduced the SIRT3 level, while *trans*-(−)-ε-viniferin treatment attenuated HTT, causing oxidative stress and cell death in an SIRT3-dependent manner.[77] In cultured neurons, lacking SIRT3 rendered cells more vulnerable to excitotoxic, oxidative and metabolic stress. Upon treatment with 3-nitropropionic acid (3-NPA; succinate dehydrogenase inhibitor), SIRT3 loss increased the vulnerability of striatal and hippocampal neurons and increased mortality.[52]

Amyotrophic lateral sclerosis (ALS) has been characterized as a disease caused by increased oxidative stress and superoxide dismutase 1 (SOD1) mutations associate tightly with ALS.[78] Although SOD1 does not localize to the mitochondrial matrix, the deleterious effect of G93A-SOD1 in astrocytes is reverted by SIRT3, but not by SIRT1 overexpression.[72] SIRT3 protein levels are also elevated in the tibialis anterior muscle and spinal cord of aged G93A-SOD1 transgenic mice.[79] Ischemia/reperfusion caused by a brain stroke also increases oxidative damage in the brain, leading to loss of neuronal viablity.[80] In this setting, ketone treatment enhanced mitochondrial function, decreased oxidative damage, and improved neurologic function after ischemia in an SIRT3-dependent manner.[81] The SIRT3 expression level has also been found to be induced by oxygen and glucose deprivation, while overexpression of SIRT3 reduced LDH release and neuronal apoptosis under these conditions.[82]

8.2.3.3 Acute kidney injury

The kidney, along with the brain and the heart, is one of the organs that consumes the most energy. Similar to the role in the heart and brain, SIRT3 function protects the kidney against stress-induced injury.[83] Cisplatin treatment decreased the SIRT3 protein level and caused acute kidney injury, while treatment with the AMPK agonist AICAR or ALCAR restored SIRT3 protein levels and improved renal function.[83] In contrast, loss of SIRT3 completely abolished these beneficial effect of AICAR and ALCAR in mice.[9] In cultured proximal tubular cells, SIRT3 ameliorated saturated palmitate-stimulated ROS production.[84] Moreover, in kidney proximal tubular cells, inhibition of poly(ADP-ribose) polymerase 1 (PARP1) improved the expression and activity of antioxidant enzymes after cisplatin-induced injury, while SIRT3 deficiency abolished these beneficial effects.[85]

8.2.3.4 Cancer

The role of SIRT3 in cancer cannot be simply defined as oncogenic or tumor suppressing. Although SIRT3 was originally identified as a tumor suppressor, elevated SIRT3 levels have been observed in some cancer types indicating that SIRT3 may possess protumorigenic functions.[86] These contrasting observations may be due to the central function of SIRT3 in mediating certain cell type-specific stress responses.

SIRT3 levels are decreased across human cancers, including testicular, glioblastoma multiforme, prostate, clear cell renal, breast and gastric cancers. In liver cancer, decreased SIRT3 levels are associated with increased recurrence, decreased survival rate, differentiation grade and portal vein tumor thrombus.[87] SIRT3 levels are also decreased in head and neck squamous cell carcinoma and further downregulated during the advanced stages of this disease.[88] Moreover, SIRT3 was significantly downregulated in metastasized ovarian cancer.[89] Next to these correlative studies, the role of SIRT3 in cancer has been examined in a mechanistic fashion in cell lines and mouse models. SIRT3 KO mice spontaneously develop tumors that resemble human luminal B breast cancer and displayed high ROS and elevated SOD2 acetylation.[90] In breast cancers, SIRT3 loss results in HIF1alpha stabilization, induction of glycolytic gene expression, and metabolic reprogramming to fuel tumor growth.[56,57] In cultured lung adenocarcinoma cells, overexpression of SIRT3 increased the BAX/BCL-2 and BAD/BCL-x/L ratios, upregulated p53 and p21 protein levels, elevated ROS levels, and induced apoptosis.[91] In gastric cancer cells, overexpression of SIRT3 suppressed proliferation and colony formation.[92] Moreover, overexpression of SIRT3 in malignant B cell lines reduced cell growth and diminished the Warburg-like phenotype probably by reducing ROS production through deacetylation of IDH2.[93] On the other hand, depletion of SIRT3 enhanced tumor cell growth and colony formation in gastric cancer cells.[92] Lack of SIRT3 also rendered malignant B cell lines more susceptible to ROS scavenger[93] and rescued human retinal pigment epithelial cell line from Bcl-2 siRNA-induced G1 cell cycle arrest.[94]

It is still not clear which mitochondrial targets of SIRT3 contribute most to its antitumor activity. Mechanistically, the antioxidant function of SIRT3 seems to involve deacetylation and activation of IDH2 and SOD2 in cancers.[90,93] Increased ROS levels caused by SIRT3 loss stabilized HIF-1α and contributed to the Warburg effect, a hallmark of many cancers.[56,57] In addition, SIRT3 loss and increased oxidative stress resulted in increased genomic instability.[55] Deacetylation by SIRT3 induced apoptosis through a BCL-2-dependent pathway.[94] SIRT3 loss also renders cancer cells more vulnerable to anticancer treatment. Overexpression of SIRT3 caused leukemia cells

K562 to be more sensitive to kaempferol[95] and rendered p53-deficient nonsmall-cell lung cancers more susceptible to minnelide/triptolide treatment by rescuing mitochondrial bioenergetics.[96] In sum, SIRT3 functions as a tumor suppressor in many different cancer types by balancing ROS levels. As numerous metabolic pathways contribute to ROS generation, it will be interesting for future studies to rigorously probe the source of ROS generation regulated by SIRT3 in cancer.

SIRT3 may also function as an oncogene in different cancer types. In nonsmall-cell lung cancer tissue, oral cancer, melanoma, and esophageal cancer, SIRT3 levels are increased.[86] Moreover, high SIRT3 levels are correlated with shorter overall survival duration, whereas reduced SIRT3 levels cause induced sensitivity to radiation and increased cisplatin sensitivity.[97–100] It remains unclear how SIRT3 functions to promote tumor growth. Interestingly, the positive effect of SIRT3 on the proliferation of esophageal cancer might be due to the ability of tumors to adapt to increased ROS levels. While all these studies show that SIRT3 plays an important role in cancer, further studies are needed to investigate the context-dependent functions of SIRT3 in tumor development, progression, and resistance to therapy.

8.3 SIRT4

Among all the members of the sirtuin family, SIRT4 has the weakest enzymatic activity in vitro. Weak ADP-ribosylation, deacetylation, deacylation, and lipoamidase activity have been reported for SIRT4.[14–17] SIRT4 is capable of binding proteins containing the more bulky modifications, lipoyl and biotinyl, and displays much higher activity towards removing these moieties from target proteins than removing acetyl modifications.[17] Recently, SIRT4 was found to possess even broader deacylation activity by removing (hydroxyl)methylglutaryl and 3-methylglutaconyl from lysine residues.[16] Many of these activities appear more robust in vivo than in vitro, suggesting there are unaccounted for layers of SIRT4 regulation.

8.3.1 SIRT4 SUBSTRATES AND PATHWAYS

The search for SIRT4 substrates and biology has been hampered by difficulties obtaining purified protein and its low enzymatic activity on a cadre of canonical sirtuin substrates. However, a number of SIRT4 substrates have been identified thusfar revealing a central role for SIRT4 in fuel choice (Fig. 8.1). The first SIRT4 substrate identified was GDH or GLUD1, a mitochondrial enzyme that converts glutamate to α-ketoglutarate and thus promotes glutamate and glutamine metabolism.[15] Through ADP-ribosyltransferase activity, SIRT4 inhibits GDH activity in insulinoma cells and pancreatic β-cells, thereby downregulating insulin secretion. A similar mechanism for SIRT4-mediated regulation of insulin secretion was described in insulin-producing INS-1E cells.[101] Through coimmunoprecipitations followed by mass spectrometry, SIRT4 was identified as a binding partner of ANT2, ANT3, and IDE, providing additional mechanisms through which SIRT4 may regulate insulin secretion.[101] This inhibitory role of SIRT4 on insulin secretion has been confirmed recently through the finding that SIRT4 binds and activates the methylcrotonyl-CoA carboxylase complex (MCCC) to promote leucine catabolism, also inhibiting GDH activity.[16] In this setting, SIRT4 acts as a deacylase and removes the inhibitory effect of (hydroxyl)methylglutaryl and 3-methylglutaconyl lysine modifications from the MCCC complex, thereby regulating insulin secretion.

SIRT4 appears to act to limit multiple arms of fuel catabolism in addition to amino acids. SIRT4 inhibits lipid catabolism and synthesis through deacetylation and subsequent inactivation of malonyl-CoA decarboxylase (MCD).[102] In addition, by removing lipoyl modifications from the PDH subunit DLAT, SIRT4 inhibits the activity of PDH and limits the metabolic flux through this complex.[17]

To identify SIRT4 binding proteins, which reveal novel SIRT4 substrates and/or regulators, an extensive mass spectrometry study was undertaken to build a mitochondrial sirtuin network.[18] This proteomic approach revealed SIRT4 shares many common binding proteins with SIRT3, suggesting they may regulate some common biological functions, but perhaps in a distinct manner. For example, while SIRT3 and SIRT4 both interact with GDH, studies have demonstrated that SIRT3 seems to activate GDH, whereas SIRT4 inhibits GDH function. Next to GDH, this strategy also identified SIRT4 binding with numerous mitochondrial transporters. Future studies will need to examine whether SIRT4 binding partners are bona fide substrates.

SIRT4 has also been reported to impact mitochondrial-derived signaling. SIRT4 reduces AMPK signaling, and downregulates PGC-1α and ERRα in an ANT2-dependent manner.[103] In addition, depletion of SIRT4 activates PPAR alpha to induce FAO through transcriptional control.[104]

In conclusion, SIRT4 activities reported to date suggest that SIRT4 is activated under fuel-rich conditions to limit excessive mitochondrial metabolism. Future studies should aim to provide the necessary mechanistic insight on how SIRT4 activity controls fuel choice and mitochondrial metabolism.

8.3.2 REGULATION OF SIRT4

A better understanding of SIRT4 expression may provide insight into when its activity is physiologically relevant. Intriguingly, conditions that induce SIRT3 expression *downregulate* SIRT4. For instance, fasting and CR increase SIRT3 protein levels but decrease SIRT4 expression in the liver.[104] In addition, SIRT3 protein level decreases with age while SIRT4 appears to increase.[105] Chemically and radiation-induced genotoxic stress significantly increased SIRT4 levels but did not affect the level of SIRT3 protein.[10,106] mTOR signaling was shown to suppress SIRT4 expression by promoting degradation of cAMP response element binding-2.[107] SIRT4 expression was also found to be suppressed by the transcriptional corepressor C-terminal-binding protein, which also plays an essential role in promoting glutaminolysis.[108] Lysine-specific demethylase 1 (LSD1) regulates histone methylation and suppresses SIRT4 transcription. Here, loss of LSD1 induced decreased glutamine metabolism, lowered oxygen consumption, and diminished mitochondrial membrane potential, which could be rescued by forced SIRT4 expression.[109] At the posttranscriptional level, two microRNAs have recently been identified to suppress the SIRT4 expression level. In cardiac hypertrophy models, overexpression of miR-497 inhibited myocardial hypertrophy and suppressed SIRT4.[110] In addition, miR-15b has been identified as a negative regulator of SIRT4 expression.[111] Taken together, these studies suggest that SIRT4 may be a key player in the metabolic response to cellular stress.

8.3.3 SIRT4-RELATED DISEASES

Numerous studies have identified changes in SIRT4 levels in cancer. SIRT4 mRNA levels are decreased in lung, breast, colorectal, and esophageal cancers.[108,112–115] SIRT4 may protect against tumorigenesis via its function as a metabolic checkpoint.[106] Genotoxic and replicative stressors induce SIRT4 levels, which in turn leads to inhibition of GDH and halts the cell cycle. Subsequently, loss of SIRT4 results in increased genome instability and causes mice to develop spontaneous lung tumors.[106] Using a separate model, Jeong et al. found that the loss of SIRT4 increased lymphomagenesis in Eμ-Myc transgenic mouse.[116] It will be interesting for future studies to identify further mechanisms by which SIRT4 controls DNA repair or cellular proliferation.

Numerous studies have linked SIRT4 with metabolic homeostasis and obesity. In addition to the control of insulin secretion (discussed above), SIRT4 regulates pathways linked with lipid metabolism.[102,117] SIRT4 expression increases with a high-fat diet and decreases under fasting conditions and positively correlates with triglyceride/lipoprotein A levels, while correlating negatively with high-density lipoprotein cholesterol.[118,119] SIRT4 KO tissues, such as skeletal muscle and white adipose, possess high MCD activity and low levels of malonyl CoA. Whole-body SIRT4 KO mice have normal body weight and fat mass, and demonstrate protection from diet-induced obesity.[102] It will be interesting for future studies to identify a tissue-specific role for SIRT4 in obesity and metabolic homeostasis.

8.4 SIRT5

In contrast to the other mitochondrial sirtuins, SIRT5 operates both in the mitochondria as well as in the cytosol, although regulation of its subcellular localization has not been examined. Proof for a cytosolic function of SIRT5 comes from its subcellular localization combined with the discovery of mitochondrial and cytosolic substrates and interaction partners.[18,120] As a member of the sirtuin family, SIRT5 was initially identified as a deacetylase.[121] However, cells lacking SIRT5 showed no significant changes in the mitochondrial acetylome, as was demonstrated for SIRT3.[13] Later studies uncovered that SIRT5 displays demalonylation, deglutarylation, and desuccinylation activity. In fact, the desuccinylation activity of SIRT5 is much higher than its deacetylation activity.[14] Proteomics studies that were designed to reveal changes in the succinylome and malonylome identified many proteins that showed hypersuccinylation (140 proteins) and hypermalonylation (120 proteins) in the absence of SIRT5.[3,12] The preference of SIRT5 for the negatively charged substrates succinyl, malonyl, and glutaryl groups may be explained by the presence of positively charged amino acid groups in the active site of SIRT5.[122,123]

8.4.1 SUBSTRATES OF SIRT5

Few SIRT5 substrates have been identified thus far (Fig. 8.1), but the progress in identification of large-scale changes in the lysine modification in the absence of SIRT5 have contributed to a better understanding of SIRT5 function. The first SIRT5 substrate identified was carbamoyl phosphate synthetase 1 (CPS1), the enzyme that catalyzes the first step in the urea cycle.[121] Mice lacking

SIRT5 display lowered CPS-1 activity and defects in the urea cycle in the liver leading to hyperammonemia.[121] Follow-up studies found that CPS1 is also modified through succinylation and glutarylation, and SIRT5 is able to remove these moieties to activate CPS1 function.[124,125] In addition, SIRT5 was also identified as a deacetylase and activator for urate oxidase; an enzyme that breaks down uric acid and is present in most animals but absent from humans.[126] The majority of SIRT5 substrates have been identified as desuccinylation targets. For instance, the majority of proteins involved in ketogenesis and FAO show hypersuccinlylation in the absence of SIRT5.[3] In addition, HMGCS2 was characterized as a direct target of SIRT5 and its desuccinylation activity.[12] In agreement with the finding that SIRT5 regulates FAO, HADHA was identified to be desuccinylated and activated by SIRT5.[127] SIRT5 has also been implicated to play a role in glucose metabolism and the TCA cycle through desuccinylation and deactivation of the PDH and SDH complexes and functions as a suppressor of cellular respiration.[120] Loss of SIRT5 decreased complex I, complex II, and ATP synthase activity, indicating SIRT5 activity regulates ETC efficiency and ATP production.[128] In agreement with these findings, complex II function and oxygen consumption are enhanced in liver mitochondria from SIRT5 KO mice. In the cytosol, SIRT5 regulates glycolysis through desuccinylation and inhibition of pyruvate kinase M2 and demalonylation and activation of glyceraldehyde 3-phosphate dehydrogenase and promotes glycolysis.[12,129]

SIRT5 also functions in ROS defense through binding and desuccinylation of SOD1.[130] In addition, SIRT5 also desuccinylates and deglutarylates IDH2 and glucose-6-phosphate dehydrogenase to increase NADPH production, thereby indirectly contributing to glutathione production and ROS defense.[131]

8.4.2 REGULATION OF SIRT5

Little is known about how SIRT5 is regulated. On one hand, SIRT5 protein levels do not significantly change with aging or exercise,[132] but PGC-1α has been found to regulate SIRT5 expression, and both fasting and CR induce SIRT5 expression.[133,134] At the posttranscriptional level, the microRNA miR-19b was reported to regulate SIRT5 transcription levels under a low-protein diet to reduce SIRT5 levels and inhibit urea cycle activity.[135]

8.4.3 SIRT5-RELATED DISEASES

There are limiting and conflicting reports connecting SIRT5 dysregulation to cancer.[136] In IDH1-mutant cancer, SIRT5 suppresses the oncogenic effect of the oncometabolite 2-hydroxyglutarate by reversing hypersuccinylation mediated by succinyl-CoA accumulation.[137]

Succinyl-CoA is the most abundant acyl-CoA molecule in the heart, suggesting that succinylation and the desuccinylation activity of SIRT5 play an important role in cardiac function. SIRT5 KO mice display decreased ATP levels under fasting conditions and developed hypertrophy.[127] Loss of SIRT5 in the heart also renders animals more vulnerable to ischemia—reperfusion injury, compared to wildtype littermates.[138] Here, SIRT5-deficient animals displayed an increase in infarct size most likely due to hypersuccinylation and activation of SDH.[138]

8.5 CONCLUDING REMARKS

Mitochondrial sirtuins play important roles in mitochondrial metabolism and homeostasis. We are now beginning to gain insights into how these enzymes coordinate the activity of multiple substrates to regulate a concerted response to very distinct stress stimuli. While loss of mitochondrial sirtuin function is not lethal under steady-state conditions, organisms are more vulnerable to a range of different stress conditions and become more susceptible to cancer and metabolic diseases. However, due to the wide range of different substrates, enzymatic activities and dynamic responses to different stress conditions research remain challenging.

Despite some important efforts to unravel the mechanisms of substrate recognition by sirtuins,[122] we still have insufficient understanding of the specificity of mitochondrial sirtuin activity, while proteomics data sets obtained from interactome and acylome studies suggest that such substrate preferences exists. However, the lack of direct experimental evidence on how SIRT3–5 recognize their substrates prevents us from gaining a definitive answer about whether they have specific targets or target the posttranslational modifications on mitochondrial proteins. Interestingly, mitochondrial sirtuins also share common interacting proteins and converge on similar mitochondrial processes. Based on the published literature it is tempting to propose that SIRT3 and SIRT4 antagonize the effect on their shared substrate proteins, suggesting they are active under distinct conditions. One challenge for future research is to unravel the interplay between the different sirtuins in regulation of the substrates and biological pathways that are controlled by SIRT3–5.

While the primary enzymatic function of SIRT3 is deacetylation, SIRT4 and SIRT5 possess more than one enzymatic activity. Based on the current literature, one could argue that SIRT4 and SIRT5 do not possess a major enzymatic activity but instead facilitate the removal of a wide range of lysine modifications. In addition, their enzymatic activities might be context-dependent and controlled by unknown signals including allosteric regulation or posttranslational modifications on mitochondrial sirtuins. Defining the enzymatic activities of SIRT4 and SIRT5 will help to better understand the biological function and regulatory spectra of these sirtuins.

REFERENCES

1. Hebert AS, Dittenhafer-Reed KE, Yu W, Bailey DJ, Selen ES, Boersma MD, et al. Calorie restriction and SIRT3 trigger global reprogramming of the mitochondrial protein acetylome. *Mol Cell* 2013;**49**:186–99.
2. Rardin MJ, Newman JC, Held JM, Cusack MP, Sorensen DJ, Li B, et al. Label-free quantitative proteomics of the lysine acetylome in mitochondria identifies substrates of SIRT3 in metabolic pathways. *Proc Natl Acad Sci USA* 2013;**110**:6601–6.
3. Rardin MJ, He W, Nishida Y, Newman JC, Carrico C, Danielson SR, et al. SIRT5 regulates the mitochondrial lysine succinylome and metabolic networks. *Cell Metab* 2013;**18**:920–33.
4. Kim SC, Sprung R, Chen Y, Xu Y, Ball H, Pei J, et al. Substrate and functional diversity of lysine acetylation revealed by a proteomics survey. *Mol Cell* 2006;**23**:607–18.
5. Baeza J, Smallegan MJ, Denu JM. Site-specific reactivity of nonenzymatic lysine acetylation. *ACS Chem Biol* 2015;**10**:122–8.
6. Rine J, Herskowitz I. Four genes responsible for a position effect on expression from HML and HMR in *Saccharomyces cerevisiae*. *Genetics* 1987;**116**:9–22.

7. Imai S, Armstrong CM, Kaeberlein M, Guarente L. Transcriptional silencing and longevity protein Sir2 is an NAD-dependent histone deacetylase. *Nature* 2000;**403**:795−800.

8. Schwer B, North BJ, Frye RA, Ott M, Verdin E. The human silent information regulator (Sir)2 homologue hSIRT3 is a mitochondrial nicotinamide adenine dinucleotide-dependent deacetylase. *J Cell Biol* 2002;**158**:647−57.

9. Michishita E, Park JY, Burneskis JM, Barrett JC, Horikawa I. Evolutionarily conserved and nonconserved cellular localizations and functions of human SIRT proteins. *Mol Biol Cell* 2005;**16**:4623−35.

10. Iwahara T, Bonasio R, Narendra V, Reinberg D. SIRT3 functions in the nucleus in the control of stress--related gene expression. *Mol Cell Biol* 2012;**32**:5022−34.

11. Scher MB, Vaquero A, Reinberg D. SirT3 is a nuclear NAD + -dependent histone deacetylase that translocates to the mitochondria upon cellular stress. *Genes Dev* 2007;**21**:920−8.

12. Nishida Y, Rardin MJ, Carrico C, He W, Sahu AK, Gut P, et al. SIRT5 regulates both cytosolic and mitochondrial protein malonylation with glycolysis as a major target. *Mol Cell* 2015;**59**:321−32.

13. Lombard DB, Alt FW, Cheng H-L, Bunkenborg J, Streeper RS, Mostoslavsky R, et al. Mammalian Sir2 homolog SIRT3 regulates global mitochondrial lysine acetylation. *Mol Cell Biol* 2007;**27**:8807−14.

14. Feldman JL, Baeza J, Denu JM. Activation of the protein deacetylase SIRT6 by long-chain fatty acids and widespread deacylation by mammalian sirtuins. *J Biol Chem* 2013;**288**:31350−6.

15. Haigis MC, Mostoslavsky R, Haigis KM, Fahie K, Christodoulou DC, Murphy AJ, et al. SIRT4 inhibits glutamate dehydrogenase and opposes the effects of calorie restriction in pancreatic beta cells. *Cell* 2006;**126**:941−54.

16. Anderson KA, Huynh FK, Fisher-Wellman K, Stuart JD, Peterson BS, Douros JD, et al. SIRT4 is a lysine deacylase that controls leucine metabolism and insulin secretion. *Cell Metab* 2017;**25**:838−855e15.

17. Mathias RA, Greco TM, Oberstein A, Budayeva HG, Chakrabarti R, Rowland EA, et al. Sirtuin 4 is a lipoamidase regulating pyruvate dehydrogenase complex activity. *Cell* 2014;**159**:1615−25.

18. Yang W, Nagasawa K, Munch C, Xu Y, Satterstrom K, Jeong S, et al. Mitochondrial sirtuin network reveals dynamic SIRT3-dependent deacetylation in response to membrane depolarization. *Cell* 2016;**167**:985−1000e21.

19. Hallows WC, Yu W, Smith BC, Devries MK, Devires MK, Ellinger JJ, et al. Sirt3 promotes the urea cycle and fatty acid oxidation during dietary restriction. *Mol Cell* 2011;**41**:139−49.

20. Hirschey MD, Shimazu T, Goetzman E, Jing E, Schwer B, Lombard DB, et al. SIRT3 regulates mitochondrial fatty-acid oxidation by reversible enzyme deacetylation. *Nature* 2010;**464**:121−5.

21. Zhang Y, Bharathi SS, Rardin MJ, Uppala R, Verdin E, Gibson BW, et al. SIRT3 and SIRT5 regulate the enzyme activity and cardiolipin binding of very long-chain acyl-CoA dehydrogenase. *PLoS One* 2015;**10**: e0122297.

22. Shimazu T, Hirschey MD, Hua L, Dittenhafer-Reed KE, Schwer B, Lombard DB, et al. SIRT3 deacetylates mitochondrial 3-hydroxy-3-methylglutaryl CoA synthase 2 and regulates ketone body production. *Cell Metab* 2010;**12**:654−61.

23. Schlicker C, Gertz M, Papatheodorou P, Kachholz B, Becker CF, Steegborn C. Substrates and regulation mechanisms for the human mitochondrial sirtuins Sirt3 and Sirt5. *J Mol Biol* 2008;**382**:790−801.

24. Jing E, O'Neill BT, Rardin MJ, Kleinridders A, Ilkeyeva OR, Ussar S, et al. Sirt3 regulates metabolic flexibility of skeletal muscle through reversible enzymatic deacetylation. *Diabetes* 2013;**62**:3404−17.

25. Liang L, Li Q, Huang L, Li D, Li X. Sirt3 binds to and deacetylates mitochondrial pyruvate carrier 1 to enhance its activity. *Biochem Biophys Res Commun* 2015;**468**:807−12.

26. Fan J, Shan C, Kang HB, Elf S, Xie J, Tucker M, et al. Tyr phosphorylation of PDP1 toggles recruitment between ACAT1 and SIRT3 to regulate the pyruvate dehydrogenase complex. *Mol Cell* 2014;**53**:534−48.

27. Yang H, Zuo XZ, Tian C, He DL, Yi WJ, Chen Z, et al. Green tea polyphenols attenuate high-fat diet-induced renal oxidative stress through SIRT3-dependent deacetylation. *Biomed Environ Sci* 2015;**28**:455−9.

28. Fernandes J, Weddle A, Kinter CS, Humphries KM, Mather T, Szweda LI, et al. Lysine acetylation activates mitochondrial aconitase in the heart. *Biochemistry* 2015;**54**:4008−18.

29. Someya S, Yu W, Hallows WC, Xu J, Vann JM, Leeuwenburgh C, et al. Sirt3 mediates reduction of oxidative damage and prevention of age-related hearing loss under caloric restriction. *Cell* 2010;**143**:802−12.

30. Finley LWS, Haas W, Desquiret-Dumas V, Wallace DC, Procaccio V, Gygi SP, et al. Succinate dehydrogenase is a direct target of sirtuin 3 deacetylase activity. *PLoS One* 2011;**6**:e23295.

31. Cui XX, Li X, Dong SY, Guo YJ, Liu T, Wu YC. SIRT3 deacetylated and increased citrate synthase activity in PD model. *Biochem Biophys Res Commun* 2017;**484**:767−73.

32. Bao J, Scott I, Lu Z, Pang L, Dimond CC, Gius D, et al. SIRT3 is regulated by nutrient excess and modulates hepatic susceptibility to lipotoxicity. *Free Radic Biol Med* 2010;**49**:1230−7.

33. Pellegrini L, Pucci B, Villanova L, Marino ML, Marfe G, Sansone L, et al. SIRT3 protects from hypoxia and staurosporine-mediated cell death by maintaining mitochondrial membrane potential and intracellular pH. *Cell Death Differ* 2012;**19**:1815−25.

34. Ahn B-H, Kim H-S, Song S, Lee IH, Liu J, Vassilopoulos A, et al. A role for the mitochondrial deacetylase Sirt3 in regulating energy homeostasis. *Proc Natl Acad Sci U S A* 2008;**105**:14447−52.

35. Rahman M, Nirala NK, Singh A, Zhu LJ, Taguchi K, Bamba T, et al. Drosophila Sirt2/mammalian SIRT3 deacetylates ATP synthase β and regulates complex V activity. *J Cell Biol* 2014;**206**:289−305.

36. Vassilopoulos A, Pennington JD, Andresson T, Rees DM, Bosley AD, Fearnley IM, et al. SIRT3 deacetylates ATP synthase F1 complex proteins in response to nutrient- and exercise-induced stress. *Antioxid Redox Signal* 2014;**21**:551−64.

37. Zhao S, Xu W, Jiang W, Yu W, Lin Y, Zhang T, et al. Regulation of cellular metabolism by protein lysine acetylation. *Science* 2010;**327**:1000−4.

38. Bell EL, Guarente L. The SirT3 divining rod points to oxidative stress. *Mol Cell* 2011;**42**:561−8.

39. van de Ven RAH, Santos D, Haigis MC. Mitochondrial sirtuins and molecular mechanisms of aging. *Trends Mol Med* 2017;**23**:320−31.

40. German NJ, Haigis MC. Sirtuins and the metabolic hurdles in cancer. *Curr Biol* 2015;**25**:R569−83.

41. Tao R, Coleman MC, Pennington JD, Ozden O, Park S-H, Jiang H, et al. Sirt3-mediated deacetylation of evolutionarily conserved lysine 122 regulates MnSOD activity in response to stress. *Mol Cell* 2010;**40**:893−904.

42. Cheng Y, Ren X, Gowda AS, Shan Y, Zhang L, Yuan YS, et al. Interaction of Sirt3 with OGG1 contributes to repair of mitochondrial DNA and protects from apoptotic cell death under oxidative stress. *Cell Death Dis* 2013;**4**:e731.

43. Hafner AV, Dai J, Gomes AP, Xiao C-Y, Palmeira CM, Rosenzweig A, et al. Regulation of the mPTP by SIRT3-mediated deacetylation of CypD at lysine 166 suppresses age-related cardiac hypertrophy. *Aging (Albany NY)* 2010;**2**:914−23.

44. Sundaresan NR, Gupta M, Kim G, Rajamohan SB, Isbatan A, Gupta MP. Sirt3 blocks the cardiac hypertrophic response by augmenting Foxo3a-dependent antioxidant defense mechanisms in mice. *J Clin Invest* 2009;**119**:2758−71.

45. Sundaresan NR, Samant SA, Pillai VB, Rajamohan SB, Gupta MP. SIRT3 is a stress-responsive deacetylase in cardiomyocytes that protects cells from stress-mediated cell death by deacetylation of Ku70. *Mol Cell Biol* 2008;**28**:6384−401.

46. Song CL, Tang H, Ran LK, Ko BC, Zhang ZZ, Chen X, et al. Sirtuin 3 inhibits hepatocellular carcinoma growth through the glycogen synthase kinase-3beta/BCL2-associated X protein-dependent apoptotic pathway. *Oncogene* 2016;**35**:631−41.

47. Yang Y, Cimen H, Han MJ, Shi T, Deng JH, Koc H, et al. NAD + -dependent deacetylase SIRT3 regulates mitochondrial protein synthesis by deacetylation of the ribosomal protein MRPL10. *J Biol Chem* 2010;**285**:7417−29.

48. Pillai VB, Sundaresan NR, Kim G, Gupta M, Rajamohan SB, Pillai JB, et al. Exogenous NAD blocks cardiac hypertrophic response via activation of the SIRT3-LKB1-AMP-activated kinase pathway. *J Biol Chem* 2010;**285**:3133−44.
49. Gibellini L, Pinti M, Beretti F, Pierri CL, Onofrio A, Riccio M, et al. Sirtuin 3 interacts with Lon protease and regulates its acetylation status. *Mitochondrion* 2014;**18**:76−81.
50. Imai S-I, Guarente L. NAD + and sirtuins in aging and disease. *Trends Cell Biol* 2014;**24**:464−71.
51. Palacios OM, Carmona JJ, Michan S, Chen KY, Manabe Y, Ward JL, et al. Diet and exercise signals regulate SIRT3 and activate AMPK and PGC-1alpha in skeletal muscle. *Aging (Albany NY)* 2009;**1**:771−83.
52. Cheng A, Yang Y, Zhou Y, Maharana C, Lu D, Peng W, et al. Mitochondrial SIRT3 mediates adaptive responses of neurons to exercise and metabolic and excitatory challenges. *Cell Metab* 2016;**23**:128−42.
53. Kendrick AA, Choudhury M, Rahman SM, McCurdy CE, Friederich M, Van Hove JL, et al. Fatty liver is associated with reduced SIRT3 activity and mitochondrial protein hyperacetylation. *Biochem J* 2011;**433**:505−14.
54. Weir HJ, Murray TK, Kehoe PG, Love S, Verdin EM, O'Neill MJ, et al. CNS SIRT3 expression is altered by reactive oxygen species and in Alzheimer's disease. *PLoS One* 2012;**7**:e48225.
55. Kim HS, Patel K, Muldoon-Jacobs K, Bisht KS, Aykin-Burns N, Pennington JD, et al. SIRT3 is a mitochondria-localized tumor suppressor required for maintenance of mitochondrial integrity and metabolism during stress. *Cancer Cell* 2010;**17**:41−52.
56. Bell EL, Emerling BM, Ricoult SJ, Guarente L. SirT3 suppresses hypoxia inducible factor 1alpha and tumor growth by inhibiting mitochondrial ROS production. *Oncogene* 2011;**30**:2986−96.
57. Finley LWS, Carracedo A, Lee J, Souza A, Egia A, Zhang J, et al. SIRT3 opposes reprogramming of cancer cell metabolism through HIF1α destabilization. *Cancer Cell* 2011;**19**:416−28.
58. Buler M, Andersson U, Hakkola J. Who watches the watchmen? Regulation of the expression and activity of sirtuins. *FASEB J* 2016;**30**:3942−60.
59. Lim JH, Gerhart-Hines Z, Dominy JE, Lee Y, Kim S, Tabata M, et al. Oleic acid stimulates complete oxidation of fatty acids through protein kinase A-dependent activation of SIRT1-PGC1alpha complex. *J Biol Chem* 2013;**288**:7117−26.
60. Liu TF, Vachharajani V, Millet P, Bharadwaj MS, Molina AJ, McCall CE. Sequential actions of SIRT1-RELB-SIRT3 coordinate nuclear-mitochondrial communication during immunometabolic adaptation to acute inflammation and sepsis. *J Biol Chem* 2015;**290**:396−408.
61. Xu H, Hertzel AV, Steen KA, Bernlohr DA. Loss of fatty acid binding protein 4/aP2 reduces macrophage inflammation through activation of SIRT3. *Mol Endocrinol* 2016;**30**:325−34.
62. Koentges C, Pfeil K, Schnick T, Wiese S, Dahlbock R, Cimolai MC, et al. SIRT3 deficiency impairs mitochondrial and contractile function in the heart. *Basic Res Cardiol* 2015;**110**:36.
63. Grillon JM, Johnson KR, Kotlo K, Danziger RS. Non-histone lysine acetylated proteins in heart failure. *Biochim Biophys Acta* 2012;**1822**:607−14.
64. Cheung KG, Cole LK, Xiang B, Chen K, Ma X, Myal Y, et al. Sirtuin-3 (SIRT3) protein attenuates doxorubicin-induced oxidative stress and improves mitochondrial respiration in H9c2 cardiomyocytes. *J Biol Chem* 2015;**290**:10981−93.
65. Chen T, Liu J, Li N, Wang S, Liu H, Li J, et al. Mouse SIRT3 attenuates hypertrophy-related lipid accumulation in the heart through the deacetylation of LCAD. *PLoS One* 2015;**10**:e0118909.
66. Porter GA, Urciuoli WR, Brookes PS, Nadtochiy SM. SIRT3 deficiency exacerbates ischemia-reperfusion injury: implication for aged hearts. *Am J Physiol Heart Circ Physiol* 2014;**306**:H1602−9.
67. Pillai VB, Samant S, Sundaresan NR, Raghuraman H, Kim G, Bonner MY, et al. Honokiol blocks and reverses cardiac hypertrophy in mice by activating mitochondrial Sirt3. *Nat Commun* 2015;**6**:6656.

68. Guo F, Liu B, Tang F, Lane S, Souslova EA, Chudakov DM, et al. Astroglia are a possible cellular substrate of angiotensin(1-7) effects in the rostral ventrolateral medulla. *Cardiovasc Res* 2010;**87**:578−84.
69. Lin MT, Beal MF. Mitochondrial dysfunction and oxidative stress in neurodegenerative diseases. *Nature* 2006;**443**:787−95.
70. Yin JX, Turner GH, Lin HJ, Coons SW, Shi J. Deficits in spatial learning and memory is associated with hippocampal volume loss in aged apolipoprotein E4 mice. *J Alzheimers Dis* 2011;**27**:89−98.
71. Yang W, Zou Y, Zhang M, Zhao N, Tian Q, Gu M, et al. Mitochondrial Sirt3 expression is decreased in APP/PS1 double transgenic mouse model of Alzheimer's disease. *Neurochem Res* 2015;**40**:1576−82.
72. Han P, Tang Z, Yin J, Maalouf M, Beach TG, Reiman EM, et al. Pituitary adenylate cyclase-activating polypeptide protects against beta-amyloid toxicity. *Neurobiol Aging* 2014;**35**:2064−71.
73. Dias V, Junn E, Mouradian MM. The role of oxidative stress in Parkinson's disease. *J Parkinsons Dis* 2013;**3**:461−91.
74. Shi H, Deng HX, Gius D, Schumacker PT, Surmeier DJ, Ma YC. Sirt3 protects dopaminergic neurons from mitochondrial oxidative stress. *Hum Mol Genet* 2017;**26**:1915−26.
75. Liu L, Peritore C, Ginsberg J, Kayhan M, Donmez G. SIRT3 attenuates MPTP-induced nigrostriatal degeneration via enhancing mitochondrial antioxidant capacity. *Neurochem Res* 2015;**40**:600−8.
76. Zhang X, Ren X, Zhang Q, Li Z, Ma S, Bao J, et al. PGC-1alpha/ERRalpha-Sirt3 pathway regulates DAergic neuronal death by directly deacetylating SOD2 and ATP synthase beta. *Antioxid Redox Signal* 2016;**24**:312−28.
77. Fu J, Jin J, Cichewicz RH, Hageman SA, Ellis TK, Xiang L, et al. trans-(-)-epsilon-Viniferin increases mitochondrial sirtuin 3 (SIRT3), activates AMP-activated protein kinase (AMPK), and protects cells in models of Huntington Disease. *J Biol Chem* 2012;**287**:24460−72.
78. Redler RL, Dokholyan NV. The complex molecular biology of amyotrophic lateral sclerosis (ALS). *Prog Mol Biol Transl Sci* 2012;**107**:215−62.
79. Salvatori I, Valle C, Ferri A, Carri MT. SIRT3 and mitochondrial metabolism in neurodegenerative diseases. *Neurochem Int* 2017;**109**:184−92.
80. Chen H, Kim GS, Okami N, Narasimhan P, Chan PH. NADPH oxidase is involved in post-ischemic brain inflammation. *Neurobiol Dis* 2011;**42**:341−8.
81. Yin J, Han P, Tang Z, Liu Q, Shi J. Sirtuin 3 mediates neuroprotection of ketones against ischemic stroke. *J Cereb Blood Flow Metab* 2015;**35**:1783−9.
82. Dai SH, Chen T, Wang YH, Zhu J, Luo P, Rao W, et al. Sirt3 attenuates hydrogen peroxide-induced oxidative stress through the preservation of mitochondrial function in HT22 cells. *Int J Mol Med* 2014;**34**:1159−68.
83. Morigi M, Perico L, Rota C, Longaretti L, Conti S, Rottoli D, et al. Sirtuin 3-dependent mitochondrial dynamic improvements protect against acute kidney injury. *J Clin Invest* 2015;**125**:715−26.
84. Koyama T, Kume S, Koya D, Araki S, Isshiki K, Chin-Kanasaki M, et al. SIRT3 attenuates palmitate-induced ROS production and inflammation in proximal tubular cells. *Free Radic Biol Med* 2011;**51**:1258−67.
85. Yoon SP, Kim J. Poly(ADP-ribose) polymerase 1 contributes to oxidative stress through downregulation of sirtuin 3 during cisplatin nephrotoxicity. *Anat Cell Biol* 2016;**49**:165−76.
86. Chen Y, Fu LL, Wen X, Wang XY, Liu J, Cheng Y, et al. Sirtuin-3 (SIRT3), a therapeutic target with oncogenic and tumor-suppressive function in cancer. *Cell Death Dis* 2014;**5**:e1047.
87. Zhang B, Qin L, Zhou CJ, Liu YL, Qian HX, He SB. SIRT3 expression in hepatocellular carcinoma and its impact on proliferation and invasion of hepatoma cells. *Asian Pac J Trop Med* 2013;**6**:649−52.
88. Lai CC, Lin PM, Lin SF, Hsu CH, Lin HC, Hu ML, et al. Altered expression of SIRT gene family in head and neck squamous cell carcinoma. *Tumour Biol* 2013;**34**:1847−54.

89. Dong X-C, Jing L-M, Wang W-X, Gao Y-X. Down-regulation of SIRT3 promotes ovarian carcinoma metastasis. *Biochem Biophys Res Commun* 2016;**475**:245−50.

90. Zou X, Santa-Maria CA, O'Brien J, Gius D, Zhu Y. Manganese superoxide dismutase acetylation and dysregulation, due to loss of SIRT3 activity, promote a luminal B-like breast carcinogenic-permissive phenotype. *Antioxid Redox Signal* 2016;**25**:326−36.

91. Xiao K, Jiang J, Wang W, Cao S, Zhu L, Zeng H, et al. Sirt3 is a tumor suppressor in lung adenocarcinoma cells. *Oncol Rep* 2013;**30**:1323−8.

92. Wang L, Wang WY, Cao LP. SIRT3 inhibits cell proliferation in human gastric cancer through down-regulation of Notch-1. *Int J Clin Exp Med* 2015;**8**:5263−71.

93. Yu W, Denu RA, Krautkramer KA, Grindle KM, Yang DT, Asimakopoulos F, et al. Loss of SIRT3 provides growth advantage for B cell malignancies. *J Biol Chem* 2016;**291**:3268−79.

94. Allison SJ, Milner J. SIRT3 is pro-apoptotic and participates in distinct basal apoptotic pathways. *Cell Cycle* 2007;**6**:2669−77.

95. Marfe G, Tafani M, Indelicato M, Sinibaldi-Salimei P, Reali V, Pucci B, et al. Kaempferol induces apoptosis in two different cell lines via Akt inactivation, Bax and SIRT3 activation, and mitochondrial dysfunction. *J Cell Biochem* 2009;**106**:643−50.

96. Kumar A, Corey C, Scott I, Shiva S, D'Cunha J. Minnelide/triptolide impairs mitochondrial function by regulating SIRT3 in P53-dependent manner in non-small cell lung cancer. *PLoS One* 2016;**11**:e0160783.

97. Alhazzazi TY, Kamarajan P, Joo N, Huang JY, Verdin E, D'Silva NJ, et al. Sirtuin-3 (SIRT3), a novel potential therapeutic target for oral cancer. *Cancer* 2011;**117**:1670−8.

98. George J, Nihal M, Singh CK, Zhong W, Liu X, Ahmad N. Pro-proliferative function of mitochondrial sirtuin deacetylase SIRT3 in human melanoma. *J Invest Dermatol* 2016;**136**:809−18.

99. Xiong Y, Wang M, Zhao J, Wang L, Li X, Zhang Z, et al. SIRT3 is correlated with the malignancy of non-small cell lung cancer. *Int J Oncol* 2017;**50**:903−10.

100. Zhao Y, Yang H, Wang X, Zhang R, Wang C, Guo Z. Sirtuin-3 (SIRT3) expression is associated with overall survival in esophageal cancer. *Ann Diagn Pathol* 2013;**17**:483−5.

101. Ahuja N, Schwer B, Carobbio S, Waltregny D, North BJ, Castronovo V, et al. Regulation of insulin secretion by SIRT4, a mitochondrial ADP-ribosyltransferase. *J Biol Chem* 2007;**282**:33583−92.

102. Laurent G, German NJ, Saha AK, de Boer VCJ, Davies M, Koves TR, et al. SIRT4 coordinates the balance between lipid synthesis and catabolism by repressing malonyl CoA decarboxylase. *Mol Cell* 2013;**50**:686−98.

103. Ho L, Titus AS, Banerjee KK, George S, Lin W, Deota S, et al. SIRT4 regulates ATP homeostasis and mediates a retrograde signaling via AMPK. *Aging (Albany NY)* 2013;**5**:835−49.

104. Laurent G, de Boer VCJ, Finley LWS, Sweeney M, Lu H, Schug TT, et al. SIRT4 represses peroxisome proliferator-activated receptor α activity to suppress hepatic fat oxidation. *Mol Cell Biol* 2013;**33**:4552−61.

105. Wong DW, Soga T, Parhar IS. Aging and chronic administration of serotonin-selective reuptake inhibitor citalopram upregulate Sirt4 gene expression in the preoptic area of male mice. *Front Genet* 2015;**6**:281.

106. Jeong SM, Xiao C, Finley LWS, Lahusen T, Souza AL, Pierce K, et al. SIRT4 has tumor-suppressive activity and regulates the cellular metabolic response to DNA damage by inhibiting mitochondrial glutamine metabolism. *Cancer Cell* 2013;**23**:450−63.

107. Csibi A, Fendt S-M, Li C, Poulogiannis G, Choo AY, Chapski DJ, et al. The mTORC1 pathway stimulates glutamine metabolism and cell proliferation by repressing SIRT4. *Cell* 2013;**153**:840−54.

108. Wang L, Zhou H, Wang Y, Cui G, Di LJ. CtBP maintains cancer cell growth and metabolic homeostasis via regulating SIRT4. *Cell Death Dis* 2015;**6**:e1620.

109. Castex J, Willmann D, Kanouni T, Arrigoni L, Li Y, Friedrich M, et al. Inactivation of Lsd1 triggers senescence in trophoblast stem cells by induction of Sirt4. *Cell Death Dis* 2017;**8**:e2631.

110. Xiao Y, Zhang X, Fan S, Cui G, Shen Z. MicroRNA-497 inhibits cardiac hypertrophy by targeting sirt4. *PLoS One* 2016;**11**:e0168078.

111. Lang A, Grether-Beck S, Singh M, Kuck F, Jakob S, Kefalas A, et al. MicroRNA-15b regulates mitochondrial ROS production and the senescence-associated secretory phenotype through sirtuin 4/SIRT4. *Aging (Albany NY)* 2016;**8**:484−505.

112. Fu L, Dong Q, He J, Wang X, Xing J, Wang E, et al. SIRT4 inhibits malignancy progression of NSCLCs, through mitochondrial dynamics mediated by the ERK-Drp1 pathway. *Oncogene* 2017;**36**:2724−36.

113. Huang G, Cheng J, Yu F, Liu X, Yuan C, Liu C, et al. Clinical and therapeutic significance of sirtuin-4 expression in colorectal cancer. *Oncol Rep* 2016;**35**:2801−10.

114. Igci M, Kalender ME, Borazan E, Bozgeyik I, Bayraktar R, Bozgeyik E, et al. High-throughput screening of Sirtuin family of genes in breast cancer. *Gene* 2016;**586**:123−8.

115. Nakahara Y, Yamasaki M, Sawada G, Miyazaki Y, Makino T, Takahashi T, et al. Downregulation of SIRT4 expression is associated with poor prognosis in esophageal squamous cell carcinoma. *Oncology* 2016;**90**:347−55.

116. Jeong SM, Lee A, Lee J, Haigis MC. SIRT4 protein suppresses tumor formation in genetic models of Myc-induced B cell lymphoma. *J Biol Chem* 2014;**289**:4135−44.

117. Nasrin N, Wu X, Fortier E, Feng Y, Bare OC, Chen S, et al. SIRT4 regulates fatty acid oxidation and mitochondrial gene expression in liver and muscle cells. *J Biol Chem* 2010;**285**:31995−2002.

118. Drew JE, Farquharson AJ, Horgan GW, Williams LM. Tissue-specific regulation of sirtuin and nicotinamide adenine dinucleotide biosynthetic pathways identified in C57Bl/6 mice in response to high-fat feeding. *J Nutr Biochem* 2016;**37**:20−9.

119. Song R, Xu W, Chen Y, Li Z, Zeng Y, Fu Y. The expression of Sirtuins 1 and 4 in peripheral blood leukocytes from patients with type 2 diabetes. *Eur J Histochem* 2011;**55**:e10.

120. Park J, Chen Y, Tishkoff DX, Peng C, Tan M, Dai L, et al. SIRT5-mediated lysine desuccinylation impacts diverse metabolic pathways. *Mol Cell* 2013;**50**:919−30.

121. Nakagawa T, Lomb DJ, Haigis MC, Guarente L. SIRT5 deacetylates carbamoyl phosphate synthetase 1 and regulates the urea cycle. *Cell* 2009;**137**:560−70.

122. Rauh D, Fischer F, Gertz M, Lakshminarasimhan M, Bergbrede T, Aladini F, et al. An acetylome peptide microarray reveals specificities and deacetylation substrates for all human sirtuin isoforms. *Nat Commun* 2013;**4**:2327.

123. Yu J, Haldar M, Mallik S, Srivastava DK. Role of the substrate specificity-defining residues of human SIRT5 in modulating the structural stability and inhibitory features of the enzyme. *PLoS One* 2016;**11**:e0152467.

124. Du J, Zhou Y, Su X, Yu JJ, Khan S, Jiang H, et al. Sirt5 is a NAD-dependent protein lysine demalonylase and desuccinylase. *Science* 2011;**334**:806−9.

125. Tan M, Peng C, Anderson KA, Chhoy P, Xie Z, Dai L, et al. Lysine glutarylation is a protein posttranslational modification regulated by SIRT5. *Cell Metab* 2014;**19**:605−17.

126. Nakamura Y, Ogura M, Ogura K, Tanaka D, Inagaki N. SIRT5 deacetylates and activates urate oxidase in liver mitochondria of mice. *FEBS Lett* 2012;**586**:4076−81.

127. Sadhukhan S, Liu X, Ryu D, Nelson OD, Stupinski JA, Li Z, et al. Metabolomics-assisted proteomics identifies succinylation and SIRT5 as important regulators of cardiac function. *Proc Natl Acad Sci* 2016;**113**:4320−5.

128. Zhang Y, Bharathi SS, Rardin MJ, Lu J, Maringer KV, Sims-Lucas S, et al. SIRT5 binds to cardiolipin and regulates the electron transport chain. *J Biol Chem* 2017;**292**(24):10239−49.

129. Xiangyun Y, Xiaomin N, Linping G, Yunhua X, Ziming L, Yongfeng Y, et al. Desuccinylation of pyruvate kinase M2 by SIRT5 contributes to antioxidant response and tumor growth. *Oncotarget* 2017;**8**:6984−93.

130. Lin Z-F, Xu H-B, Wang J-Y, Lin Q, Ruan Z, Liu F-B, et al. SIRT5 desuccinylates and activates SOD1 to eliminate ROS. *Biochem Biophys Res Commun* 2013;**441**:191−5.

131. Zhou L, Wang F, Sun R, Chen X, Zhang M, Xu Q, et al. SIRT5 promotes IDH2 desuccinylation and G6PD deglutarylation to enhance cellular antioxidant defense. *EMBO Rep* 2016;**17**:811−22.

132. Karvinen S, Silvennoinen M, Vainio P, Sistonen L, Koch LG, Britton SL, et al. Effects of intrinsic aerobic capacity, aging and voluntary running on skeletal muscle sirtuins and heat shock proteins. *Exp Gerontol* 2016;**79**:46−54.

133. Geng YQ, Li TT, Liu XY, Li ZH, Fu YC. SIRT1 and SIRT5 activity expression and behavioral responses to calorie restriction. *J Cell Biochem* 2011;**112**:3755−61.

134. Li L, Zhang P, Bao Z, Wang T, Liu S, Huang F. PGC-1alpha promotes ureagenesis in mouse periportal hepatocytes through SIRT3 and SIRT5 in response to glucagon. *Sci Rep* 2016;**6**:24156.

135. Sun RP, Xi QY, Sun JJ, Cheng X, Zhu YL, Ye DZ, et al. In low protein diets, microRNA-19b regulates urea synthesis by targeting SIRT5. *Sci Rep* 2016;**6**:33291.

136. Osborne B, Cooney GJ, Turner N. Are sirtuin deacylase enzymes important modulators of mitochondrial energy metabolism? *Biochim Biophys Acta* 2014;**1840**:1295−302.

137. Li F, He X, Ye D, Lin Y, Yu H, Yao C, et al. NADP(+)-IDH mutations promote hypersuccinylation that impairs mitochondria respiration and induces apoptosis resistance. *Mol Cell* 2015;**60**:661−75.

138. Boylston JA, Sun J, Chen Y, Gucek M, Sack MN, Murphy E. Characterization of the cardiac succinylome and its role in ischemia-reperfusion injury. *J Mol Cell Cardiol* 2015;**88**:73−81.

MULTITASKING ROLES OF THE MAMMALIAN DEACETYLASE SIRT6

9

Sylvana Hassanieh[1,2,3] and Raul Mostoslavsky[1,2,3]

[1]*The Massachusetts General Hospital Cancer Center, Boston, MA, United States* [2]*Harvard Medical School, Boston, MA, United States* [3]*Broad Institute of Harvard and MIT, Cambridge, MA, United States*

9.1 INTRODUCTION

Sirtuins (SIRT) are highly conserved proteins first described as nicotinamide adenine dinucleotide (NAD^+)-dependent type III histone deacetylases. They are the mammalian homologs of the SIR2 gene found in yeast.[1] Seven homologs (SIRT1−7) have been described, all sharing an NAD^+ binding catalytic domain comprising 275 amino acids and variable N and C termini.[2] This difference in their terminal domains dictates distant roles and cellular localization of the members of the sirtuin family. The roles and features of other sirtuins have been described elsewhere[3] and in other chapters in this book.

In this review, we will focus on SIRT6, its structure, enzymatic activity, and regulation. We will discuss its role in genomic stability and DNA repair, metabolism, and development. Finally, we will review SIRT6's involvement in several diseases.

9.2 SIRT6 STRUCTURE, ENZYMATIC ACTIVITY, AND REGULATION

SIRT6 is an NAD^+-dependent deacetylase with 355 amino acids. It transfers the acetyl group from the lysine side chain of a protein or peptide substrate to the cofactor NAD^+, generating deacetylated lysine, nicotinamide, and the intermediate 2′-*O*-acetyl-ADP-ribose.[4] The resulting nicotinamide allosterically inhibits SIRT6 activity (Fig. 9.1). Different from other sirtuins, SIRT6 can still bind NAD^+ in the absence of acetylated substrate, suggesting that it might be acting as an NAD^+ sensor. The structures of SIRT6 reveal unique features, including a splayed zinc-binding domain and the absence of a helix bundle that in other sirtuin structures connects the zinc-binding motif and Rossmann fold domain. SIRT6 also lacks the conserved, highly flexible, NAD^+-binding loop and instead contains a stable single helix.[5] In a series of studies, several years ago it was demonstrated that SIRT6 acts as a specific histone deacetylase for histone H3 lysine K9 and K56.[6−8] Although other activities were described later (see below) these remain the main targets for SIRT6, explaining its function as a gene repressor.

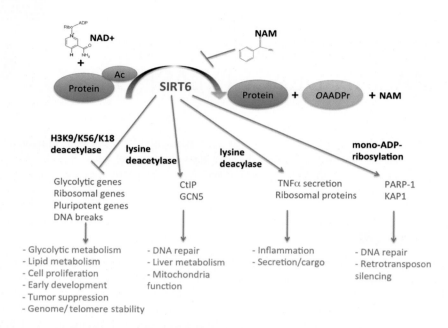

FIGURE 9.1 Enzymatic and biological activities for SIRT6.

A diagram depicting the different enzymatic and consequent biological functions for SIRT6.

Although originally characterized as a lysine deacetylase, recent studies indicate that SIRT6 can also remove long-chain fatty acyl groups (myristoyl and palmitoyl) from lysine residues generating myristoyl-adenosine 5'-diphosphoribose and the deacylated substrate.[9] In vivo, SIRT6 is able to remove acyl groups from TNFα and ribosomal proteins, influencing their secretion,[9,10] suggesting that while the deacetylase activity may influence gene expression, the deacylase activity may be important for protein secretion.

Additionally, SIRT6 has been reported to catalyze mono-ADP-ribosylation to poly ADP-ribose polymerase 1 (PARP1),[11] yet this activity was shown in vitro, therefore its physiological significance remains to be determined (Fig. 9.1).

In recent years, research has been focused on studying how SIRT6 is regulated (Fig. 9.2). Calorie restriction and physical exercise can effectively modulate the activity of sirtuins, particularly SIRT6. It has been demonstrated that exercise training regulates sirtuin levels in skeletal muscle of aged rats and slows down the aging process.[12] Recently, it has been reported that SIRT6 activation by calorie restriction suppresses NFκB signaling and delays aging.[13] SIRT6 half-life can be reduced due to increased proteasome-mediated degradation in cells deficient in carboxyl terminus of Hsp70-interacting protein (CHIP). CHIP is a protein that exhibits both cochaperone and ubiquitin ligase activities, and is an integral component of protein quality control. Cells overexpressing CHIP showed an increased SIRT6 protein expression without affecting SIRT6 transcription. CHIP noncanonically ubiquitinates SIRT6 at lysine 170 (K170), which stabilizes SIRT6

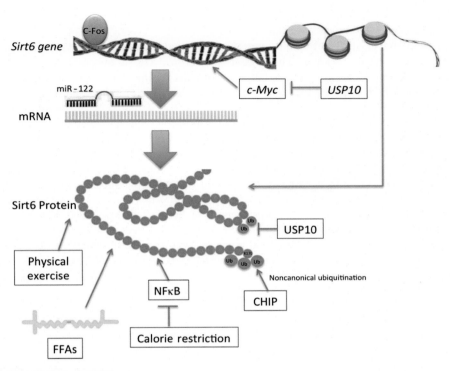

FIGURE 9.2 Regulation of SIRT6.

Multiple modulators of SIRT6 have been described. c-Fos induces SIRT6 transcription by binding to an AP-1 binding site (TAAGTCA) at position 208 base pairs in the SIRT6 promoter. In addition, the USP10 ubiquitin-specific peptidase antagonizes the transcriptional activity of the c-Myc oncogene through SIRT6, as well as p53, to inhibit cell cycle progression, cancer cell growth, and tumor formation. In addition, SIRT6 deacetylase activity is significantly enhanced in the context of nucleosomes, while the microRNA MiR-122 binds to three sites on the SIRT6 3′ UTR and reduces its levels. Exercise training regulates sirtuin levels in skeletal muscle, slowing down the aging process. FFA at physiological concentrations can increase up to 35-fold the catalytic activity of SIRT6, while SIRT6 activation by calorie restriction suppresses NFκB signaling and delays aging. The factor CHIP (C-terminus domain of Hsp70-interacting protein) noncanonically ubiquitinates SIRT6 at lysine 170 (K170), stabilizing SIRT6 and preventing SIRT6 canonical ubiquitination by other ubiquitin ligases, thus increasing SIRT6 stability without affecting SIRT6 transcription. In this context, the deubiquitinase USP10 suppresses SIRT6 ubiquitination and protects SIRT6 from proteasomal degradation.

and prevents SIRT6 canonical ubiquitination by other ubiquitin ligases[14] (Fig. 9.2). Moreover, a ubiquitin-specific peptidase USP10 has been identified as one of the SIRT6-interacting proteins. USP10 suppresses SIRT6 ubiquitination to protect SIRT6 from proteasomal degradation. USP10 antagonizes the transcriptional activity of the c-Myc oncogene through SIRT6, as well as p53, to inhibit cell cycle progression, cancer cell growth, and tumor formation.[15]

In addition, it has been shown that SIRT6 deacetylase activity is significantly enhanced in the context of nucleosomes (compared to naked histones).[16]

As mentioned above, SIRT6 efficiently removes long-chain fatty acyl groups, such as myristoyl and palmitoyl, from lysine residues. The crystal structure of SIRT6 reveals a large hydrophobic pocket that can accommodate long-chain fatty acyl groups. Using that information, the Denu lab demonstrated that SIRT6 can bind free fatty acids (FFAs), and such binding significantly stimulates SIRT6 deacetylase activity[17] in vitro. It has been demonstrated that several biologically relevant FFAs (including myristic, oleic, and linoleic acids) at physiological concentrations induce up to a 35-fold increase in catalytic efficiency of SIRT6 but not SIRT1. The activation mechanism is consistent with fatty acid inducing a conformation that binds acetylated H3 with greater affinity. Binding of long-chain FFA and myristoylated H3 peptide is mutually exclusive.[17] Both the nucleosome dependency and the FFA dependency can explain why SIRT6 exhibits poor deacetylase activity in vitro, while exhibiting robust histone deacetylase activity in vivo.

Recently, pyrrolo[1,2-*a*]quinoxaline derivatives have been synthesized and screened, yielding the first synthetic Sirt6 activators. Biochemical assays show direct, substrate-independent compound binding to the SIRT6 catalytic core and potent activation of Sirt6-dependent deacetylation of peptide substrates and complete nucleosomes. Crystal structures of SIRT6/activator complexes reveal that the compounds bind to a SIRT6-specific acyl channel pocket and identify key interactions.[18] Whether such activators work in vivo remained to be established, but they could represent important therapeutic tools for diseases where SIRT6 activity is decreased, as described in detail below.

9.3 SIRT6 FUNCTIONS IN GENOMIC STABILITY AND DNA REPAIR

9.3.1 TELOMERE AND CENTROMERE MAINTENANCE

Early studies have shown that SIRT6 deacetylates H3K9[6] and H3K56[7,8] at telomeric histones, and such activity was required for the stable association of WRN, the factor that is mutated in Werner syndrome, a premature aging disorder. SIRT6 depletion leads to telomere dysfunction with end-to-end chromosomal fusions and premature cellular senescence. Moreover, SIRT6-depleted cells exhibit abnormal telomere structures that resemble defects observed in Werner syndrome. SIRT6 contributes to the propagation of a specialized chromatin state at mammalian telomeres, which in turn is required for proper telomere metabolism and function.[6] The role of SIRT6 and telomerase has been studied in a mouse model of transverse aortic constriction (TAC)-induced heart failure. SIRT6, telomerase reverse transcriptase (TERT), and telomere repeat binding factor (TRF)-1 were significantly downregulated in TAC mice compared with their expression in sham-operated mice. Overexpression of SIRT6 upregulated TERT and TRF1 and increased the survival of mice after TAC. SIRT6 protects the myocardium against damage and this effect may be mediated by the modulation of telomeres.[19] Additionally, SIRT6 was shown to bind, upon replication stress, to TRF2, a factor involved in telomere maintenance and DNA damage response, in a DNA-independent manner.[20] Knockdown of SIRT6 upregulates TRF2 protein levels and counteracts its downregulation during the DNA damage response, leading to cell survival. Moreover, SIRT6 deacetylates in vivo the TRFH domain of TRF2, which in turn, is ubiquitylated in vivo, activating the ubiquitin-dependent proteolysis. Notably, overexpression of the TRF2cT mutant failed to be stabilized by SIRT6 depletion, demonstrating that the TRFH domain is required for its posttranscriptional modification.[20]

Recent studies have uncovered a novel role for SIRT6 in maintaining heterochromatin in pericentromeric regions. Specifically, Chua and colleagues identified a new enzymatic activity, in which SIRT6 deacetylates H3K18 specifically in pericentromeric regions, and such deacetylation protected cells against mitotic errors and senescence.[21]

9.4 BASE EXCISION REPAIR

SIRT6 was hypothesized to have a role in base excision repair (BER) because SIRT6-deficient cells experience increased sensitivity to MMS and H_2O_2, both damage repaired by BER, and those could be rescued by overexpression of the rate-limiting enzymes in BER.[22] Recent studies indicate that SIRT6 regulates BER in a PARP1-dependent manner.[23] Moreover, it has been shown that SIRT6 interacts with and stimulates MYH glycosylase, APE1, two key BER enzymes, and the Rad9-Rad1-Hus1 (9-1-1) checkpoint clamp, which promotes cell cycle checkpoint signaling and DNA repair. These interactions are enhanced following oxidative stress, and play important roles to maintain genomic and telomeric integrity in mammalian cells.[24]

9.5 DOUBLE-STRAND BREAKS REPAIR

Several studies have shown that SIRT6 is involved in double-strand break (DSB) repair. SIRT6 is recruited to the sites of DNA DSBs and stimulates repair, through nonhomologous end joining (NHEJ) and homologous recombination (HR). Specifically, SIRT6 associates with PARP1 and mono-ADP-ribosylates PARP1 on lysine residue 521 stimulating PARP1 poly-ADP-ribosylase activity and enhancing DSB repair, yet this was only observed under oxidative stress, suggesting that such roles for SIRT6 on PARP1 may represent a unique mechanism of DNA repair following oxidative stress.[11]

Interestingly, SIRT6 has been shown to deacetylate and enhance the activity of CtIP, the first nonhistone substrate discovered for SIRT6 deacetylase activity.[25] CtIP is crucial for DSB end resection, resulting in single-stranded DNA (ssDNA), which is consequently bound by replication protein A, leading to the formation of an ssDNA-RAD51 nucleoprotein filament that mediates HR.[26] In this context, upon depletion of SIRT6 there was impaired accumulation of RPA and ssDNA at DNA damage sites along with reduced rates of HR.[25] In addition, SIRT6 forms a macromolecular complex with the DNA DSB repair factor DNA-PK (DNA-dependent protein kinase), promoting NHEJ repair.[27] These studies were performed in vitro, thus its significance remains to be determined in vivo (particularly since lymphocytes, which depend on NHEJ, develop normally in SIRT6-deficient mice). Interestingly, SIRT6 recruits the ATP-dependent chromatin remodeler SNF2H to DSBs and deacetylates histone H3K56 that are both required for proper DSB repair. Chromatin remodeling is impaired by a lack of SIRT6 and SNFH2, affecting downstream recruitment of DNA repair pathway factors.[28] The role of SIRT6 in this context is further explained by

H2AX stabilization, resulting from the inhibition of proteasome-mediated degradation.[29] Such protecting roles for SIRT6 and SNF2H have recently proved critical in protecting against neuronal degeneration in vivo, suggesting roles for SIRT6 in neurodegenerative diseases. Brain-specific SIRT6-deficient mice survive, but present behavioral defects with major learning impairments by 4 months of age. Moreover, the brains of these mice show increased signs of DNA damage, cell death, and hyperphosphorylated Tau—a critical mark in several neurodegenerative diseases.[30]

SIRT6 is also a powerful repressor of L1 retrotransposons, which are an abundant class of transposable elements that pose a threat to genome stability. Specifically, SIRT6 binds to the 5'-UTR of L1 loci leading to its silencing, where it mono-ADP ribosylates the nuclear corepressor protein, KAP1, and facilitates KAP1 interaction with the heterochromatin factor, HP1α. Upon DNA damage, it was found that SIRT6 is depleted from L1 loci, allowing the activation of these previously silenced L1.[31]

Lastly, it has been found that Sirt6 is downregulated in induced pluripotent stem cells (iPSCs) from old mice. Sirt6 directly binds to Ku80 and facilitates interaction between Ku80 and DNA-PKcs, leading to DNA-PKcs phosphorylation and consequently efficient NHEJ. Introducing a combination of Sirt6 and the Yamanaka factors during reprogramming significantly promotes DNA DSB repair by activating NHEJ in iPSCs derived from old mice.[32] In this context, it has been found that lncPRESS1, a p53 regulated long noncoding RNA (lncRNA), physically interacts with SIRT6 and prevents SIRT6 chromatin localization by sequestering it from targeted chromatin interactions, maintaining high levels of histone H3K56 and H3K9 acetylation at promoters of pluripotency genes. This is the first known interaction of SIRT6 with lncRNAs affecting its function and genome stability.[33] Even though all these studies suggest that SIRT6 modulates early embryogenesis through its roles in DNA repair, a separate study demonstrated that SIRT6 directly silences the expression of pluripotent genes during early embryo differentiation.[34] Embryonic stem (ES) cells deficient for SIRT6 retain high levels of pluripotent genes upon differentiation, exhibiting a default towards neuroectoderm lineage. Such an effect depends on a failure to silence the Tet dyoxygenase enzymes, in turn sustaining high levels of 5-OhC on neuroectoderm genes.[34] In summary, these multiple roles for SIRT6 exemplify how this critical enzyme evolved to maintain cellular homeostasis, not only by guarding against genomic instability, but also acting as a key modulator of metabolism, as detailed below.

9.6 METABOLIC FUNCTIONS OF SIRT6

9.6.1 GLUCOSE METABOLISM

SIRT6-deficient mice die at 4 weeks due to age-degenerative processes after developing normally for the first 2 weeks. Defects in those mice include hypoglycemia, low levels of serum insulin growth factor receptor-1 (IGF-1), a key factor regulating lifespan, in addition to loss of subcutaneous fat, a curved spine, lymphopenia, and other metabolic disorders.[22] These mice show an increase in glucose uptake both in skeletal muscle and brown adipose tissue, a phenotype correlated with an increase in membrane glucose transporter-1 (GLUT1) expression, enhanced glycolysis, and a decreased mitochondrial respiration. Interestingly, SIRT6 functions as a corepressor of the transcription factor hypoxia-inducible factor 1-alpha (HIF1α), which is an important regulator of nutrients in stress conditions. Such an interaction leads to repression of several Hif1-a target glycolytic genes including pyruvate dehydrogenase kinase-1 (PDK1), lactate dehydrogenase (LDH), phosphofructokinase-1 (PFK1), and GLUT1[35] (Fig. 9.1).

In separate studies, SIRT6 was shown to control gluconeogenesis in the liver. It does so by regulating peroxisome proliferator-activated receptor-γ coactivator 1-α (PGC-1α), which increases the expression of gluconeogenic enzymes.[36] Sirt6 induces PGC-1α acetylation, inactivating it and suppressing hepatic glucose production. SIRT6 interacts with and modifies GCN5, enhancing its activity. Consequently, GCN5 acetylates PGC-1α which dictates its relocalization away from gluconeogenic enzyme target genes.[37] Finally, it has been shown that the transcription factor E2F1 enhances glycolysis by inhibiting SIRT6 expression. E2F1 directly binds to the SIRT6 promoter and suppresses SIRT6 promoter activity under both normoxic and hypoxic culture conditions.[38]

Although all the previous functions are thought to be dependent on nutrient conditions and independent of insulin, SIRT6 also has a critical role in glucose-stimulated insulin secretion (GSIS). Mice lacking Sirt6 in pancreatic β cells developed glucose intolerance with severely impaired GSIS. In this case, SIRT6 deacetylates forkhead box protein O1 (FoxO1) to trigger its nuclear export and releases its transcriptional repression of key glucose-sensing genes such as Pdx1 and Glut2.[39]

9.7 LIPID METABOLISM

Emerging data indicate important roles for SIRT6 in lipid metabolism as well (Fig. 9.1). In one study it was found that another sirtuin, SIRT1, forms a complex with FOXO3a and NRF1 on the SIRT6 promoter and positively regulates expression of SIRT6, which in turn deacetylates H3K9 in the promoter of multiple genes involved in β-oxidation and TG synthesis. The absence of SIRT6 results in an increase in triglycerides (TG) leading to liver disease.[40] To explore the role of SIRT6 in metabolic stress, wildtype and transgenic mice overexpressing SIRT6 were fed a high-fat diet. SIRT6 transgenic mice accumulated significantly less visceral fat, LDL-cholesterol, and triglycerides and reduction in the expression of PPARγ dependent genes, which are important in lipid metabolism.[41] Moreover, SIRT6 has been shown to have protective effects in the context of rosiglitazone (RGZ), a thiazolidinedione that acts as an agonist of PPARγ and is a treatment for hepatic steatosis. RGZ treatment ameliorated hepatic lipid accumulation and increased expression of SIRT6, PGC-1α, FoxO1, and AMP-activated protein kinase (AMPK) phosphorylation.[42]

In addition, SIRT6 and microRNA miR-122 negatively regulate each other to control various aspects of liver physiology. SIRT6 downregulates miR-122 by deacetylating H3K56 in the promoter region. MiR-122 binds to three sites on the SIRT6 3′ UTR and reduces its levels. This is implicated in the regulation of a set of metabolic genes and fatty acid β-oxidation affecting lipid metabolism.[43] SIRT6 has also been deleted specifically in adipose tissue. Fat-specific Sirt6 knockout (FKO) sensitized mice to high-fat diet-induced obesity leading to adipocyte hypertrophy. This resulted in decreased expression of adipose triglyceride lipase (ATGL), a key lipolytic enzyme. The suppression of ATGL in FKO mice was accounted for by the increased phosphorylation and acetylation of FoxO1. Moreover, decreased SIRT6 and ATGL expression was observed in obese patients. Thus, FKO sensitizes to high-fat diet-induced obesity and insulin resistance by inhibiting lipolysis.[44]

SIRT6 has also been associated with cholesterol homeostasis. SIRT6 plays a critical role in the regulation of proprotein convertase subtilisin/kexin type 9 (PCSK) gene expression in mice. Knockdown of SIRT6 in the liver leads to elevated gene expression of PCSK9 and LDL-cholesterol, where Sirt6 is recruited by forkhead transcription factor (FoxO3) to the promoter region of the PCSK9 gene and deacetylates H3K9 and H3K56, suppressing gene expression.[45] Additionally, SIRT6 negatively influences the lipogenic transcription factors SREBP1 and SREBP2. SIRT6 represses SREBP1 and SREBP2 by several mechanisms: SIRT6 represses the transcription levels of SREBP1/SREBP2 and that of their target genes, it inhibits the cleavage of SREBP1/SREBP2 into their active form (although it remains unknown how SIRT6 modulates cleavage of these proteins), and it activates AMPK, which promotes phosphorylation and inhibition of SREBP1 by AMPK.[43]

Finally, it has been shown that SIRT6 deficiency in preadipocytes blocks their differentiation into mature adipocytes. SIRT6 negatively regulates KIF5C, which belongs to the kinesin family. KIF5C interacts with casein kinase 2 alpha (CK2α), a catalytic subunit of CK2 blocking its nuclear translocation and CK2 kinase activity, consequently inhibiting mitotic clonal expansion during adipogenesis.[46] Together, these studies reveal a crucial role for SIRT6 in lipid metabolism and provide potential therapeutic targets for lipid metabolism-related diseases.

9.8 CANCER

In previous sections, we have discussed biological and biochemical roles for SIRT6 in the context of normal cell function (Fig. 9.1). In this section, we will describe roles for SIRT6 in disease. One of the main diseases SIRT6 has been proven critical to protect us from is cancer. Multiple studies have proven beyond doubt that SIRT6 acts as a potent tumor suppressor.[47] One of the hallmarks of cancer cells is their switch towards increased glycolysis in order to increase biomass for their cellular needs, a phenomenon known as the Warburg effect. Notably, SIRT6 deficiency contributed to tumor formation even in the absence of known oncogene activation. Injecting wildtype immortalized mouse embryonic fibroblasts (MEFs) into immunodeficient mice does not generate tumors (although they lack a p53/p19 checkpoint, you need to transform these MEFs by activating an oncogene), yet SIRT6 KO MEFs readily formed tumors,[48] and such a phenotype was fully dependent on increased glycolytic activity, as described before in normal cells.[39] Further, in an in vivo model of colon adenocarcinoma, SIRT6 acted as a potent tumor suppressor by inhibiting this switch towards aerobic glycolysis.[48] In addition to its effects on glycolysis, SIRT6 also functions as a regulator of ribosome metabolism by corepressing MYC transcriptional activity, which regulates cell proliferation, ribosome biogenesis, and protein synthesis by controlling the transcription and assembly of ribosome components.[48] In separate studies, it was found that in various breast cancer cell lines, SIRT6 was phosphorylated at Ser[338] by the kinase AKT1, which induced the interaction and ubiquitination of SIRT6 by MDM2, targeting SIRT6 for protease-dependent degradation. The survival of breast cancer patients is positively correlated with an abundance of SIRT6 and inversely correlated with the phosphorylation of SIRT6 at Ser[338]. In a panel of breast tumor biopsies, SIRT6 abundance was inversely correlated with an abundance of phosphorylated AKT. Inhibiting AKT or preventing SIRT6 phosphorylation by mutating Ser338 prevented the degradation of SIRT6 mediated by MDM2, suppressed the proliferation of breast cancer cells in culture, and inhibited the growth of breast tumor xenografts in mice.[49] Moreover in the context of breast cancer, researchers have described RUNX2 transcription factor as a mediator of breast cancer metastasis to bone. RUNX2 expression in luminal breast cancer cells correlated with lower estrogen receptor-α levels, anchorage-independent growth, expression of glycolytic genes, increased glucose uptake, and sensitivity to glucose starvation, but not to inhibitors of oxidative phosphorylation. Interestingly, SIRT6 was a critical regulator of these RUNX2-mediated metabolic changes. RUNX2 expression resulted in elevated pAkt, HK2, and PDHK1 glycolytic protein levels that were reduced by ectopic expression of SIRT6.[50]

Separate studies using a genetic mouse model specific for liver cancer initiation established that survival of cancer-initiating cells is controlled by c-Jun, independently of p53, through suppression of c-Fos-mediated apoptosis, a mechanism inhibited by SIRT6. Mechanistically, c-Fos induced SIRT6 transcription, which repressed survivin by reducing histone H3K9 acetylation and NF-κB activation. Overexpression of SIRT6 in this model impaired cancer development.[51] In addition, it has been shown that SIRT6 binds to and deacetylates nuclear pyruvate kinase M2 (PKM2), which is a glycolytic enzyme with nuclear oncogenic functions. SIRT6-mediated deacetylation results in PKM2 nuclear export, abolishing its nuclear protein kinase and transcriptional coactivator functions. Thus, SIRT6 suppresses PKM2 oncogenic functions, resulting in reduced cell proliferation, migration potential, and invasiveness.[52]

In recent studies using a pancreatic ductal adenocarcinoma (PDAC) model, SIRT6 was shown to act as a potent tumor suppressor. SIRT6 inactivation accelerated PDAC progression and increased metastasis formation via upregulation of Lin28b, a negative regulator of the let-7 microRNA family. SIRT6 loss results in histone hyperacetylation at the Lin28b promoter, Myc recruitment, and pronounced induction of Lin28b and downstream let-7 target genes, including HMGA2, IGF2BP1, and IGF2BP3. Thus, in addition to its other tumor suppression roles, SIRT6 has been identified as an important PDAC tumor suppressor and this work has uncovered the Lin28b pathway as a potential therapeutic target for this highly lethal malignancy.[53]

Despite all the overwhelming data supporting SIRT6 roles as a tumor suppressor, few studies appear to indicate that SIRT6 could function as an oncogene in the context of specific cancers. For instance, patients with chronic lymphocytic leukemia (CLL) show a fourfold increase in SIRT6 expression compared to healthy volunteers,[54] suggesting that CLL cancers may benefit from upregulation of SIRT6. Such roles seem to depend on the ability of SIRT6 to protect against genomic instability,[22] since inhibition of SIRT6 in related tumors (multiple myeloma) renders them more sensitive to chemotherapy.[55] In addition, SIRT6 expression was significantly higher in hepatocellular carcinoma (HCC) cell lines and HCC tissues from 138 patients compared to immortalized hepatocyte cell line and nontumor tissues. SIRT6 silencing significantly prevented the growth of HCC cell lines by inducing cellular senescence in a p16/Rb- and p53/p21-pathway independent manner.[56] These results contradict the studies of Wagner and colleagues[51] discussed above, and it remains to be determined what the differences are between these two studies.

Taken together, these results indicate that, although SIRT6 acts as a strong tumor suppressor in many cancers, there are few instances where upregulation of SIRT6 may benefit the tumor, therefore future therapies designed to tackle SIRT6 must be considered on a cancer-type basis.

9.9 INFLAMMATION

As mentioned earlier, SIRT6 efficiently removes long-chain fatty acyl groups, such as myristoyl and palmitoyl, from lysine residues. SIRT6 promotes the secretion of TNF-α by removing fatty acyl groups on K19/K20 of TNF-α. TNF-α is a proinflammatory cytokine involved in systemic inflammation, suggesting that SIRT6 activity on TNFα is proinflammatory.[9] In addition, SIRT6 limits the expression of NF-κB target genes, acting as a corepressor of NF-κB. SIRT6 interacts with the Rel-A subunit of NF-κB and destabilizes it, while at the same time SIRT6 deacetylates H3K9 on the promoter of the target genes.[57] NF-κB is a stress-responsive transcription factor that controls DNA transcription and cytokine production. Therefore, in this context SIRT6 appears to have antiinflammatory immunosuppressive functions. Similarly, it has been shown that calorie restriction-induced SIRT6 activation suppresses NF-κB signaling. SIRT6 overexpression is sufficient to delay the replicative senescence of human fibroblasts WI38 by attenuating NF-κB signaling, while SIRT6 knockdown results in accelerated cell senescence and overactive NF-κB signaling.[13]

SIRT6 also plays roles in vascular inflammation. TNF-α treatment of vascular adventitial fibroblasts decreased the expression of SIRT6 and SIRT1. In contrast, TNF-α significantly increased the expression of monocyte chemotactic protein 1 (MCP-1) and interleukin (IL-6). In addition, knockdown of SIRT1 and SIRT6, respectively, augmented TNF-α-induced generation of reactive oxygen species (ROS) and phosphorylation of protein kinase B (Akt),[58] indicating protective roles for SIRT6 in vessels.

SIRT6 has also shown antiinflammatory properties in the liver. Hepatocyte-specific Sirt6 knockout (KO) mice fed a high-fat and high-fructose (HFHF) diet for 16 weeks exhibited increased hepatic steatosis, inflammation, aggravated glucose intolerance, and insulin resistance compared with wildtype mice. In the livers of KO mice, nuclear factor erythroid 2-related factor 2 (Nrf2) was downregulated; conversely, BTB domain and CNC homolog 1 (Bach1), a nuclear repressor of Nrf2, was upregulated. It has been shown that Sirt6 promotes Nrf2 binding to the antioxidant response element (ARE) in response to oxidative stimuli.[59]

9.10 CARDIOVASCULAR DISEASES

SIRT6 appears to have important roles in the heart. In neonatal rat cardiomyocytes, overexpression of SIRT6 significantly attenuated angiotensin II (Ang-II)-induced cardiac hypertrophy.[60] Additionally, SIRT6 functions at the level of chromatin to directly inhibit IGF-Akt signaling. Sirt6 binds to and suppresses the promoter of IGF signaling-related genes by interacting with c-Jun and deacetylating H3K9. Lack of SIRT6 increased expression of the genes, leading to cardiac hypertrophy and heart failure.[61] Moreover, it has been shown that a novel poly PARP1 inhibitor, AG-690/11026014, prevents Ang II-induced cardiomyocyte hypertrophy by reversing the depletion of cellular NAD + and SIRT6 deacetylase activity.[62]

9.11 LIFESPAN

Different members of the sirtuin family of proteins have been implicated in lifespan extension in yeast, nematodes, fruitflies, and mice.[63] Specifically in mice, a study targeting SIRT6 has shown that male, but not female, transgenic mice overexpressing SIRT6 have a significantly longer lifespan than wildtype mice.[64] Gene expression analysis revealed significant differences between male SIRT6-transgenic mice and male wildtype mice: transgenic males displayed lower serum levels of IGF1, higher levels of IGF-binding protein 1, and altered phosphorylation levels of major components of IGF1 signaling, providing a potential explanation for the sexual dimorphism in the lifespan extension phenotype.[64] Separate studies found that human dermal fibroblasts (HDFs) from older human subjects were more resistant to reprogramming by classic Yamanaka factors than those from younger human subjects. However, upon the addition of SIRT6 to the older HDFs, the efficiency of reprogramming improved dramatically. In this context, SIRT6 regulates transcription of the microRNA miR-766 via a feedback regulatory loop, although how such regulation influences reprogramming of aging cells remains to be defined.[65] Recently, in a cohort study of 43 healthy males and 92 male control subjects who had died of natural causes, single nucleotide polymorphisms (SNPs) in the exons and their surroundings of the *SIRT6* were studied. The SNP rs117385980 (C > T) was found in heterozygous form in 1/43 longer-living healthy men and in 9/92 controls. These results suggest an inverse association between the T allele of rs117385980 and longevity.[66] These results need to be confirmed in a larger study and for it to be determined whether rs117385980 itself has a causal effect or whether it rather represents a mere genetic marker.

In summary, we have tried in this chapter to bring together the different studies that indicate the unique and broad biological functions that this chromatin deacetylase has evolved to perform, acting as a multitasking factor maintaining cellular homeostasis through its roles in development, metabolism, gene transcription, and DNA repair. Although much has been learnt in recent years, it is likely that future studies will keep uncovering important functions for this unique enzyme, as well as better understanding on how these different functions are coordinated and regulated in the cell.

REFERENCES

1. Imai S, Armstrong CM, Kaeberlein M, Guarente L. Transcriptional silencing and longevity protein Sir2 is an NAD-dependent histone deacetylase. *Nature* 2000;**403**(6771):795–800.
2. Mei Z, Zhang X, Yi J, Huang J, He J, Tao Y. Sirtuins in metabolism, DNA repair and cancer. *J Exp Clin Cancer Res* 2016;**35**(1):182–96.
3. Michishita E, Park JY, Burneskis JM, Barrett JC, Horikawa I. Evolutionarily conserved and nonconserved cellular localizations and functions of human SIRT proteins. *Mol Biol Cell* 2005;**16**(10):4623–35.
4. Jackson MD, Denu JM. Structural identification of 2'- and 3'-O-acetyl-ADP-ribose as novel metabolites derived from the Sir2 family of beta -NAD + -dependent histone/protein deacetylases. *J Biol Chem* 2002;**277**(21):18535–44.
5. Pan PW, Feldman JL, Devries MK, Dong A, Edwards AM, Denu JM. Structure and biochemical functions of SIRT6. *J Biol Chem* 2011;**286**(16):14575–87.
6. Michishita E, McCord RA, Berber E, Kioi M, Padilla-Nash H, Damian M, et al. SIRT6 is a histone H3 lysine 9 deacetylase that modulates telomeric chromatin. *Nature* 2008;**452**(7186):492–6.

7. Michishita E, McCord RA, Boxer LD, Barber MF, Hong T, Gozani O, et al. Cell cycle-dependent deacetylation of telomeric histone H3 lysine K56 by human SIRT6. *Cell Cycle* 2009;**8**(16):2664–6.
8. Yang B, Zwaans BMM, Eckersdorff M, Lombard D. The sirtuin SIRT6 deacetylates H3 K56Ac in vivo to promote genomic stability. *Cell Cycle* 2009;**8**(16):2662–3.
9. Jiang H, Khan S, Wang Y, Charron G, He B, Sebastian C, et al. SIRT6 regulates TNF-alpha secretion through hydrolysis of long-chain fatty acyl lysine. *Nature* 2013;**496**(7443):110–13.
10. Zhang X, Khan S, Jiang H, Antonyak MA, Chen X, Spiegelman NA, et al. Identifying the functional contribution of the defatty-acylase activity of SIRT6. *Nat Chem Biol* 2016;**12**(8):614–20.
11. Mao Z, Hine C, Tian X, Van Meter M, Au M, Vaidya A, et al. SIRT6 promotes DNA repair under stress by activating PARP1. *Science* 2011;**332**(6036):1443–6.
12. Koltai E, Szabo Z, Atalay M, Boldogh I, Naito H, Goto S, et al. Exercise alters SIRT1, SIRT6, NAD and NAMPT levels in skeletal muscle of aged rats. *Mech Ageing Dev* 2010;**131**(1):21–8.
13. Zhang N, Li Z, Mu W, Li L, Liang Y, Lu M, et al. Calorie restriction-induced SIRT6 activation delays aging by suppressing NF-κB signaling. *Cell Cycle* 2016;**15**(7):1009–18.
14. Ronnebaum SM, Wu Y, McDonough H, Patterson C. The Ubiquitin ligase CHIP prevents SirT6 degradation through noncanonical ubiquitination. *Mol Cell Biol* 2013;**33**(22):4461–72.
15. Lin Z, Yang H, Tan C, Li J, Liu Z, Quan Q, et al. USP10 antagonizes c-Myc transcriptional activation through SIRT6 stabilization to suppress tumor formation. *Cell Rep* 2013;**5**(6):1639–49.
16. Gil R, Barth S, Kanfi Y, Cohen HY. SIRT6 exhibits nucleosome-dependent deacetylase activity. *Nucl Acids Res* 2013;**41**(18):8537–45.
17. Feldman JL, Baeza J, Denu JM. Activation of the protein deacetylase SIRT6 by long-chain fatty acids and widespread deacylation by mammalian sirtuins. *J Biol Chem* 2013;**288**(43):31350–6.
18. You W, Rotili D, Li TM, Kambach C, Meleshin M, Schutkowski M, et al. Structural basis of sirtuin 6 activation by synthetic small molecules. *Angew Chem Int Ed Eng* 2017;**56**(4):1007–11.
19. Li Y, Meng X, Wang W, Liu F, Hao Z, Yang Y, et al. Cardioprotective effects of SIRT6 in a mouse model of transverse aortic constriction-induced heart failure. *Front Physiol* 2017;**8**:394.
20. Rizzo A, Iachettini S, Salvati E, Zizza P, Maresca C, D'Angelo C, et al. SIRT6 interacts with TRF2 and promotes its degradation in response to DNA damage. *Nucl Acids Res* 2017;**45**(4):1820–34.
21. Taselli L, Xi Y, Zheng W, Tenen RI, Odrowaz Z, Simeoni F, et al. Silencing of pericentromeric heterochromatin associated with SIRT6-dependent H3K18 deacetylation protects against mitotic errors and cellular senescence. *Nat Struct Mol Biol* 2016;**23**(5):434–40.
22. Mostoslavsky R, Chua KF, Lombard DB, Pang WW, Fischer MR, Gellon L, et al. Genomic instability and aging-like phenotype in the absence of mammalian SIRT6. *Cell* 2006;**124**(2):315–29.
23. Xu Z, Zhang L, Zhang W, Meng D, Zhang H, Jiang Y, et al. SIRT6 rescues the age related decline in base excision repair in a PARP1-dependent manner. *Cell Cycle* 2015;**14**(2):269–76.
24. Hwang B-J, Jin J, Gao Y, Shi G, Madabushi A, Yan A, et al. SIRT6 protein deacetylase interacts with MYH DNA glycosylase, APE1 endonuclease, and Rad9–Rad1–Hus1 checkpoint clamp. *BMC Mol Biol* 2015;**16**:12–28.
25. Kaidi A, Weinert BT, Choudhary C, Jackson SP. Human SIRT6 promotes DNA end resection through CtIP deacetylation. *Science* 2010;**329**:1348–53.
26. Sartori AA, Lukas C, Coates J, Mistrik M, Fu S, Bartek J, et al. Human CtIP promotes DNA end resection. *Nature* 2007;**450**(7169):509–14.
27. McCord RA, Michishita E, Hong T, Berber E, Boxer LD, Kusumoto R, et al. SIRT6 stabilizes DNA-dependent Protein Kinase at chromatin for DNA double-strand break repair. *Aging* 2009;**1**(1):109–21.
28. Toiber D, Erdel F, Bouazoune K, Silberman DM, Zhong L, Mulligan P, et al. SIRT6 recruits SNF2H to sites of DNA breaks, preventing genomic instability through chromatin remodeling. *Mol Cell* 2013;**51**(4):454–68.

29. Atsumi Y, Minakawa Y, Ono M, Dobashi S, Shinohe K, Shinohara A, et al. ATM and SIRT6/SNF2H mediate transient H2AX stabilization when DSBs form by blocking HUWE1 to allow efficient gammaH2AX foci formation. *Cell Rep* 2015;**13**(12):2728−40.

30. Kaluski S, Portillo M, Besnard A, Stein D, Einav M, Zhong L, et al. Neuroprotective functions for the histone deacetylase SIRT6. *Cell Rep* 2017;**18**(13):3052−62.

31. Van Meter M, Kashyap M, Rezazadeh S, Geneva AJ, Morello TD, Seluanov A, et al. SIRT6 represses LINE1 retrotransposons by ribosylating KAP1 but this repression fails with stress and age. *Nat Commun* 2014;**5**:5011.

32. Chen W, Liu N, Zhang H, Zhang H, Qiao J, Jia W, et al. Sirt6 promotes DNA end joining in iPSCs derived from old mice. *Cell Rep* 2017;**18**(12):2880−92.

33. Jain AK, Xi Y, McCarthy R, Allton K, Akdemir KC, Patel LR, et al. LncPRESS1 is a p53-regulated LncRNA that safeguards pluripotency by disrupting SIRT6-mediated de-acetylation of histone H3K56. *Mol Cell* 2016;**64**(5):967−81.

34. Etchegaray J-P, Chavez L, Huang Y, Ross KN, Choi J, Martinez-Pastor B, et al. The histone deacetylase Sirt6 controls embryonic stem cell fate via tet-mediated production of 5-hydroxymethylcytosine. *Nat Cell Biol* 2015;**17**(5):545−57.

35. Zhong L, D'Urso A, Toiber D, Sebastian C, Henry RE, Vadysirisack DD, et al. The histone deacetylase SIRT6 regulates glucose homeostasis via Hif1α. *Cell* 2010;**140**(2):280.

36. Puigserver P, Rhee J, Donovan J, Walkey CJ, Yoon JC, Oriente F, et al. Insulin-regulated hepatic gluco-neogenesis through FOXO1-PGC-1alpha interaction. *Nature* 2003;**423**(6939):550−5.

37. Dominy JE, Lee Y, Jedrychowski MP, Chim H, Jurczak MJ, Camporez JP, et al. The deacetylase Sirt6 activates the acetyltransferase GCN5 and suppresses hepatic gluconeogenesis. *Mol Cell* 2012;**48**(6):900−13.

38. Wu M, Seto E, Zhang J. E2F1 enhances glycolysis through suppressing Sirt6 transcription in cancer cells. *Oncotarget* 2015;**6**(13):11252−63.

39. Song M-Y, Wang J, Ka S-O, Bae EJ, Park B-H. Insulin secretion impairment in Sirt6 knockout pancreatic β cells is mediated by suppression of the FoxO1-Pdx1-Glut2 pathway. *Sci Rep* 2016;**6**:30321.

40. Kim H-S, Xiao C, Wang R-H, Lahusen T, Xu X, Vassilopoulos A, et al. Hepatic specific disruption of SIRT6 in mice results in fatty liver formation due to enhanced glycolysis and triglyceride synthesis. *Cell Metab* 2010;**12**(3):224−36.

41. Kanfi Y, Peshti V, Gil R, Naiman S, Nahum L, Levin E, et al. SIRT6 protects against pathological dam-age caused by diet-induced obesity. *Aging Cell* 2010;**9**(2):162−73.

42. Yang SJ, Choi JM, Chae SW, Kim WJ, Park SE, Rhee EJ, et al. Activation of peroxisome proliferator-activated receptor gamma by rosiglitazone increases sirt6 expression and ameliorates hepatic steatosis in rats. *PLoS One* 2011;**6**(2):e17057.

43. Elhanati S, Kanfi Y, Varvak A, Roichman A, Carmel-Gross I, Barth S, et al. Multiple regulatory layers of SREBP1/2 by SIRT6. *Cell Rep* 2013;**4**(5):905−12.

44. Kuang J, Zhang Y, Liu Q, Shen J, Pu S, Cheng S, et al. Fat-specific Sirt6 ablation sensitizes mice to high-fat diet-induced obesity and insulin resistance by inhibiting lipolysis. *Diabetes* 2017;**66**(5):1159−71.

45. Tao R, Xiong X, DePinho RA, Deng C-X, Dong XC. FoxO3 transcription factor and sirt6 deacetylase regulate low density lipoprotein (LDL)-cholesterol homeostasis via control of the proprotein convertase subtilisin/kexin type 9 (*Pcsk9*) gene expression. *J Biol Chem* 2013;**288**(41):29252−9.

46. Chen Q, Hao W, Xiao C, Wang R, Xu X, Lu H, et al. SIRT6 is essential for adipocyte differentiation by regulating mitotic clonal expansion. *Cell Rep* 2017;**18**(13):3155−66.

47. Kugel S, Mostoslavsky R. Chromatin and beyond: the multitasking roles for SIRT6. *Trends Biochem Sci* 2014;**39**(2):72−81.

48. Sebastian C, Zwaans BM, Silberman DM, Gymrek M, Goren A, Zhong L, et al. The histone deacetylase SIRT6 is a tumor suppressor that controls cancer metabolism. *Cell* 2012;**151**(6):1185−99.
49. Thirumurthi U, Shen J, Xia W, LaBaff AM, Wei Y, Li C-W, et al. MDM2-mediated degradation of SIRT6 phosphorylated by AKT1 promotes tumorigenesis and trastuzumab resistance in breast cancer. *Sci Signaling* 2014;**7**(336):ra71.
50. Choe M, Brusgard JL, Chumsri S, Bhandary L, Zhao XF, Lu S, et al. The RUNX2 transcription factor negatively regulates SIRT6 expression to alter glucose metabolism in breast cancer cells. *J Cell Biochem* 2015;**116**(10):2210−26.
51. Min L, Ji Y, Bakiri L, Qiu Z, Cen J, Chen X, et al. Liver cancer initiation is controlled by AP-1 through SIRT6-dependent inhibition of survivin. *Nat Cell Biol* 2012;**14**(11):1203−11.
52. Bhardwaj A, Das S. SIRT6 deacetylates PKM2 to suppress its nuclear localization and oncogenic functions. *Proc Nat Acad Sci U S A* 2016;**113**(5):E538−47.
53. Kugel S, Sebastian C, Fitamant J, Ross KN, Saha SK, Jain E, et al. SIRT6 suppresses pancreatic cancer through control of Lin28b. *Cell* 2016;**165**(6):1401−15.
54. Wang JC, Kafeel MI, Avezbakiyev B, Chen C, Sun Y, Rathnasabapathy C, et al. Histone deacetylase in chronic lymphocytic leukemia. *Oncology* 2011;**81**(5-6):325−9.
55. Cea M, Cagnetta A, Adamia S, Acharya C, Tai Y-T, Fulciniti M, et al. Evidence for a role of the histone deacetylase SIRT6 in DNA damage response of multiple myeloma cells. *Blood* 2016;**127**(9):1138−50.
56. Lee N, Ryu HG, Kwon JH, Kim DK, Kim SR, Wang HJ, et al. SIRT6 depletion suppresses tumor growth by promoting cellular senescence induced by DNA damage in HCC. *PLoS One* 2016;**11**(11):e0165835.
57. Kawahara TLA, Michishita E, Adler AS, Damian M, Berber E, Lin M, et al. SIRT6 links histone H3 lysine 9 deacetylation to control of NF-κB dependent gene expression and organismal lifespan. *Cell* 2009;**136**(1):62−74.
58. He Y, Xiao Y, Yang X, Li Y, Wang B, Yao F, et al. SIRT6 inhibits TNF-alpha-induced inflammation of vascular adventitial fibroblasts through ROS and akt signaling pathway. *Exp Cell Res* 2017;**357**(1):88−97.
59. Ka SO, Bang IH, Bae EJ, Park BH. Hepatocyte-specific sirtuin 6 deletion predisposes to nonalcoholic steatohepatitis by up-regulation of Bach1, an Nrf2 repressor. *FASEB J* 2017;**31**(9):3999−4010.
60. Cai Y, Yu SS, Chen SR, Pi RB, Gao S, Li H, et al. Nmnat2 protects cardiomyocytes from hypertrophy via activation of SIRT6. *FEBS Lett* 2012;**586**(6):866−74.
61. Sundaresan NR, Vasudevan P, Zhong L, Kim G, Samant S, Parekh V, et al. The sirtuin SIRT6 blocks IGF-Akt signaling and development of cardiac hypertrophy by targeting c-Jun. *Nat Med* 2012;**18**(11):1643−50.
62. Liu M, Li Z, Chen GW, Li ZM, Wang LP, Ye JT, et al. AG-690/11026014, a novel PARP-1 inhibitor, protects cardiomyocytes from AngII-induced hypertrophy. *Mol Cell Endocrinol* 2014;**392**:14−22.
63. Imai S, Guarente L. NAD$^+$ and sirtuins in aging and disease. *Trends Cell Biol* 2014;**24**(8):464−71.
64. Kanfi Y, Naiman S, Amir G, Peshti V, Zinman G, Nahum L, et al. The sirtuin SIRT6 regulates lifespan in male mice. *Nature* 2012;**483**(7388):218−21.
65. Sharma A, Diecke S, Zhang WY, Lan F, He C, Mordwinkin NM, et al. The role of SIRT6 protein in aging and reprogramming of human induced pluripotent stem cells. *J Biol Chem* 2013;**288**(25):18439−47.
66. Hirvonen K, Laivuori H, Lahti J, Strandberg T, Eriksson JG, Hackman P. SIRT6 polymorphism rs117385980 is associated with longevity and healthy aging in Finnish men. *BMC Med Genet* 2017;**18**(41).

CHROMATIN AND NUCLEAR SIGNALING: SIRT7 FUNCTION IN THE NUCLEOLUS AND BEYOND

10

Maria Angulo-Ibanez[1,2] and Katrin F. Chua[1,2]

[1]Stanford University School of Medicine, Stanford, CA, United States [2]Veterans Affairs Palo Alto Health Care System, Palo Alto, CA, United States

Sirtuins are NAD^+-dependent enzymes related to yeast silent information regulator-2, a histone deacetylase that prevents premature aging through chromatin silencing. Mammalian sirtuins protect against diverse aging-related pathologic states, from cancer to metabolic and neurodegenerative disease. In addition to chromatin regulation, however, mammalian sirtuins also govern myriad cellular processes through a growing list of enzymatic activities and substrates. Nucleolar-enriched SIRT7 is among the less well-understood mammalian sirtuins, with only a handful of known substrates. Exciting new work has expanded knowledge of SIRT7's biochemical targets and physiologic functions, and uncovered unique links between novel nuclear/nucleolar functions of SIRT7, cellular homeostasis, and aging and disease biology. Here, we review these discoveries, with an emphasis on biochemical insights into SIRT7 catalytic activities and substrates, dynamic functions of SIRT7 in cellular and stress signaling, and emerging insights into the roles of SIRT7 in aging and disease.

10.1 CATALYTIC ACTIVITIES OF SIRT7: STRUCTURE, ENZYMATIC MODULATION, AND SUBSTRATES

Mammalian sirtuin family members share a conserved NAD^+-dependent catalytic domain, but are differentially involved in specific physiologic and disease pathways. Hence, elucidation of structural features and protein substrates that are specific to each family member may be key for developing pharmacologic strategies that selectively target individual sirtuins. Unique among mammalian sirtuins, SIRT7 is enriched in nucleoli, where it deacetylates substrates involved in ribosome biogenesis. In response to stress, however, SIRT7 can also relocate out of nucleoli to deacetylate nucleoplasmic targets. The conserved NAD^+-dependent catalytic domain of SIRT7 is flanked by unique N- and C-terminal extensions, which contain nuclear and nucleolar localization signals.[1,2] The N-terminus of SIRT7 can also bind nucleic acids, and it has been proposed that interactions of this domain with specific RNA and DNA species may allow localized activation of SIRT7 to deacetylate substrates where the corresponding nucleic acids are present, such as RNA processing, transcription, or chromatin regulatory machinery.[2−4]

Among mammalian sirtuins, SIRT7 is most closely related to SIRT6. As discussed below, these sirtuins have been particularly challenging to study at the enzymatic level in vitro, and in early

Introductory Review on Sirtuins in Biology, Aging, and Disease. DOI: https://doi.org/10.1016/B978-0-12-813499-3.00010-1

studies, were proposed to lack the deacetylase activity characteristic of the sirtuin family. More recently, findings from numerous groups have shown that SIRT6 and SIRT7 are highly selective deacetylase enzymes with stringent substrate requirements. Only a handful of SIRT7 substrates, including sites on histone and nonhistone proteins, have been well characterized (Fig. 10.1A). However, it is likely that the list of SIRT7 substrates will expand considerably with new proteomics and functional studies. Indeed, a recent proteomic analysis identified over 250 candidate SIRT7 deacetylation sites where acetylation was increased in SIRT7-deficient cells.[5] The SIRT7 protein interactome includes many factors involved in chromatin remodeling and transcription regulation, such as B-WICH, NoRC, and SWI/SNF, which could include SIRT7 substrates.[6,7] SIRT7 also associates with a number of proteins that lack known acetylation sites, and it is possible that these SIRT7 interactions may serve nonenzymatic or scaffolding roles.

SIRT7 was first shown to have highly selective histone deacetylase activity on lysine K18 of histone H3 (H3K18Ac), but not numerous other histone acetylation sites.[8] In vitro, SIRT7 deacetylates H3K18Ac on nucleosome substrates but not free histones, highlighting the importance of the chromatin context for SIRT7 biochemical activity. In vivo, SIRT7 deacetylates H3K18Ac at promoters of target genes involved in various tumor-suppressive and stress-responsive pathways.[8−10] Consistent with observations of low H3K18Ac levels in tumor tissues, SIRT7-dependent H3K18 deacetylation contributes to maintaining malignant features of cancer cells,[8] and increased SIRT7 levels correlate with tumor aggressiveness and poor patient prognosis in multiple types of cancers.[11,12] SIRT7 also deacetylates H3K18Ac at sites of DNA damage, where it is important for DNA repair.[13] This function of SIRT7 may further contribute to maintaining cancer cell viability in the face of endogenous or therapeutic DNA damage agents, but may also have positive effects in primary cells through enhancement of genomic stability.

In addition to deacetylating histones at chromatin, SIRT7 also has nonhistone targets involved in various aspects of cellular homeostasis and stress responses. SIRT7 is enriched in nucleoli, where it promotes RNA polymerase PolI binding to rDNA promoters and deacetylates the PAF53 subunit of the PolI complex, which promotes pre-rRNA transcription.[14,15] In addition, SIRT7 deacetylates U3-55K, a subunit of the U3 snoRNP, to promote pre-rRNA processing and maturation.[16] Through these functions, SIRT7 facilitates ribosome biogenesis, and consequently, cell growth and proliferation. Beyond the nucleolus, SIRT7 also promotes transcriptional elongation of PolII-dependent transcription of mRNAs and snoRNAs, through deacetylation of CDK9 kinase, a component of the P-TEF-b elongation factor.[2] As described below, in response to nucleolar and genotoxic stress, SIRT7 relocates from nucleoli to nucleoplasm, leading to reduced deacetylation of its nucleolar substrates and potential enhancement of activity on nucleoplasmic targets.[15] Indeed, several nucleoplasmic substrates of SIRT7 are implicated in nuclear signaling and stress responses. For example, SIRT7 influences mitochondrial functions—essential for cellular energy homeostasis—through multiple mechanisms. It deacetylates GABPβ1, a master regulator of nuclear-encoded mitochondrial factors,[17] and represses expression of both cytosolic and mitochondrial translation machinery through H3K18Ac deacetylation.[8,9] SIRT7 also deacetylates the transcription factor FOXO3 to regulate apoptosis in macrophages,[18] and regulates Akt-dependent cell survival pathways through deacetylation of the Akt-scaffolding protein, FKBP51.[19] SIRT7 was also reported to deacetylate p53 in vitro, and thus could have effects on p53-dependent stress responses.[20]

An emerging area of sirtuin biology focuses on identification of NAD^+-dependent catalytic activities beyond deacetylation. This work has already expanded the list of substrates through

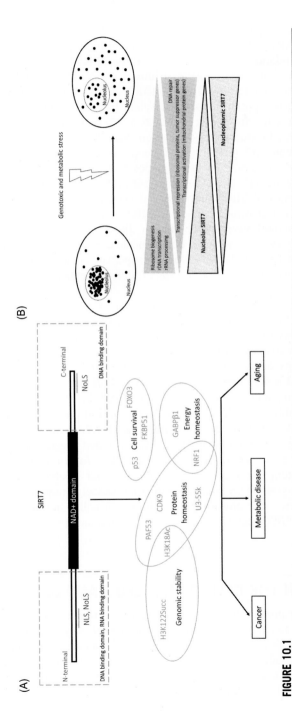

FIGURE 10.1

(A) Schematic of SIRT7 structure showing the NAD$^+$-binding catalytic domain, nuclear localization signal (NLS), nucleolar localization signals (NoLS), and proposed DNA and RNA binding domains. SIRT7 interacts with several proteins involved in different cellular processes, leading to diverse pathophysiological outcomes. (B) Schematic of dynamic subnuclear movements of SIRT7 in response to stress and associated nuclear and nucleolar functions. Black dots indicate relative levels of SIRT7 in nucleoli versus nucleoplasm under baseline or stress conditions.

which several mammalian SIRTs impact on cellular functions. SIRT6 is reported to promote mono-ADP-ribosylation of the PARP1 chromatin regulatory factor and KAP1 transcriptional corepressor,[21,22] and given the close relationship of SIRT7 to SIRT6, SIRT7 might also have ribosylation targets. Recent work has revealed that several sirtuins have NAD$^+$-dependent deacylation activity beyond deacetylation, and can remove lysine modifications involving long-chain fatty-acyl groups. For example, SIRT6 is reported to demyristoylate lysine residues of TNFα, and can defatty-acylate several histone sites in vitro (reviewed in Ref. [23]). Recent studies provide evidence that SIRT7 can also catalyze histone defatty-acylation. Specifically, SIRT7 was reported to desuccinylate lysine K122 of histone H3 (H3K122succ) to promote chromatin condensation and facilitate access to DNA repair machinery.[24] In vitro studies have also revealed that SIRT7 can demyristoylate lysine K9 of histone H3 (H3K9myr), a substrate that is shared with SIRT6.[3] Nonhistone defatty-acylase substrates of SIRT7 have not yet been uncovered. Notably, a majority of known acylated proteins are cytoplasmic,[25] whereas SIRT7 (and SIRT6) are predominantly nuclear proteins. Future work should investigate whether these sirtuins target novel nuclear acylated proteins, or have defatty-acylase activity on nonnuclear proteins. Interestingly, SIRT6 has a pocket that can bind long fatty-acid chains, but this pocket is not conserved in SIRT7. Thus, the structural underpinnings of the reported SIRT7 defatty-acylation activity remain to be clarified.

The H3K18 deacetylase activity of SIRT7 in vitro is relatively weak when assayed on histone peptide or free histone substrates, but is much more efficient on nucleosomes, suggesting that physiologic deacetylation by SIRT7 may require specific conditions, such as posttranslational modifications of SIRT7 or interaction with cofactors. In the case of SIRT6, in vitro deacetylase activity can be increased $\sim$35-fold by free fatty acids.[26] While activation of SIRT7 by free fatty acids has not yet been observed, both DNA and RNA molecules can stimulate both deacetylation and deacylase activity of SIRT7 in vitro.[3,27] Thus, addition of DNA to in vitro SIRT7 reactions uncovered weak deacetylation of H3K9Ac, and addition of rRNAs or tRNAs not only stimulated histone deacetylation by SIRT7, but also uncovered previously undetected histone H3K9myr demyristoylation activity. High-throughput sequencing of SIRT7-bound RNAs revealed enrichment for rRNAs, suggesting a physiologic role for these RNAs in modulating SIRT7 function in the cell. Indeed, approximately 30% of SIRT7 protein interactions, as well its nucleolar localization, are dependent on RNA.[2] Future studies should investigate how RNAs affect the deacetylation or defatty-acylation activity of SIRT7 on its substrates under physiologic conditions in cells, and aim to uncover yet other mechanisms through which SIRT7 catalytic activity can be modulated.

10.2 SIRT7 DYNAMICS IN THE CELL: AT NUCLEOLI AND BEYOND

Originally described as a nucleolar protein, SIRT7 is now understood to partition between nucleolar and nucleoplasmic compartments, and shuttles dynamically out of nucleoli in response to metabolic or genotoxic stress. Thus, SIRT7 both participates in nucleolar stress-sensing mechanisms, and triggers transcriptional and nuclear signaling responses that oppose the instigating stress (Fig. 10.1B). Intriguingly, SIRT7 also affects cellular homeostasis beyond the nucleus, contributing to nuclear–mitochondrial signaling and, potentially, regulation of proteins in the cytoplasm.

10.2.1 STRESS-DEPENDENT PARTITIONING OF NUCLEOLAR AND NUCLEAR FUNCTIONS BY SIRT7

Under baseline conditions, SIRT7 promotes ribosome biogenesis in nucleoli through multiple mechanisms. In addition to augmenting rDNA transcription and rRNA processing,[14−16] SIRT7 also impacts on the translation machinery by activating PolIII-dependent transcription of tRNAs.[28] Interestingly, anchoring of SIRT7 in nucleoli is itself dependent on rRNA binding and is disrupted by rDNA transcription inhibition, suggesting a self-reinforcing feedback loop.[2,15] Moreover, SIRT7 also facilitates PolII-mediated transcriptional elongation through deacetylation of CDK9 kinase.[2] Thus, SIRT7 has global transcriptional activating effects on all three polymerases. These functions of SIRT7 facilitate cell growth and proliferation, and, accordingly, high SIRT7 expression is observed in highly metabolic tissues and in many cancers.[29−32]

In addition to its global effects on polymerase activity, SIRT7 also acts locally to regulate transcription of specific protein-coding gene networks. SIRT7 represses the expression of ribosomal proteins (RPs) and translation machinery genes through histone deacetylation,[8,9] whereas it augments transcription of nuclear-encoded mitochondrial genes through deacetylation of GABPβ1.[17] How these disparate functions of SIRT7 are balanced and contribute to cellular homeostasis likely depends critically on conditions that trigger dynamic relocalization of SIRT7 from nucleolus to nucleoplasm, or to specific chromatin features or genomic locations.

Under conditions of metabolic stress such as glucose deprivation, movement of SIRT7 out of nucleoli leads to loss of ribosome biogenesis activities, with reduced rDNA transcription and rRNA processing,[15] potentially serving as a brake on cell proliferation in response to the energetic restriction. Correspondingly, increased nucleoplasmic SIRT7 levels could lead to increased SIRT7-dependent H3K18Ac deacetylation and reduced transcription of RP genes. Thus, reciprocal loss of nucleolar SIRT7 and increased nucleoplasmic SIRT7 could provide a mechanism for functionally coupling levels of the RNA and protein components of the ribosome. Increased nucleoplasmic SIRT7 under glucose deprivation could also promote GABPβ1-dependent augmentation of mitochondrial function and ATP energy generation. Such a mechanism could contribute to a mitochondrial adaptation response to physiologic stresses of fasting or starvation.[17] Strategies to activate the coordinated effects of SIRT7 shuttling out of nucleoli could also be beneficial in approaches to combat cancer cell proliferation, by putting a brake on both ribosome biogenesis and mitochondrial ATP production.

SIRT7 also translocates out of nucleoli in response to DNA damage agents, and is important for chromatin changes that promote efficient DNA repair.[1,13,24] SIRT7 is recruited dynamically to chromatin surrounding DNA double-strand breaks (DSBs), where it is proposed to promote chromatin compaction via deacetylation of H3K18Ac or desuccinylation of H3K122succ.[13,24] In turn, these chromatin changes are important for recruitment of DNA repair factors to DNA DSBs, and efficient DSB repair via nonhomologous end joining, and possibly homologous recombination. Accordingly, SIRT7-deficient mice show signs of DNA repair defects, including increased DNA damage sensitivity, replication stress, and impaired class switch recombination of immunoglobulin genes. Unanswered questions are whether SIRT7-dependent chromatin changes are important for repair of DNA damage at nucleolar rDNA, and if so, whether DNA damage-induced mobilization of SIRT7 to genome-wide DSBs comes at the expense of increased nucleolar DNA instability.

10.2.2 SIRT7 SIGNALING FROM THE NUCLEUS: TO MITOCHONDRIA AND BEYOND

Growing evidence suggests that SIRT7 has far-ranging effects on cellular homeostasis beyond its chromatin and nuclear/nucleolar signaling activities, most notably in controlling mitochondrial homeostasis. A majority of mitochondrial proteins are encoded in the nucleus, and transcriptional activation of their genes by SIRT7-dependent deacetylation of GABPβ1 is important for maintaining mitochondrial functions and energy homeostasis.[17] Accordingly, SIRT7-deficient mice exhibit metabolic alterations due to mitochondrial dysfunction, particularly in the contexts of cardiac hypertrophy, hepatic steatosis, and hearing defects. In addition, SIRT7 also interacts with NRF1, another key transcriptional regulator of nuclear-encoded mitochondrial genes.[10] However, whereas SIRT7 activates GABPβ1-dependent transcription, the SIRT7−NRF1 interaction represses transcription of mitochondrial RP and translation factor genes. Through this pathway, SIRT7 is proposed to protect against cellular dysfunction due to mitochondrial protein folding stress (PFSmt). Indeed, SIRT7 protein levels are reduced in aging hematopoietic stem cells (HSCs), associated with deregulated PFSmt, and loss of HSC regenerative potential.

Finally, recent findings suggest potential functional effects of SIRT7 in cytoplasmic regulatory pathways. In certain cell types and conditions, SIRT7 is observed to interact with proteins that are predominantly cytoplasmic, and SIRT7 itself can be detected in cytoplasmic biochemical extracts.[1,19] Dicer, a microRNA-processing enzyme implicated in heterochromatin formation in lower organisms, is proposed to hold a portion of SIRT7 in the cytoplasm.[33] Upon DNA damage, increased Dicer expression leads to reduced chromatin-associated SIRT7, and a resultant increase in H3K18Ac acetylation. Thus, cytoplasmic sequestration of SIRT7 by Dicer may be a mechanism for regulating its chromatin and nuclear signaling functions. A potential function for SIRT7 in the cytoplasm has been proposed in studies showing that SIRT7 regulates the PI3K−Akt pathway through modulation of the FKBP51−Akt−PHLPP axis.[19] Mechanistically, SIRT7 deacetylates the scaffolding protein FKBP51, and appears to colocalize with FKBP51 in both nuclear and cytoplasmic compartments. This facilitates interaction of FKBP51 with Akt and PHLPP, and inhibits Akt activation. Thus, the low levels of SIRT7 that are observed in many cancers could contribute to overactivation of Akt signaling and consequent increased chemotherapy resistance.[19] Future work should seek to further investigate mechanisms underlying the dynamic modulation of SIRT7 function in multiple cellular compartments, and elucidate their functional consequences in health and disease.

10.3 SIRT7 FUNCTION IN AGING AND DISEASE

10.3.1 AGING-LIKE PHENOTYPES OF SIRT7-DEFICIENT MICE

Chromatin silencing by yeast silent information regulator-2, the founding member of the sirtuin family, protects against genomic instability and replicative senescence in yeast. Mammalian sirtuins are implicated in many pathways relevant for aging biology. SIRT7-deficient mice have significantly shortened lifespan, genomic instability, and phenotypes that overlap with aging-associated pathologies. These include cardiac cardiomyopathy, lipodystrophy, reduced fat stores, kyphosis, reduced IGF-1 levels, and loss of HSC regenerative potential.[9,13,20] It will be important to characterize the molecular underpinnings of these phenotypes, and investigate whether this constellation of phenotypes is driven by a global "aging program." In this context, defective DNA repair and increased genomic instability in the absence of SIRT7 were proposed to be major contributors to cellular dysfunction and aging-related phenotypes in SIRT7-deficient mice.[13] Notably, SIRT7 levels are reduced in aged HSCs, and overexpression of SIRT7 can rescue the aging-associated functional decline of HSCs.[10] Similarly, in NIH3T3 cells, overexpression of SIRT7 protects against cellular senescence induced by doxorubicin.[34] These observations suggest potential beneficial effects of SIRT7 activation. Future studies should examine whether SIRT7 is also downregulated in other tissues and cell types during aging.

10.3.2 SIRT7 REGULATION OF METABOLIC DISEASE

Sirtuin family members are implicated in regulating many aspects of metabolism, and nuclear sirtuins are ideally suited to link chromatin regulation and genome maintenance to cellular metabolic homeostasis. Much evidence indicates that SIRT7 has central functions in metabolic regulation, although independent studies of SIRT7-deficient mice have yielded some discrepancies in this mechanism. Shin and colleagues observed that SIRT7-deficient mice develop fatty liver hepatosteatosis associated with abnormal lipid handling.[9] Moreover, overexpression of SIRT7 protected against fatty liver pathology in obese, high-fat diet (HFD)-fed mice. Mechanistically, SIRT7 was shown to counteract the pathologic effects of ER (endoplasmic reticulum) stress, a critical instigator of fatty liver pathology, through repression of RP gene expression. In an independent study by Ryu and colleagues, fatty liver pathology observed in SIRT7-deficient mice was further linked to mitochondrial dysfunction due to deregulation of GAPBPβ1 transcriptional programs.[17] Unexpectedly, however, Yoshizawa and collaborators observed contrasting effects of SIRT7 on liver pathology. Their SIRT7-deficient mice showed decreased, rather than increased, susceptibility to hepatosteatosis under HFD, associated with impaired transcription of lipid metabolism genes controlled by the TR4 nuclear receptor.[35] It is likely that these discrepancies reflect complex interactions in the mouse genetic background and SIRT7 biology, and future studies should aim to disentangle the signaling networks through which SIRT7 impacts on metabolic stress challenges.

In addition to regulating lipid handling in the liver, SIRT7 also has an important role in adipose tissue, and decreased fat tissue stores are observed in SIRT7-deficient mice.[9,13,36] Indeed, SIRT7 appears to be a major driver of in vivo adipogenesis, by promoting adipocyte differentiation and maturation.[36] SIRT7 inactivation in cells and mice attenuates in vitro adipogenesis and in vivo adiposity, suggesting that SIRT7 could play a role in promoting obesity.

10.3.3 SIRT7 IN CANCER: EPIGENETIC AND ENERGETIC HOMEOSTASIS

Many clinicopathologic observations of SIRT7 expression in cancer, coupled with diverse mechanistic links of SIRT7 to tumor-promoting pathways, have converged to support the oncogenic activity of SIRT7.[37–39] In many tumor types, high levels of SIRT7 correlate with tumor aggressiveness, metastasis, and poor patient survival. Strikingly, depletion of SIRT7 can reverse the essential features of oncogenic transformation in cancer cells, and attenuate both primary tumor growth and metastasis in vivo.[8,12,31,40,41] Mechanistically, SIRT7 coordinates multiple pathways that favor cancer cell proliferation and tumor growth. SIRT7 is important for epigenetic maintenance of the transformed state of cancer cells through its repression of tumor-suppressive gene networks.[8] In addition, SIRT7 also specifically augments the metastatic potential of cancer cells through regulation of genes involved in epithelial—mesenchymal-like transition, independent of cancer cell proliferative capacity.[12] The effects of SIRT7 on enhancing ribosome biogenesis and mitochondrial function may also be essential contributors to supporting the high metabolic needs of cancer cells. Indeed, tumor-suppressive effects of Wnt5a, a member of the noncanonical WNT pathway, are proposed to result from its displacement of SIRT7 from the PolI machinery, and consequent reduction in rDNA transcription and ribosome biogenesis.[42]

It is important to note, however, that while SIRT7 has tumor-promoting effects in the context of cancer cells, increasing SIRT7 levels in noncancer primary cell lines do not appear to promote oncogenic transformation.[8] Thus, SIRT7 may have different effects on the initiation versus progression of cancer. In this context, the function of SIRT7 in preventing genomic instability and promoting DNA repair may have distinct effects in conditions of cancer versus normal physiology.[13,24] In noncancerous cells, SIRT7 protects against genomic instability and mutagenesis, which can have protective effects against aging-related pathologies and potentially oncogenic mutations. In tumor cells, however, SIRT7 could confer resistance to genotoxic chemotherapeutic strategies.

10.4 CONCLUDING REMARKS AND FUTURE PERSPECTIVES

In recent years, SIRT7 has emerged as a key modulator of gene expression and cellular homeostasis. The establishment of SIRT7 as a physiologic deacetylase enzyme in vivo, and characterization of potential deacylation activity, sets the stage for studies to delineate the full proteome of SIRT7 substrates. Moreover, a growing body of work highlights the dynamic relocalization of SIRT7 from nucleoli to nucleoplasm, and even cytoplasm, in response to genotoxic, metabolic, or other types of stress. These findings suggest intriguing models for how subnuclear translocation of SIRT7 can be important for synchronized, reciprocal regulation of nucleolar versus nuclear targets of SIRT7. Moreover, suggestions that SIRT7 could function to deacylate fatty-acylated proteins, a majority of which are cytoplasmic, lend credence to the notion that the cellular reach of SIRT7 may extend to cytosolic pathways as well. At the same time, the dynamic subcellular mobility of SIRT7 underscores the importance of the molecular context for SIRT7 catalytic function. For example, the stage is set to test models in which interactions of SIRT7 with specific DNA and RNA species, regulatory proteins such as Dicer, or potentially free fatty acids, might target SIRT7 to select subcellular or genomic locations and substrates. Structural studies of SIRT7 in the context of these or other interactions can inform the development of chemical modulators of selective SIRT7 activities. The diverse cellular functions of SIRT7 that continue to be elucidated are matched by its equally diverse functions in mammalian physiology, aging, and disease. A challenge for the future will be to assign, if possible, specific molecular activities and substrates of SIRT7 to particular cellular and organismal processes. Such mechanistic untangling can be instrumental for efforts to selectively activate beneficial SIRT7 functions without triggering potential deleterious additional consequences. The clear relevance for SIRT7 in cancer, metabolism, and aging biology, together with its unique interplay with chromatin regulation and nuclear stress signaling, nominate this protein as a promising target for therapeutic interventions.

REFERENCES

1. Kiran S, Chatterjee N, Singh S, Kaul SC, Wadhwa R. Intracellular distribution of human SIRT7 and mapping of the nuclear/nucleolar localization signal. *FEBS J* 2013;**280**:3451−66.
2. Blank MF, et al. SIRT7-dependent deacetylation of CDK9 activates RNA polymerase II transcription. *Nucleic Acids Res* 2017;**45**:2675−86.
3. Tong Z, et al. SIRT7 is an RNA-activated protein lysine deacylase. *ACS Chem Biol* 2016. Available from: https://doi.org/10.1021/acschembio.6b00954.
4. Priyanka A, Solanki V, Parkesh R, Thakur KG. Crystal structure of the N-terminal domain of human SIRT7 reveals a three-helical domain architecture. *Proteins Struct Funct Bioinforma* 2016;**84**:1558−63.
5. Zhang C, et al. Quantitative proteome-based systematic identification of SIRT7 substrates. *Proteomics* 2017;**17**:13−14.
6. Tsai Y-C, Greco TM, Boonmee A, Miteva Y, Cristea IM. Functional proteomics establishes the interaction of SIRT7 with chromatin remodeling complexes and expands its role in regulation of RNA polymerase I transcription. *Mol Cell Proteomics* 2012;**11**. M111.015156-M111.015156.
7. Lee N, et al. Comparative interactomes of SIRT6 and SIRT7: implication of functional links to aging. *Proteomics* 2014;**14**:1610−22.
8. Barber MF, et al. SIRT7 links H3K18 deacetylation to maintenance of oncogenic transformation. *Nature* 2012;**487**:1−7.
9. Shin J, et al. SIRT7 represses myc activity to suppress er stress and prevent fatty liver disease. *Cell Rep* 2013;**5**:654−65.
10. Mohrin M, et al. A mitochondrial UPR-mediated metabolic checkpoint regulates hematopoietic stem cell aging. *Science* 2015;**347**:1374−7.
11. Zhang S, et al. Sirt7 promotes gastric cancer growth and inhibits apoptosis by epigenetically inhibiting miR-34a. *Sci Rep* 2015;**5**:9787.
12. Malik S, et al. SIRT7 inactivation reverses metastatic phenotypes in epithelial and mesenchymal tumors. *Sci Rep* 2015;**5**:9841.
13. Vazquez BN, et al. SIRT7 promotes genome integrity and modulates non-homologous end joining DNA repair. *EMBO J* 2016;**35**:1488−503.
14. Ford E, et al. Mammalian Sir2 homolog SIRT7 is an activator of RNA polymerase I transcription. *Genes Dev* 2006;**20**:1075−80. Available from: https://doi.org/10.1101/gad.1399706.
15. Chen S, et al. Repression of RNA polymerase I upon stress is caused by inhibition of RNA-dependent deacetylation of PAF53 by SIRT7. *Mol Cell* 2013;**52**:303−13.
16. Chen S, et al. SIRT7-dependent deacetylation of the U3-55k protein controls pre-rRNA processing. *Nat Commun* 2016;**7**:10734.
17. Ryu D, et al. A SIRT7-dependent acetylation switch of GABPβ1 controls mitochondrial function. *Cell Metab* 2014;**20**:856−69.
18. Li Z, Bridges B, Olson J, Weinman SA. The interaction between acetylation and serine-574 phosphorylation regulates the apoptotic function of FOXO3. *Oncogene* 2016;**36**:1−12.
19. Yu J, et al. Regulation of serine-threonine kinase akt activation by NAD + -dependent deacetylase SIRT7. *Cell Rep* 2017;**18**:1229−40.
20. Vakhrusheva O, et al. Sirt7 increases stress resistance of cardiomyocytes and prevents apoptosis and inflammatory cardiomyopathy in mice. *Circ Res* 2008;**102**:703−10.
21. Mao Z, et al. SIRT6 promotes DNA repair under stress by activating PARP1. *Science* 2011;**332**:1443−6.
22. Van Meter M, et al. SIRT6 represses LINE1 retrotransposons by ribosylating KAP1 but this repression fails with stress and age. *Nat Commun* 2014;**5**:5011.

23. Tasselli L, Zheng W, Chua KF. SIRT6: novel mechanisms and links to aging and disease. *Trends Endocrinol Metab* 2017;**28**:168−85.

24. Li L, et al. SIRT7 is a histone desuccinylase that functionally links to chromatin compaction and genome stability. *Nat Commun* 2016;**7**:12235.

25. Resh MD. Fatty acylation of proteins: the long and the short of it. *Prog Lipid Res* 2016;**63**:120−31.

26. Feldman JL, Baeza J, Denu JM. Activation of the protein deacetylase SIRT6 by long-chain fatty acids and widespread deacylation by Mammalian Sirtuins. *J Biol Chem* 2013;**288**:31350−6.

27. Tong Z, et al. SIRT7 is activated by DNA and deacetylates histone H3 in the chromatin context. *ACS Chem Biol* 2016;**11**:742−7.

28. Tsai Y-C, Greco TM, Cristea IM. Sirtuin 7 plays a role in ribosome biogenesis and protein synthesis. *Mol Cell Proteomics* 2014;**13**:73−83.

29. Michishita E, Park JY, Burneskis JM, Barrett JC, Horikawa I. Evolutionarily conserved and nonconserved cellular localizations and functions of human SIRT proteins. *Genetics* 1990;**125**:351−69.

30. Ashraf N, et al. Altered sirtuin expression is associated with node-positive breast cancer. *Br J Cancer* 2006;**95**:1056−61.

31. Kim JK, et al. Sirtuin7 oncogenic potential in human hepatocellular carcinoma and its regulation by the tumor suppressors MiR-125a-5p and MiR-125b. *Hepatology* 2013;**57**:1055−67.

32. Frye R. 'SIRT8' expressed in thyroid cancer is actually SIRT7. *Br J Cancer* 2002;**87**:1479.

33. Zhang PY, et al. Dicer interacts with SIRT7 and regulates H3K18 deacetylation in response to DNA damaging agents. *Nucleic Acids Res* 2016;**44**:3629−42.

34. Kiran S, Oddi V, Ramakrishna G. Sirtuin 7 promotes cellular survival following genomic stress by attenuation of DNA damage, SAPK activation and p53 response. *Exp Cell Res* 2015;**331**:123−41.

35. Yoshizawa T, et al. SIRT7 controls hepatic lipid metabolism by regulating the ubiquitin-proteasome pathway. *Cell Metab* 2014;**19**:712−21.

36. Cioffi M, et al. MiR-93 controls adiposity via inhibition of Sirt7 and Tbx3. *Cell Rep* 2015;**12**:1594−605.

37. Kim JR, et al. Expression of SIRT1 and DBC1 is associated with poor prognosis of soft tissue sarcomas. *PLoS One* 2013;**8**:e74738.

38. Wang H-L, et al. SIRT7 exhibits oncogenic potential in human ovarian cancer cells. *Asian Pac J Cancer Prev* 2015;**16**:3573−7.

39. Shi H, Ji Y, Zhang D, Liu Y, Fang P. MicroRNA-3666-induced suppression of SIRT7 inhibits the growth of non-small cell lung cancer cells. *Oncol Rep* 2016;**36**:3051−7.

40. Yu H, et al. Overexpression of Sirt7 exhibits oncogenic property and serves as a prognostic factor in colorectal cancer. *Clin Cancer Res* 2014;**20**:3434−45.

41. Zhao L, Wang W. miR-125b suppresses the proliferation of hepatocellular carcinoma cells by targeting Sirtuin7. *Int J Clin Exp Med* 2015;**8**:18469−75.

42. Dass RA, et al. Wnt5a signals through DVL1 to repress ribosomal DNA transcription by RNA polymerase I. *PLoS Genet* 2016;**12**:1−19.

MAMMALIAN SIRTUINS, CELLULAR ENERGY REGULATION, AND METABOLISM, AND CARCINOGENESIS

11

Athanassios Vassilopoulos[1], Rui-Hong Wang[2] and David Gius[1]

[1]Northwestern University, Chicago, IL, United States [2]University of Macau, Macau SAR, China

11.1 INTRODUCTION

11.1.1 SIRTUINS IN AGING, CELLULAR METABOLISM, AND METABOLIC REPROGRAMMING

One of the fundamental observations in oncology is that the rate of malignancies increases significantly with age.[1,2] In fact, the single strongest prognostic variable that predicts the incidence of cancer is increasing age, and this is especially true for breast malignancies.[3] As such, breast cancer, together with many other types of cancer, is an aging-related disease,[4,5] and this exponential increase suggests a fundamental link between longevity and carcinogenesis. When the incidence of developing a solid malignancy is analyzed as a function of age, the data clearly show that the risk of developing a tumor begins with an early slope that is gradual and flat. However, as the age increases, an inflection point with a steep slope occurs after 40 years old.[6] Beyond this inflection point, a steep slope is presented in a logarithmic scale, which indicates a late, exponential increase of human cancer incidence rate.

Over 60 years ago, Otto Warburg described that tumor cells tend to have aberrant mitochondrial metabolism. Specifically, cancer cells always exhibit a higher level of glucose consumption (i.e., glycolysis) when compared to their normal counterparts.[7] However, tumor cells also exhibit a wide range of changes in cellular metabolism that is often referred to as metabolic reprogramming, which results in a cellular phenotype that favors all of the necessary demands required for cellular division. These demands include increasing cellular protein, lipids, and other essential organelles that must be increased in size and number to support cellular division. In addition, rapid proliferation also results in the acute accumulation of cellular metabolites, i.e., reactive oxygen species (ROS), which alter cell metabolism.

In this regard, over the last 15 years multiple mice genetically altered to delete specific sirtuin genes have been made as a means to determine if there is a mechanistic connection between this gene family and human illness that is closely associated with increasing age. In fact, several of

Introductory Review on Sirtuins in Biology, Aging, and Disease. DOI: https://doi.org/10.1016/B978-0-12-813499-3.00011-3

these sirtuin knockout mice develop tumors as well as exhibit dysregulated cellular and mitochondrial ROS, and this has been proposed, at least in some significant part, in the mechanism establishing an in vivo tumor-permissive phenotype. Thus, it is suggested that the loss of and/or dysregulation of one of the seven sirtuins may result in a physiological mismatch on cells that direct energy metabolism and/or ROS, and this allows the gradual accumulation of cellular damage that, under specific conditions, results in an in vivo tumor-permissive phenotype. Based on these observations, it is proposed that sirtuins may function as fidelity proteins that maintain energy and metabolic homeostasis during aging and the loss of these cellular processes allows metabolic damage and a tumor-permissive phenotype.

11.1.2 SIRTUINS CAN FUNCTION AS METABOLISM-FIDELITY OR TUMOR-SUPPRESSOR PROTEINS

Mammalian cells express proteins that protect against endogenous and exogenous forms of genotoxic stresses that induce genomic instability.[8–10] These proteins monitor the integrity of cellular metabolism, as well as respond to stressful conditions by activating compensatory pathways.[11–13] An extension of this observation would be that the loss of function or genetic mutation of these fidelity proteins creates a cellular environment that is permissive for the development of tumors,[14–17] suggesting that these proteins also function as tumor-suppressor genes (TSG) in the context of aging and metabolism.[18] Unlike proteins that promote transformation, i.e., oncogenes, tumor-suppressor (TSs) require that both alleles that code for a particular protein must be affected before an effect is manifested. This would suggest a model for carcinogenesis whereby if one allele is damaged, the second can still produce the correct protein, suggesting that TSs exhibit a recessive genetic phenotype while, in contrast, mutant oncogene alleles would exhibit dominant genetics. Since it is unlikely that evolutionary pressure selected for proteins in mammalian cells to prevent carcinogenesis, these proteins are more likely fidelity proteins that have evolved over time to protect specific organelles from damage caused by agents that induce genotoxic stress.

While there are many definitions of a TS gene or protein, in an overview, it should be a gene that protects a cell from one step on the path to transformation and/or carcinogenesis. As such, it has been proposed that a TS gene mutation and/or a loss or reduction in function can alter a cellular phenotype, usually in combination with other genetic changes, which may result in an altered cellular environment permissive for proliferation and progression to cancer. In this regard, it has been suggested that the loss of TS genes may be more important than protooncogene/oncogene activation for the formation of many kinds of human cancer cells.

TSG can be grouped into categories including fidelity or watchdog/caretaker genes, and/or gatekeeper genes that are proposed to continuously sense changes in environmental cellular conditions, such as DNA damage, nutrient status, etc. and initiate specific physiological processes that address these conditions to maintain cell homeostatic poise or equilibrium. In this regard, it seems clear that sirtuins, which are NAD-dependent enzymes, respond to changes in metabolic, genotoxic, oxidative, and osmotic stresses and most importantly, their stress responses appear to link aging,[19–21] oxidative stress, and thus, may function as TS genes/proteins.[22]

11.1.3 MITOCHONDRIAL SIRTUINS AND CARCINOGENESIS

There is a potential mechanistic connection between mitochondrial function and carcinogenesis. In this regard, it is well established that the mitochondria of tumor cells exhibit aberrant ROS, and this observation has been suggested to account for the high degree of genomic instability demonstrated in cancer cells.[22,23] A fundamental hypothesis in cancer research is that increased or aberrant cellular mitochondrial ROS is an early event in cell damage that results in the generation of genomic instability that, under specific cellular conditions, can result in dedifferentiation and carcinogenesis.[24–26] There are three mammalian sirtuins, SIRT3, 4, and 5, as well as four others (SIRT1, 2, 6, and 7), which regulate multiple downstream targets, via varying types of posttranslational modifications, which direct mitochondrial functions. While the dysregulation of the various mitochondrial sirtuins, and their downstream targets, vary, the common theme is aberrant ROS levels.

SIRT3 appears to be the primary mitochondrial deacetylase that directs the mitochondrial acetylome.[27,28] In this regard, *Sirt3*-deficient cells exhibit an in vitro transformation permissive phenotype, as compared to control cells, and mice genetically altered to have a deletion of the *Sirt3* gene develop mammary tumors with a long latency.[29] In addition, *SIRT3* shRNA knockdown in human cancer cells increased xenograft tumor size and reduced latency while, in contrast, *SIRT3*-enforced expression decreased xenograft tumorigenicity.[30] Finally, SIRT3 levels are decreased in breast malignancies,[29,31] and *SIRT3* is deleted in 40% of human breast and ovarian cancers, further supporting a tumor-suppressor role for this protein in human tumors.[32,33]

While it seems likely that the TS function of SIRT3 is a result of the dysregulation of multiple downstream targets, it appears that one common theme is that SIRT3 suppresses the production of ROS via deacetylation to activate an antioxidant enzyme, such as MnSOD (SOD2),[34–36] or other mitochondrial enzymes involved in energy generation, such as acetyl-CoA synthetase 2,[37] IDH2,[38,39] glutamate dehydrogenase,[40,41] succinate dehydrogenase,[42] mitochondrial ribosome subunit MRPL10,[43] and FoxO3a.[44,45] In this regard, our laboratory has shown that there appears to be a subgroup of women with estrogen receptor positive (ER+), luminal B tumors that exhibit a loss of SIRT3-MnSOD-Ac signature that display a more aggressive form of tumor than observed in women with ER +, luminal A malignancies. Based on our data, and those of others, it has been proposed that cells lacking SIRT3 exhibit increased ROS levels promoting nuclear and mitochondrial genome instability, increased HIF1-alpha levels which reprogram and/or result in a mismatch of mitochondrial energy generation, due to the aberrant acetylation of downstream proteins, resulting in a cellular phenotype permissive for tumor development and carcinogenesis.[29,32,33,35]

It has also been proposed that SIRT4 functions as an in vitro and in vivo TS that was first published in a seminal manuscript from the Haigis laboratory.[46] This work demonstrated a mechanistic link between cellular oxidative stress and the DNA damage response (DDR) by directing mitochondrial glutamine metabolism with important implications for the DDR and tumorigenesis.[47,48] In this regard, it was shown that SIRT4 is induced by genotoxic stress and subsequently induces a mitochondrial glutamine signaling program that directs maintenance of genomic integrity in response to DNA damage. Thus, this group proposed that cells lacking *Sirt4* exhibit the dysregulation of this DDR program that results in a cell phenotype permissive for accumulation of DNA damage.

SIRT4 deacetylates multiple downstream targets. In low-nutrient environmental conditions, malonyl CoA decarboxylase (MCD) is one such target. MCD produces acetyl CoA from malonyl CoA, the latter providing a carbon skeleton for lipogenesis under nutrient-rich conditions.[49] When deacetylated by SIRT4, MCD functions less efficiently, and animals lacking SIRT4 present with increased MCD activity, dysregulated lipid metabolism, and protection against diet-induced obesity. Therefore, SIRT4 opposes fatty acid oxidation, promoting lipid anabolism by regulating MCD function/malonyl CoA levels.[49] Similarly, in both myocytes and hepatocytes, loss of SIRT4 increased fatty acid oxidation gene expression and cell respiration.[50] *SIRT4* mRNA expression is reduced in several malignancies, including breast, colon, bladder, gastric, ovarian, and thyroid cancers, though *SIRT4* loss was particularly pronounced in lung cancer patients. These results point to SIRT4 as a TS, and its downregulation may serve to facilitate the progression of several human cancers and this may be due to the loss of or the dysregulation of glutamine/glutamate metabolism.

One characteristic that distinguishes SIRT5, compared to the other mitochondrial sirtuins, is its function in carrying out additional posttranslational modifications other than acetylation. In this regard, SIRT5 has been found to have a much greater affinity to remove negatively charged acyl groups from proteins, including malonyl, succinyl, and glutaryl groups,[51–53] which can be explained considering the presence of two positively charged amino acid groups in its active site. These findings, together with proteomics-based analyses,[54,55] have established SIRT5 as a critical player in regulating diverse metabolic pathways. Given that impaired metabolism is a hallmark of cancer, it would be reasonable to suggest a functional involvement in tumorigenesis.

Studies have shown that SIRT5 deacetylates cytochrome c, which plays a role in both oxidative phosphorylation and apoptosis.[41] This was consistent with the presence of both the deacetylase and the substrate in the mitochondrial intermembrane space, suggesting that subcellular localization may contribute to the specificity of biological functions regulated by individual family members. As SIRT5 can be found in the matrix as well, it was reported later that it interacts with and deacetylates carbamoyl phosphate synthetase 1 (CPS1), an enzyme essential for ammonia detoxification and disposal. Deacetylation results in increased enzymatic activity, which is further supported by in vivo studies showing that $Sirt5^{-/-}$ mice fail to upregulate CPS1 activity and control blood ammonia during nutrient stress conditions such as fasting and calorie restriction.[56]

In contrast to other mammalian sirtuins, no tumor development has been observed in mice lacking *Sirt5*.[28,57] Although this suggests that SIRT5 might not be a bona fide tumor suppressor, there is some scientific evidence supporting specific tumor-suppressive properties. Specifically, *IDH1* mutation or *SDH* inactivation resulted in mitochondrial dysfunctions, driven by hypersuccinylation, including respiration inhibition and altered metabolism with concomitant BCL-2-mediated apoptosis resistance and enhanced tumor growth. However, *SIRT5* overexpression reversed these phenotypes, suggesting that SIRT5 can function as a tumor suppressor, at least in the context of IDH mutant tumors.[58] Furthermore, *SIRT5* expression has been reported as low in different types of cancer, such as head and neck squamous cell carcinoma[59] and endometrial carcinoma.[60]

In contrast to these previous findings, a tumorigenic role for SIRT5 has been described in the context of ROS management. In this regard, it has been shown that Cu/Zn superoxide dismutase (SOD1) can induce tumorigenesis,[61,62] and its oncogenic properties may be attributed to SIRT5 through desuccinylation and activation of SOD1.[63] *SIRT5* is overexpressed in human nonsmall-cell lung cancer, and high expression predicts poor survival; however, *SIRT5* knockdown represses lung cancer cell growth and makes lung cancer cells more sensitive to drug treatment.[64] In line with this observation, SIRT5 was described as a negative regulator of the tumor suppressor SUN2, which further supports a potential oncogenic role for SIRT5 in lung cancer. These results strongly suggest that the mitochondrial sirtuins are a family of fidelity proteins that function, at least in some part, to prevent dysregulated energy metabolism as well as to prevent tumorigenesis.

11.1.4 NONMITOCHONDRIAL SIRTUINS AND CARCINOGENESIS

Sirt1, the most conserved mammalian homolog of yeast Sir2, localizes in both cytoplasm and nucleus, and plays a pivotal role in the regulation of the cell cycle, cell metabolism, stress response, circadian rhythm, and genome stability.[65,66] The role of Sirt1 in cancer is controversial, and it has been proposed to function as a tumor suppressor or an oncogene. In 2008, by using different *Sirt1* knockout mouse models, two independent studies demonstrated that SIRT1 can function as a tumor suppressor.[67,68] For example, Sirt1 haplo-insufficiency led to carcinoma, sarcoma, and lymphoma in the background of p53 insufficiency due to loss of heterochromatin formation and DNA damage repair.[67] In Apc$^{min/+}$ mice, SIRT1 deacetylated beta-catenin and suppressed its ability to activate transcription.[68] Both studies demonstrated that a certain population of human breast cancer, hepatocellular carcinoma (HCC), and colon cancer patients contained a low level of SIRT1.

In contrast, SIRT1 displayed an oncogenic effect in a Pten$^{+/-}$ mouse model via a proposed mechanism, whereby SIRT1 increased c-Myc transcriptional programs promoting prostate and thyroid tumors.[69] This report has been accompanied with a plethora of in vitro studies to demonstrate that many types of human cancer cell lines contained a high level of SIRT1, and SIRT1 antagonists can inhibit their proliferation.[70–77] The most compelling evidence about SIRT1's oncogenic effects came from studies with triple negative breast cancer (TNBC) samples. This team found that SIRT1 levels were increased in TNBC and were positively correlated with lymph node metastasis.[78] SIRT1 was also found deregulated in human gliomas;[79] and could serve as a marker for poor prognosis in advanced colorectal cancer.[80] These results indicate the obvious fact that SIRT1 is a double-edged sword during tumorigenesis and suggests a scientific manifestation that SIRT1 participates in pleiotropic pathways to maintain body homeostasis that manifests its effects on carcinogenesis in a tissue-specific context.

SIRT1 modulates multiple factors that are important for tumorigenesis through deacetylation or transcriptional regulation.[81] The Sirt1–p53 axis was the initial connection between SIRT1 and cancer. SIRT1 deacetylates p53 at multiple lysine sites, and p53 acetylation attenuates p53-mediated apoptosis under DNA damage and oxidative stress.[82] Transcription factors, such as E2F1, FOXO1, HIC1, BCL6, PGC1a, AR, RB, and NF-κB can also be deacetylated by SIRT1.[83] In addition, it has also been shown that SIRT1 maintains genome stability by promoting heterochromatin formation, via conversion of H4K9ac to an H3K9me3 heterochromatin mark, which functions to maintain genomic stability as well as enhancing DNA damage repair pathways.[67,84,85] SIRT1 forms a reciprocal feedback loop with the circadian clock where SIRT1 deacetylates BMAL1 to affect CLOCK-BMAL1-PER2 transcriptional activity,[86,87] at the same time, Per2 regulates the SIRT1 transcriptional level.[84] Finally, SIRT1 plays a key role in glucose and lipid metabolism[66] via the regulation of signaling factors such as FOXO1-mediated transactivation through deacetylation,[88] mTORC2 to affect FOXO1 activity,[89] and loop formation of p53-microRNA suggesting and confirming the complex nature of the role of SIRT1 in the process of transformation and tumor suppression.

SIRT2 is similar in sequence to yeast Hst2p,[90] and both proteins are located in the cytoplasm, which distinguishes SIRT2 from other members of the sirtuin family localized in either the nucleus or mitochondria. After the first study showing that it colocalizes with microtubules and specifically deacetylates lysine-40 of alpha-tubulin both in vitro and in vivo, several studies have identified other SIRT2-specific deacetylation targets, therefore implicating it in a variety of cellular processes. Interestingly, among the first functions attributed to the mainly cytoplasmic sirtuin was its role in regulating mitosis. During this phase of the cell cycle SIRT2 relocates to the nucleus where it directs histone H4 lysine 16 (H4K16)-mediated chromatin condensation[91] and mitotic exit.[92] Furthermore, several studies revealed a role for SIRT2 as a mitotic checkpoint protein by blocking entry to chromosome condensation and regulating chronic mitotic arrest under mitotic stress.[93,94]

In accordance with the previously assigned functions on cell cycle and mitosis, SIRT2 was found to associate with several mitotic structures including the centrosome, mitotic spindle, and mid-body.[95] Genetic evidence in several mouse models deficient in *Sirt2* suggests a tumor-suppressive role for this member of the sirtuin family as well.[96–98] Considering that impaired mitotic progression may trigger genomic instability, it comes as no surprise that SIRT2 was described to play a critical role in maintaining genome integrity. In this regard, *Sirt2*$^{-/-}$ mice developed tumors in multiple tissues and the tumor incidence slowly increased with increasing mouse age, providing strong genetic evidence for a tumor-suppressor function of *Sirt2*.[96] Following a proteomics-based approach to identify SIRT2-interacting proteins, the APC/C complex coactivators CDH1 and CDC20 were identified as downstream deacetylation targets.

More specifically, it was shown that increased acetylation upon *Sirt2* loss decreases protein–protein interactions between the coactivators and CDC27, resulting in reduced APC/C complex activity and subsequent mitotic defects. Moreover, *SIRT2* depletion resulted in hypersensitivity to replication stress and spontaneous induction of DNA damage based on its function as a specific deacetylase for both ATR interacting protein (ATRIP) and CDK9.[99,100] Together with the regulatory role of SIRT2 on mitotic deposition of H4K20me1,[97] all these pathways could provide the underlying mechanism for tumor development in mice with *Sirt2* deletion mainly due to impaired DDR and increased genomic instability. The tumor-suppressive role for SIRT2 is further supported by decreased expression in a number of cancers including breast, liver, and glioma, endometrial, as well as basal and gastric carcinomas.[60,96,101–103] Consistently, somatic *SIRT2* mutations can be detected across multiple cancer types and, more importantly, can contribute to DDR impairment and genomic instability by decreasing its deacetylase activity.[104]

Given the significance of tissue, cell-type and genetic specific context, it is noteworthy to mention that several studies have also highlighted oncogenic properties for SIRT2. This justifies the research efforts to explore SIRT2 inhibition as a potential therapeutic anticancer strategy. In this regard, SIRT2 was shown to be critical for survival of wildtype p53 cancer cells, as evidenced by increased cell death upon SIRT2 inhibition which depends on functional p53.[105,106] The dual functionality of SIRT2 in tumorigenesis is also suggested based on recent evidence showing that it upregulates Myc protein expression,[107] whereas genetic and pharmacologic SIRT2 inhibition exerts the opposite effect by upregulating several E3 enzymes and promoting ubiquitin-mediated degradation of Myc.[107,108] In HCC, *SIRT2* expression was upregulated and positively correlated with EMT through deacetylating and activating protein kinase B to target the glycogen synthase kinase (Akt/GSK)3-β/β-catenin signaling pathway.[109] Consistent with its tumorigenic role, a recently published study showed that SIRT2 maintains Slug protein levels through deacetylation-mediated increased protein stability. Furthermore, it was shown that elevated Slug protein caused by *SIRT2* overexpression corresponded to stronger repression of the Slug transcriptional targets, epithelial cell adhesion molecule, and E-cadherin, implying that SIRT2 regulates EMT-related phenotypes such as aggressiveness and invasion specifically in basal-like breast cancer.[110]

SIRT6 is considered as a watchdog for genome integrity due to its major histone targets, H3K9ac, H3K56ac, and H3K18ac, as well as its role in faithful mitosis.[111] In addition, SIRT6 plays key roles in glucose metabolism[112] and lipid metabolism,[113,114] all of which are dysregulated in tumorigenesis. The evidence that SIRT6 might serve as a tumor suppressor came from the observation that immortalized Sirt6$^{-/-}$ MEF cells generated xenograft tumors without oncogenic mutations.[112] In this study, researchers found that Sirt6 is deeply involved in regulating glycolytic and ribosomal genes together with Myc, hence inducing a Warburg effect. SIRT6's impact on cancer biology has also been manifested by other mechanisms. In 2015, it was shown that SIRT6 loss-of-function point mutations were detected in 12 types of human cancer samples, and all these mutations affected either the stability or catalytic activity of SIRT6.[115] It has also been shown that in both mouse and human PDAC tumors, loss of *Sirt6* caused hyperacetylation of Lin28b promotor and recruitment of Myc. Upregulated level of Lin28b reduced microRNA Let-7 and led to pancreas tumor progression and metastasis.[116] SIRT6 deacetylates PKM2, and inhibits AP-1 activity to suppress HCC formation/progression.[117,118] However, recent investigations also uncovered the contexts that inhibiting SIRT6 was beneficial for treating some types of cancers.[119]

11.2 CONCLUSIONS

Overall, these results suggest that sirtuins are the watchdog or fidelity proteins that direct physiological and cellular metabolism and, by extension, it is proposed that the dysregulation of these sirtuins results in a damage-permissive phenotype (Fig. 11.1). In this regard, and to varying degrees, there are both murine and human data suggesting that the sirtuin protein family can, depending on the tissue-specific context, function as tumor-suppressor proteins. In addition, it is also likely that the loss of function may also result in a permissive phenotype for other aging-related illnesses such as insulin resistance, cardiovascular disease, and neurodegeneration. While the underlying mechanism is not entirely clear it seems reasonable to suggest that the dysregulation of energy metabolism, as well as aberrant ROS levels, are fundamental to the observed tumor-permissive phenotype.

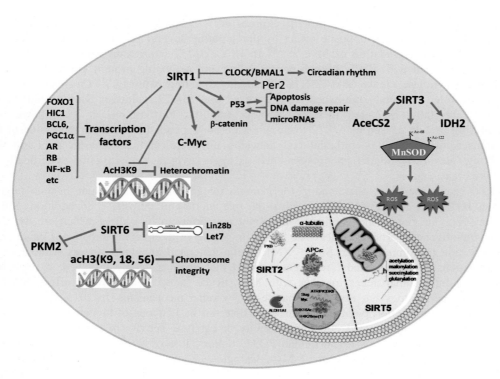

FIGURE 11.1

Sirtuins and potential downstream deacetylation targets that may play a role in the regulation of normal cellular integrity, which when dysregulated, may direct a tumor permissive phenotype.

REFERENCES

1. Ershler WB, Longo DL. The biology of aging: the current research agenda. *Cancer* 1997;**80**(7):1284—93.
2. Ershler WB, Longo DL. Aging and cancer: issues of basic and clinical science. *J Natl Cancer Inst* 1997;**89**(20):1489—97.
3. Longo VD, Fontana L. Calorie restriction and cancer prevention: metabolic and molecular mechanisms. *Trends Pharmacol Sci* 2010;**31**(2):89—98.
4. Bidoli E, Fratino L, Bruzzone S, Pappagallo M, De Paoli P, Tirelli U, et al. Time trends of cancer mortality among elderly in Italy, 1970—2008: an observational study. *BMC Cancer* 2012;**12**(1):443.
5. Barginear MF, Muss H, Kimmick G, Owusu C, Mrozek E, Shahrokni A, et al. Breast cancer and aging: results of the U13 conference breast cancer panel. *Breast Cancer Res Treat* 2014;**146**(1):1—6.
6. Zhu Y, Yan Y, Principe DR, Zou X, Vassilopoulos A, Gius D. SIRT3 and SIRT4 are mitochondrial tumor suppressor proteins that connect mitochondrial metabolism and carcinogenesis. *Cancer Metab* 2014;**2**:15.
7. Warburg O. On the origin of cancer cells. *Science* 1956;**123**(3191):309—14.
8. Slane BG, Aykin-Burns N, Smith BJ, Kalen AL, Goswami PC, Domann FE, et al. Mutation of succinate dehydrogenase subunit C results in increased O2.-, oxidative stress, and genomic instability. *Cancer Res* 2006;**66**(15):7615—20.

9. Aykin-Burns N, Ahmad IM, Zhu Y, Oberley LW, Spitz DR. Increased levels of superoxide and H2O2 mediate the differential susceptibility of cancer cells versus normal cells to glucose deprivation. *Biochem J* 2009;**418**(1):29−37.

10. Du C, Gao Z, Venkatesha VA, Kalen AL, Chaudhuri L, Spitz DR, et al. Mitochondrial ROS and radiation induced transformation in mouse embryonic fibroblasts. *Cancer Biol Ther* 2009;**8**(20).

11. Mallakin A, Sugiyama T, Taneja P, Matise LA, Frazier DP, Choudhary M, et al. Mutually exclusive inactivation of DMP1 and ARF/p53 in lung cancer. *Cancer Cell* 2007;**12**(4):381−94.

12. Uren AG, Kool J, Matentzoglu K, de Ridder J, Mattison J, van Uitert M, et al. Large-scale mutagenesis inp19(ARF)- and p53-deficient mice identifies cancer genes and their collaborative networks. *Cell* 2008;**133**(4):727−41.

13. Vousden KH, Prives C. Blinded by the light: the growing complexity of p53. *Cell* 2009;**137**(3):413−31.

14. Greger V, Passarge E, Hopping W, Messmer E, Horsthemke B. Epigenetic changes may contribute to the formation and spontaneous regression of retinoblastoma. *Hum Genet* 1989;**83**(2):155−8.

15. Baker SJ, Markowitz S, Fearon ER, Willson JK, Vogelstein B. Suppression of human colorectal carcinoma cell growth by wild-type p53. *Science* 1990;**249**(4971):912−15.

16. Feinberg AP, Johnson LA, Law DJ, Kuehn SE, Steenman M, Williams BR, et al. Multiple tumor suppressor genes in multistep carcinogenesis. *Tohoku J Exp Med* 1992;**168**(2):149−52.

17. Sherr CJ, McCormick F. The RB and p53 pathways in cancer. *Cancer Cell* 2002;**2**(2):103−12.

18. Hunter T. Oncoprotein networks. *Cell* 1997;**88**(3):333−46.

19. Chua KF, Mostoslavsky R, Lombard DB, Pang WW, Saito S, Franco S, et al. Mammalian SIRT1 limits replicative life span in response to chronic genotoxic stress. *Cell Metabol* 2005;**2**(1):67−76.

20. Droge W, Schipper HM. Oxidative stress and aberrant signaling in aging and cognitive decline. *Aging Cell* 2007;**6**(3):361−70.

21. Scher MB, Vaquero A, Reinberg D. SirT3 is a nuclear NAD + -dependent histone deacetylase that translocates to the mitochondria upon cellular stress. *Genes Dev* 2007;**21**(8):920−8.

22. Harman D. Aging: a theory based on free radical and radiation chemistry. *J Gerontol* 1956;**11**(3):298−300.

23. Gemma C, Vila J, Bachstetter A, Bickford PC. Oxidative stress and the aging brain: from theory to prevention. In: Riddle DR, editor. *Brain aging: models, methods, and mechanisms.* Boca Raton, FL: CRC Press; 2007. p. 353−74.

24. Clutton SM, Townsend KM, Walker C, Ansell JD, Wright EG. Radiation-induced genomic instability and persisting oxidative stress in primary bone marrow cultures. *Carcinogenesis* 1996;**17**(8):1633−9.

25. Spitz DR, Azzam EI, Li JJ, Gius D. Metabolic oxidation/reduction reactions and cellular responses to ionizing radiation: a unifying concept in stress response biology. *Cancer Metastasis Rev* 2004;**23** (3-4):311−22.

26. Dayal D, Martin SM, Limoli CL, Spitz DR. Hydrogen peroxide mediates the radiation-induced mutator phenotype in mammalian cells. *Biochem J* 2008;**413**(1):185−91.

27. Ahn BH, Kim HS, Song S, Lee IH, Liu J, Vassilopoulos A, et al. A role for the mitochondrial deacetylase Sirt3 in regulating energy homeostasis. *Proc Natl Acad Sci U S A* 2008;**105**(38):14447−52.

28. Lombard DB, Alt FW, Cheng HL, Bunkenborg J, Streeper RS, Mostoslavsky R, et al. Mammalian Sir2 homolog SIRT3 regulates global mitochondrial lysine acetylation. *Mol Cell Biol* 2007;**27**(24):8807−14.

29. Kim HS, Patel K, Muldoon-Jacobs K, Bisht KS, Aykin-Burns N, Pennington JD, et al. SIRT3 is a mitochondria-localized tumor suppressor required for maintenance of mitochondrial integrity and metabolism during stress. *Cancer Cell* 2010;**17**(1):41−52.

30. Bell EL, Emerling BM, Ricoult SJ, Guarente L. SirT3 suppresses hypoxia inducible factor 1α and tumor growth by inhibiting mitochondrial ROS production. *Oncogene* 2011;**30**:2986−96.

31. Desouki MM, Doubinskaia I, Gius D, Abdulkadir SA. Decreased mitochondrial SIRT3 expression is a potential molecular biomarker associated with poor outcome in breast cancer. *Hum Pathol* 2014;**45** (5):1071−7.

32. Haigis MC, Deng CX, Finley LW, Kim HS, Gius D. SIRT3 is a mitochondrial tumor suppressor: a scientific tale that connects aberrant cellular ROS, the warburg effect, and carcinogenesis. *Cancer Res* 2012;**72** (10):2468−72.

33. Finley LW, Carracedo A, Lee J, Souza A, Egia A, Zhang J, et al. SIRT3 opposes reprogramming of cancer cell metabolism through HIF1alpha destabilization. *Cancer Cell* 2011;**19**(3):416−28.

34. Qiu X, Brown K, Hirschey MD, Verdin E, Chen D. Calorie restriction reduces oxidative stress by SIRT3-mediated SOD2 activation. *Cell Metabol* 2010;**12**(6):662−7.

35. Tao R, Coleman MC, Pennington JD, Ozden O, Park SH, Jiang H, et al. Sirt3-mediated deacetylation of evolutionarily conserved lysine 122 regulates MnSOD activity in response to stress. *Mol Cell* 2010;**40** (6):893−904.

36. Tao R, Vassilopoulos A, Parisiadou L, Yan Y, Gius D. Regulation of MnSOD enzymatic activity by Sirt3 connects the mitochondrial acetylome signaling networks to aging and carcinogenesis. *Antioxid Redox Signal* 2014;**20**(10):1646−54.

37. Schwer B, Bunkenborg J, Verdin RO, Andersen JS, Verdin E. Reversible lysine acetylation controls the activity of the mitochondrial enzyme acetyl-CoA synthetase 2. *Proc Natl Acad Sci U S A* 2006;**103** (27):10224−9.

38. Someya S, Yu W, Hallows WC, Xu J, Vann JM, Leeuwenburgh C, et al. Sirt3 mediates reduction of oxidative damage and prevention of age-related hearing loss under caloric restriction. *Cell* 2010;**143** (5):802−12.

39. Zou X, Zhu Y, Park SH, Liu G, O'Brien J, Jiang H, et al. SIRT3-mediated dimerization of IDH2 directs cancer cell metabolism and tumor growth. *Cancer Res* 2017;**77**(15):3990−9.

40. Hallows WC, Lee S, Denu JM. Sirtuins deacetylate and activate mammalian acetyl-CoA synthetases. *Proc Natl Acad Sci U S A* 2006;**103**(27):10230−5.

41. Schlicker C, Gertz M, Papatheodorou P, Kachholz B, Becker CF, Steegborn C. Substrates and regulation mechanisms for the human mitochondrial sirtuins Sirt3 and Sirt5. *J Mol Biol* 2008;**382**(3):790−801.

42. Cimen H, Han MJ, Yang Y, Tong Q, Koc H, Koc EC. Regulation of succinate dehydrogenase activity by SIRT3 in mammalian mitochondria. *Biochemistry* 2010;**49**(2):304−11.

43. Yang Y, Cimen H, Han MJ, Shi T, Deng JH, Koc H, et al. NAD + -dependent deacetylase SIRT3 regulates mitochondrial protein synthesis by deacetylation of the ribosomal protein MRPL10. *J Biol Chem* 2010;**285**(10):7417−29.

44. Jacobs KM, Pennington JD, Bisht KS, Aykin-Burns N, Kim HS, Mishra M, et al. SIRT3 interacts with the daf-16 homolog FOXO3a in the mitochondria, as well as increases FOXO3a dependent gene expression. *Int J Biol Sci* 2008;**4**(5):291−9.

45. Sundaresan NR, Gupta M, Kim G, Rajamohan SB, Isbatan A, Gupta MP. Sirt3 blocks the cardiac hypertrophic response by augmenting Foxo3a-dependent antioxidant defense mechanisms in mice. *J Clin Invest* 2009.

46. Jeong SM, Xiao C, Finley LW, Lahusen T, Souza AL, Pierce K, et al. SIRT4 has tumor-suppressive activity and regulates the cellular metabolic response to DNA damage by inhibiting mitochondrial glutamine metabolism. *Cancer Cell* 2013;**23**(4):450−63.

47. Jeong SM, Lee A, Lee J, Haigis MC. SIRT4 protein suppresses tumor formation in genetic models of Myc-induced B cell lymphoma. *J Biol Chem* 2014;**289**(7):4135−44.

48. Son J, Lyssiotis CA, Ying H, Wang X, Hua S, Ligorio M, et al. Glutamine supports pancreatic cancer growth through a KRAS-regulated metabolic pathway. *Nature* 2013;**496**(7443):101−5.

49. Laurent G, German NJ, Saha AK, de Boer VC, Davies M, Koves TR, et al. SIRT4 coordinates the balance between lipid synthesis and catabolism by repressing malonyl CoA decarboxylase. *Mol Cell* 2013;**50**(5):686–98.
50. Nasrin N, Wu X, Fortier E, Feng Y, Bare OC, Chen S, et al. SIRT4 regulates fatty acid oxidation and mitochondrial gene expression in liver and muscle cells. *J Biol Chem* 2010;**285**(42):31995–2002.
51. Peng C, Lu Z, Xie Z, Cheng Z, Chen Y, Tan M, et al. The first identification of lysine malonylation substrates and its regulatory enzyme. *Mol Cel Proteom* 2011;**10**(12). M111 012658.
52. Du J, Zhou Y, Su X, Yu JJ, Khan S, Jiang H, et al. Sirt5 is a NAD-dependent protein lysine demalonylase and desuccinylase. *Science* 2011;**334**(6057):806–9.
53. Tan M, Peng C, Anderson KA, Chhoy P, Xie Z, Dai L, et al. Lysine glutarylation is a protein posttranslational modification regulated by SIRT5. *Cell Metabol* 2014;**19**(4):605–17.
54. Nishida Y, Rardin MJ, Carrico C, He W, Sahu AK, Gut P, et al. SIRT5 regulates both cytosolic and mitochondrial protein malonylation with glycolysis as a major target. *Mol Cell* 2015;**59**(2):321–32.
55. Park J, Chen Y, Tishkoff DX, Peng C, Tan M, Dai L, et al. SIRT5-mediated lysine desuccinylation impacts diverse metabolic pathways. *Mol Cell* 2013;**50**(6):919–30.
56. Nakagawa T, Lomb DJ, Haigis MC, Guarente L. SIRT5 deacetylates carbamoyl phosphate synthetase 1 and regulates the urea cycle. *Cell* 2009;**137**(3):560–70.
57. Yu J, Sadhukhan S, Noriega LG, Moullan N, He B, Weiss RS, et al. Metabolic characterization of a Sirt5 deficient mouse model. *Sci Rep* 2013;**3**:2806.
58. Li F, He X, Ye D, Lin Y, Yu H, Yao C, et al. NADP(+)-IDH mutations promote hypersuccinylation that impairs mitochondria respiration and induces apoptosis resistance. *Mol Cell* 2015;**60**(4):661–75.
59. Lai CC, Lin PM, Lin SF, Hsu CH, Lin HC, Hu ML, et al. Altered expression of SIRT gene family in head and neck squamous cell carcinoma. *Tumour Biol* 2013;**34**(3):1847–54.
60. Bartosch C, Monteiro-Reis S, Almeida-Rios D, Vieira R, Castro A, Moutinho M, et al. Assessing sirtuin expression in endometrial carcinoma and non-neoplastic endometrium. *Oncotarget* 2016;**7**(2):1144–54.
61. Papa L, Hahn M, Marsh EL, Evans BS, Germain D. SOD2 to SOD1 switch in breast cancer. *J Biol Chem* 2014;**289**(9):5412–16.
62. Somwar R, Erdjument-Bromage H, Larsson E, Shum D, Lockwood WW, Yang G, et al. Superoxide dismutase 1 (SOD1) is a target for a small molecule identified in a screen for inhibitors of the growth of lung adenocarcinoma cell lines. *Proc Natl Acad Sci U S A* 2011;**108**(39):16375–80.
63. Lin ZF, Xu HB, Wang JY, Lin Q, Ruan Z, Liu FB, et al. SIRT5 desuccinylates and activates SOD1 to eliminate ROS. *Biochem Biophys Res Commun* 2013;**441**(1):191–5.
64. Lu W, Zuo Y, Feng Y, Zhang M. SIRT5 facilitates cancer cell growth and drug resistance in non-small cell lung cancer. *Tumour Biol* 2014;**35**(11):10699–705.
65. Ren NS, Ji M, Tokar EJ, Busch EL, Xu X, Lewis D, et al. Haploinsufficiency of SIRT1 enhances glutamine metabolism and promotes cancer development. *Curr Biol* 2017;**27**(4):483–94.
66. Donmez G, Guarente L. Aging and disease: connections to sirtuins. *Aging Cell* 2010;**9**(2):285–90.
67. Wang RH, Sengupta K, Li C, Kim HS, Cao L, Xiao C, et al. Impaired DNA damage response, genome instability, and tumorigenesis in SIRT1 mutant mice. *Cancer Cell* 2008;**14**(4):312–23.
68. Firestein R, Blander G, Michan S, Oberdoerffer P, Ogino S, Campbell J, et al. The SIRT1 deacetylase suppresses intestinal tumorigenesis and colon cancer growth. *PLoS One* 2008;**3**(4):e2020.
69. Herranz D, Maraver A, Canamero M, Gomez-Lopez G, Inglada-Perez L, Robledo M, et al. SIRT1 promotes thyroid carcinogenesis driven by PTEN deficiency. *Oncogene* 2013;**32**(34):4052–6.
70. Carnevale I, Pellegrini L, D'Aquila P, Saladini S, Lococo E, Polletta L, et al. SIRT1-SIRT3 axis regulates cellular response to oxidative stress and etoposide. *J Cell Physiol* 2017;**232**(7):1835–44.
71. Cho EH, Dai Y. SIRT1 controls cell proliferation by regulating contact inhibition. *Biochem Biophys Res Commun* 2016;**478**(2):868–72.

72. Ferrer CM, Lu TY, Bacigalupa ZA, Katsetos CD, Sinclair DA, Reginato MJ. O-GlcNAcylation regulates breast cancer metastasis via SIRT1 modulation of FOXM1 pathway. *Oncogene* 2017;**36**(4):559−69.

73. Gomes AR, Yong JS, Kiew KC, Aydin E, Khongkow M, Laohasinnarong S, et al. Sirtuin1 (SIRT1) in the acetylation of downstream target proteins. *Methods Mol Biol (Clifton, NJ)*. 2016;**1436**:169−88.

74. Igci M, Kalender ME, Borazan E, Bozgeyik I, Bayraktar R, Bozgeyik E, et al. High-throughput screening of Sirtuin family of genes in breast cancer. *Gene*. 2016;**586**(1):123−8.

75. Liu L, Liu C, Zhang Q, Shen J, Zhang H, Shan J, et al. SIRT1-mediated transcriptional regulation of SOX2 is important for self-renewal of liver cancer stem cells. *Hepatology (Baltimore, Md)* 2016;**64**(3):814−27.

76. Poulsen MM, Jorgensen JO, Jessen N, Richelsen B, Pedersen SB. Resveratrol in metabolic health: an overview of the current evidence and perspectives. *Ann N Y Acad Sci* 2013;**1290**:74−82.

77. Donadini A, Rosano C, Felli L, Ponassi M. Human sirtuins: an overview of an emerging drug target in age-related diseases and cancer. *Curr Drug Targets* 2013;**14**(6):653−61.

78. Chung SY, Jung YY, Park IA, Kim H, Chung YR, Kim JY, et al. Oncogenic role of SIRT1 associated with tumor invasion, lymph node metastasis, and poor disease-free survival in triple negative breast cancer. *Clin Exp Meta* 2016;**33**(2):179−85.

79. Dali-Youcef N, Froelich S, Moussallieh FM, Chibbaro S, Noel G, Namer IJ, et al. Gene expression mapping of histone deacetylases and co-factors, and correlation with survival time and 1H-HRMAS metabolomic profile in human gliomas. *Sci Rep* 2015;**5**:9087.

80. Jiang K, Lyu L, Shen Z, Zhang J, Zhang H, Dong J, et al. Overexpression of SIRT1 is a poor prognostic factor for advanced colorectal cancer. *Chinese Med J* 2014;**127**(11):2021−4.

81. Buler M, Andersson U, Hakkola J. Who watches the watchmen? Regulation of the expression and activity of sirtuins. *Faseb J* 2016;**30**(12):3942−60.

82. Cheng HL, Mostoslavsky R, Saito S, Manis JP, Gu Y, Patel P, et al. Developmental defects and p53 hyperacetylation in Sir2 homolog (SIRT1)-deficient mice. *Proc Natl Acad Sci U S A* 2003;**100**(19):10794−9.

83. Liu X, Wang X, Zhang J, Lam EK, Shin VY, Cheng AS, et al. Warburg effect revisited: an epigenetic link between glycolysis and gastric carcinogenesis. *Oncogene* 2010;**29**(3):442−50.

84. Wang RH, Zhao T, Cui K, Hu G, Chen Q, Chen W, et al. Negative reciprocal regulation between Sirt1 and Per2 modulates the circadian clock and aging. *Sci Rep* 2016;**6**:28633.

85. Oberdoerffer P, Michan S, McVay M, Mostoslavsky R, Vann J, Park SK, et al. SIRT1 redistribution on chromatin promotes genomic stability but alters gene expression during aging. *Cell* 2008;**135**(5):907−18.

86. Masri S. Sirtuin-dependent clock control: new advances in metabolism, aging and cancer. *Curr Opin Clin Nutr Metab Care* 2015;**18**(6):521−7.

87. Masri S, Rigor P, Cervantes M, Ceglia N, Sebastian C, Xiao C, et al. Partitioning circadian transcription by SIRT6 leads to segregated control of cellular metabolism. *Cell* 2014;**158**(3):659−72.

88. Daitoku H, Hatta M, Matsuzaki H, Aratani S, Ohshima T, Miyagishi M, et al. Silent information regulator 2 potentiates Foxo1-mediated transcription through its deacetylase activity. *Proc Natl Acad Sci U S A* 2004;**101**(27):10042−7.

89. Wang RH, Kim HS, Xiao C, Xu X, Gavrilova O, Deng CX. Hepatic Sirt1 deficiency in mice impairs mTorc2/Akt signaling and results in hyperglycemia, oxidative damage, and insulin resistance. *J Clin Invest* 2011;**121**(11):4477−90.

90. Afshar G, Murnane JP. Characterization of a human gene with sequence homology to Saccharomyces cerevisiae SIR2. *Gene* 1999;**234**(1):161−8.

91. Vaquero A, Scher MB, Lee DH, Sutton A, Cheng HL, Alt FW, et al. SirT2 is a histone deacetylase with preference for histone H4 Lys 16 during mitosis. *Genes Dev* 2006;**20**(10):1256−61.

92. Dryden SC, Nahhas FA, Nowak JE, Goustin AS, Tainsky MA. Role for human SIRT2 NAD-dependent deacetylase activity in control of mitotic exit in the cell cycle. *Mol Cell Biol* 2003;**23**(9):3173−85.

93. Inoue T, Hiratsuka M, Osaki M, Yamada H, Kishimoto I, Yamaguchi S, et al. SIRT2, a tubulin deacetylase, acts to block the entry to chromosome condensation in response to mitotic stress. *Oncogene* 2007;**26**(7):945−57.

94. Inoue T, Nakayama Y, Yamada H, Li YC, Yamaguchi S, Osaki M, et al. SIRT2 downregulation confers resistance to microtubule inhibitors by prolonging chronic mitotic arrest. *Cell Cycle* 2009;**8**(8):1279−91.

95. North BJ, Verdin E. Interphase nucleo-cytoplasmic shuttling and localization of SIRT2 during mitosis. *PLoS One* 2007;**2**(8):e784.

96. Kim HS, Vassilopoulos A, Wang RH, Lahusen T, Xiao Z, Xu X, et al. SIRT2 maintains genome integrity and suppresses tumorigenesis through regulating APC/C activity. *Cancer Cell* 2011;**20**(4):487−99.

97. Serrano L, Martinez-Redondo P, Marazuela-Duque A, Vazquez BN, Dooley SJ, Voigt P, et al. The tumor suppressor SirT2 regulates cell cycle progression and genome stability by modulating the mitotic deposition of H4K20 methylation. *Genes Dev* 2013;**27**(6):639−53.

98. Song HY, Biancucci M, Kang HJ, O'Callaghan C, Park SH, Principe DR, et al. SIRT2 deletion enhances KRAS-induced tumorigenesis in vivo by regulating K147 acetylation status. *Oncotarget* 2016;**7**(49):80336−49.

99. Zhang H, Park SH, Pantazides BG, Karpiuk O, Warren MD, Hardy CW, et al. SIRT2 directs the replication stress response through CDK9 deacetylation. *Proc Natl Acad Sci U S A* 2013;**110**(33):13546−51.

100. Zhang H, Head PE, Daddacha W, Park SH, Li X, Pan Y, et al. ATRIP deacetylation by SIRT2 drives ATR checkpoint activation by promoting binding to RPA-ssDNA. *Cell Rep* 2016;**14**(6):1435−47.

101. Hiratsuka M, Inoue T, Toda T, Kimura N, Shirayoshi Y, Kamitani H, et al. Proteomics-based identification of differentially expressed genes in human gliomas: down-regulation of SIRT2 gene. *Biochem Biophys Res Commun* 2003;**309**(3):558−66.

102. Peters CJ, Rees JR, Hardwick RH, Hardwick JS, Vowler SL, Ong CA, et al. A 4-gene signature predicts survival of patients with resected adenocarcinoma of the esophagus, junction, and gastric cardia. *Gastroenterology* 2010;**139**(6):1995−2004 e15.

103. Temel M, Koc MN, Ulutas S, Gogebakan B. The expression levels of the sirtuins in patients with BCC. *Tumour Biol* 2016;**37**(5):6429−35.

104. Head PE, Zhang H, Bastien AJ, Koyen AE, Withers AE, Daddacha WB, et al. Sirtuin 2 mutations in human cancers impair its function in genome maintenance. *J Biol Chem* 2017;**292**(24):9919−31.

105. Peck B, Chen CY, Ho KK, Di Fruscia P, Myatt SS, Coombes RC, et al. SIRT inhibitors induce cell death and p53 acetylation through targeting both SIRT1 and SIRT2. *Mol Cancer Ther* 2010;**9**(4):844−55.

106. Hoffmann G, Breitenbucher F, Schuler M, Ehrenhofer-Murray AE. A novel sirtuin 2 (SIRT2) inhibitor with p53-dependent pro-apoptotic activity in non-small cell lung cancer. *J Biol Chem* 2014;**289**(8):5208−16.

107. Liu PY, Xu N, Malyukova A, Scarlett CJ, Sun YT, Zhang XD, et al. The histone deacetylase SIRT2 stabilizes Myc oncoproteins. *Cell Death Diff* 2013;**20**(3):503−14.

108. Jing H, Hu J, He B, Negron Abril YL, Stupinski J, Weiser K, et al. A SIRT2-selective inhibitor promotes c-Myc oncoprotein degradation and exhibits broad anticancer activity. *Cancer Cell* 2016;**29**(3):297−310.

109. Chen J, Chan AW, To KF, Chen W, Zhang Z, Ren J, et al. SIRT2 overexpression in hepatocellular carcinoma mediates epithelial to mesenchymal transition by protein kinase B/glycogen synthase kinase-3beta/beta-catenin signaling. *Hepatology (Baltimore, Md)* 2013;**57**(6):2287−98.

110. Zhou W, Ni TK, Wronski A, Glass B, Skibinski A, Beck A, et al. The SIRT2 deacetylase stabilizes slug to control malignancy of basal-like breast cancer. *Cell Rep* 2016;**17**(5):1302−17.

111. Pastor BM, Mostoslavsky R. SIRT6: a new guardian of mitosis. *Nat Struct Mol Biol* 2016;**23**(5):360−2.

112. Sebastian C, Zwaans BM, Silberman DM, Gymrek M, Goren A, Zhong L, et al. The histone deacetylase SIRT6 is a tumor suppressor that controls cancer metabolism. *Cell* 2012;**151**(6):1185−99.

113. Kim HS, Xiao C, Wang RH, Lahusen T, Xu X, Vassilopoulos A, et al. Hepatic-specific disruption of SIRT6 in mice results in fatty liver formation due to enhanced glycolysis and triglyceride synthesis. *Cell Metabol* 2010;**12**(3):224−36.

114. Kugel S, Mostoslavsky R. Chromatin and beyond: the multitasking roles for SIRT6. *Trends Biochem Sci* 2014;**39**(2):72−81.

115. Kugel S, Feldman JL, Klein MA, Silberman DM, Sebastian C, Mermel C, et al. Identification of and molecular basis for SIRT6 loss-of-function point mutations in cancer. *Cell Rep* 2015;**13**(3):479−88.

116. Kugel S, Sebastian C, Fitamant J, Ross KN, Saha SK, Jain E, et al. SIRT6 suppresses pancreatic cancer through control of Lin28b. *Cell* 2016;**165**(6):1401−15.

117. Bhardwaj A, Das S. SIRT6 deacetylates PKM2 to suppress its nuclear localization and oncogenic functions. *Proc Natl Acad Sci U S A* 2016;**113**(5):E538−47.

118. Min L, Ji Y, Bakiri L, Qiu Z, Cen J, Chen X, et al. Liver cancer initiation is controlled by AP-1 through SIRT6-dependent inhibition of survivin. *Nat Cell Biol* 2012;**14**(11):1203−11.

119. Lerrer B, Gertler AA, Cohen HY. The complex role of SIRT6 in carcinogenesis. *Carcinogenesis* 2016;**37**(2):108−18.

ROLES FOR SIRTUINS IN CARDIOVASCULAR BIOLOGY

Adam B. Stein, William Giblin, Angela H. Guo and David B. Lombard

University of Michigan, Ann Arbor, MI, United States

12.1 INTRODUCTION

There is a tremendous amount of scientific interest in harnessing the salubrious effects of sirtuin deacylases to promote healthy aging. In the industrialized world, cardiovascular (CV) disease exerts major negative impacts on healthy aging and overall longevity. In the United States, CV disease is responsible for about 1 out of every 3 deaths, and claims a total of about 801,000 lives per year.[1] Heart failure (HF) is the common clinical endpoint of many manifestations of CV disease, and represents a leading cause of overall morbidity and mortality. Roughly 5.7 million Americans currently suffer from HF, and there are an estimated 915,000 new HF diagnoses every year in the United States.[2] One in five individuals who reach the age of 40 will ultimately develop HF, and the incidence of HF doubles with every decade after the age of 65.[1] Specifically, there is a 14%−22% incidence in those aged 75−84 and a 30%−43% incidence in those aged 85−94.[3] As the population ages, this HF epidemic will become even more costly, both in terms of dollars and lives. Unfortunately, despite current treatments, HF remains a chronic progressive disease that contributes to over 300,000 deaths/year in the United States, and it is estimated that the overall cost of treating HF will rise to $69 billion/year by 2030, highlighting the urgent need for new mechanism-based HF therapies.[1]

What is HF? HF refers to the inability of the heart to meet the oxygen demands of the body. Ineffective cardiac pump function results in the development of a constellation of symptoms including shortness of breath with exertion, edema in the lower extremities, and poor energy levels. HF can be broadly divided into two categories, termed HF with reduced ejection fraction (HFrEF), and HF with preserved ejection fraction (HFpEF). HFrEF refers to a condition in which the heart's ability to contract or squeeze is diminished. Clinically, the heart's pumping ability is measured by the ejection fraction (EF), the percentage of the blood volume the left ventricle (LV) pumps with each beat. The pumping ability of the heart can be impaired by stressors and injuries such as myocardial infarction (MI), hypertension, valvular disease, genetic causes, viral infection, or toxins such as alcohol. The natural history of HFrEF results in a process of adverse remodeling, defined by LV chamber dilation, hypertrophy of the cardiac myocytes, and scar/fibrosis formation.[4] Medications currently used that improve morbidity and mortality in HFrEF attenuate this adverse remodeling. HFpEF refers to a condition in which the ventricles contract adequately, but the LV does not relax normally. As a result of abnormal relaxation, the heart requires elevated pressures to fill the ventricles with blood. These elevated filling pressures result in fluid retention and the symptoms of HF. HFpEF is a much more recently recognized clinical entity than HFrEF. HFpEF is particularly prevalent in the aging population, and the underlying causative mechanisms are only beginning to be investigated.[5] In contrast to HFrEF, no known therapies improve morbidity and mortality associated with HFpEF.

The diverse molecular mechanisms that lead to the development of HFpEF and HFrEF include alterations in signaling pathways, gene expression profiles, calcium homeostasis, energy metabolism, oxidative stress, mitochondrial dysfunction, and cell death. These pathways are accompanied, and driven, by the neurohormonal activation of catecholamine signaling and the renin—angiotensin—aldosterone pathway. Current HFrEF therapies include beta-blockers, angiotensin-converting enzyme inhibitors, and aldosterone antagonists.[6] Despite these therapies, HF remains a chronic progressive disorder with high associated morbidity and mortality,[7] and it is important to elucidate the underlying pathologic mechanisms so that the disease can be treated more effectively.

Sirtuins are a conserved group of NAD^+-dependent deacylase/deacetylase enzymes classified as class III histone deacetylases. However, sirtuins also modify a plethora of other protein substrates in addition to histones.[8−10] In mammals, there are seven distinct sirtuin genes, termed *SIRT1−7*. They share a conserved NAD^+-binding catalytic domain, but differ in their amino and carboxy regions. Mammalian sirtuins vary in their subcellular localization, expression patterns, enzymatic activity, and biologic functions. For example, SIRT1, 6, and 7 are found mainly in the nucleus; SIRT7 is largely nucleolar; SIRT3, 4, and 5 are predominantly found in the mitochondrial matrix; and SIRT2 is largely cytoplasmic.[11] Importantly, these patterns are far from absolute, and sirtuin proteins can show localization to multiple subcellular compartments, in a cell- and context-specific manner. As a result of their different localization patterns and numerous targets, collectively, members of the sirtuin family regulate many diverse facets of cellular biology and function. Research interest in sirtuins originally stemmed largely from their ability to extend replicative lifespan in yeast, and, more recently, for specific mammalian sirtuins, their ability to promote lifespan and healthy aging in mammals.[12−15] Given that sirtuins modulate stress responses by modifying histones and many other proteins, it is not surprising that sirtuins have been found to play prominent roles in the regulation of cardiac phenotypes. In this chapter, we discuss the intersection of sirtuins with cardiac biology, focusing mainly on recent studies linking sirtuins to cardiac function and health in mouse models. Sirtuins may prove to be promising targets for interventions to suppress the deleterious effects of adverse remodeling and HF, an area currently lacking effective targeted therapies.

12.2 **SIRT1**

Sirt1 the closest homolog to yeast *SIR2*, regulates many major cellular pathways by deacetylating a large number of proteins, including a core group of transcription factors detailed below. Pathways of importance in cardiac development and disease that are impacted by SIRT1 include cell cycle regulation, apoptosis, autophagy, reactive oxygen species (ROS) management, and energy homeostasis. Insight into the role that SIRT1 plays in cardiac biology comes mostly from mouse studies. The expression patterns of SIRT1 and its subcellular localization vary significantly with age and stress status during cardiac development. At embryonic day 12.5, SIRT1 localizes to the nucleus in cardiac myocytes. However, in adult murine tissue, SIRT1 is distributed in both the nucleus and the cytoplasm.[16] PI3K signaling promotes SIRT1 nuclear localization.[16] Changes in both SIRT1 expression and subcellular localization suggest that SIRT1 biological functions differ between developing and adult myocardium, since nuclear SIRT1 is more likely to regulate acetylation and activity of transcription factors.[16] Whether SIRT1 might show different localization patterns in hearts from young versus old animals has not been studied in depth.

In the adult murine heart, oxidative stress, pressure overload, exercise, ischemic preconditioning, and aging have all been reported to upregulate SIRT1 expression.[17,18] SIRT1 levels decline in response to ischemia reperfusion injury and ischemic stress in aged murine hearts results in SIRT1 shuttling from the nucleus to the cytoplasm.[19] In 24−26-month-old murine hearts, SIRT1 levels in both the cytoplasmic and nuclear fraction decrease compared to young adult mice.[20,21] Additionally, intracellular levels of NAD^+, the cofactor of sirtuin catalytic activity, decrease with age in myocardium.[22] Interestingly, cardiac SIRT1 levels actually increase with age in monkeys; NAD^+ levels and SIRT1 activity were not measured in this study.[17]

SIRT1 plays important roles in cardiac development. *Sirt1* null mice are born at a sub-Mendelian ratio, and rarely survive to adulthood.[23,24] Studies of *Sirt1* null embryos revealed significant cardiac structural defects, including atrial and ventricular septal defects,[23,24] a phenotype potentially related to p53 hyperacetylation and increased apoptotic sensitivity occurring in the setting of SIRT1 deficiency.[23]

In order to assess effects of SIRT1 on cardiac disease states, Alcendor et al. created a cardiac-specific SIRT1 overexpressing (OE) mouse strain.[17] Using three different transgenic lines, they found that mice with low and modest SIRT1 overexpression (2.5- and 7.5-fold, respectively) showed no baseline phenotype at 6 months of age. However, a murine line with a high (12.5-fold) level of SIRT1 expression demonstrated pathologic remodeling, including LV hypertrophy, LV chamber dilation, a decrease in LV function, and increased fibrosis when compared to littermate controls at 6 months of age. When SIRT1 overexpressors were allowed to age to 18 months, mice with low or moderate SIRT1 overexpression displayed an attenuation of age-associated cardiac decline, defined by smaller LV chamber size with improved LV function, when compared to aged littermates. Moderate overexpression of SIRT1 was associated with protection from paraquat-induced oxidative stress, lower cardiac apoptosis levels, elevated expression of the antiapoptotic protein BCL2, and enhanced citrate synthetase activity. Levels of biological aging markers, including $p15^{INK4b}$ and $p19^{ARF}$, were attenuated in low and moderate SIRT1 overexpressors. High levels of SIRT1 overexpression resulted in enhanced apoptosis, decreased ATP levels, increased oxidative stress, and reduced mitochondrial biosynthesis.[17]

Kawashima et al. also generated SIRT1 OE mice.[25] They also found adverse cardiac remodeling in their high overexpressing group consistent with the development of HFrEF. However, they found that moderate overexpression of SIRT1 resulted in no overt baseline phenotype, but treatment with dobutamine induced diastolic dysfunction, the pathophysiologic underpinning of HFpEF. Moreover, mice with moderate cardiac SIRT1 overexpression showed decreased fatty acid uptake and impaired mitochondrial respiration. Pressure overload induced by transverse aortic constriction (TAC) is a commonly used technique that induces the development of cardiac hypertrophy and adverse cardiac remodeling.[26] Mice with low levels of SIRT1 overexpression exhibited no baseline phenotype, but responded abnormally when subjected to TAC pressure overload.

Sundaresan et al. also studied SIRT1 OE mice.[27] They found that protein kinase B (AKT) signaling is regulated by SIRT1-mediated deacetylation. Specifically, SIRT1 deacetylates AKT, rendering it active. Mice with fourfold overexpression of SIRT1 showed enhanced AKT activity. These mice developed significant cardiac hypertrophy under basal conditions. These results are consistent with prior studies that show that persistent activation of AKT results in development of cardiac hypertrophy and failure.[28,29] However, these results contradict findings by Kawashima and Alcendor that show no significant hypertrophy or protection from hypertrophy in mice with a similar degree of SIRT1 overexpression. Overall, these three studies exploring the overexpression of SIRT1 in the heart arrived at discrepant conclusions regarding protective versus deleterious impacts of cardiac SIRT1 overexpression.[17,27,30]

SIRT1 deficiency has also been studied in the context of CV disease. Planavila et al. analyzed heterozygous *Sirt1* mice and *Sirt1* null mice.[30] Heterozygous *Sirt1* mice survive into adulthood. However, after 5 months, heterozygous mice develop a dilated cardiomyopathy (HFrEF), characterized by LV chamber dilation and depressed LV function. Although *Sirt1* null mice survive at sub-Mendelian ratios, Planavila and colleagues found that the few *Sirt1* null mice that survive into adulthood also developed a dilated cardiomyopathy phenotype. Electron microscopy and molecular analysis in these SIRT1-deficient models revealed defective mitochondria and abnormal mitochondrial function. They found that SIRT1-deficient mice had altered acetylation of the MEF2 family of transcription factors that regulate cardiac gene expression and mitochondrial biogenesis.[31] In this regard, a small-molecule inhibitor of MEF2 acetylation was recently shown to provide protection in three distinct mouse models of cardiac hypertrophy.[32]

Cardiomyocyte metabolism is critical to the functioning of the normal heart and to cardiac stress responses.[33] Peroxisome proliferator-activator receptor alpha (PPARα) is a member of a nuclear receptor superfamily that is activated by fatty acids, and induces the expression of genes important for fatty acid uptake and oxidation.[34] PPARα can both positively and negatively influence cardiac function, through mechanisms that remain incompletely understood. In a study examining the PPARα- and SIRT1-mediated regulation of mitochondrial function in the heart, Oka et al. utilized *Sirt1* heterozygous mice.[35] In contrast to the work by Planavila et al., Oka et al. found that *Sirt1* heterozygous mice are resistant to pressure overload-induced cardiac hypertrophy and failure. They found that SIRT1 interacts with PPARα to mediate the transcriptional suppression of a set of PPARα genes important for mitochondrial and cardiac function through SIRT1-mediated histone deacetylation.

Sirt1 null mice were also characterized by Sundaresan et al. in the context of elucidating the relationship between AKT and SIRT1.[27] *Sirt1* null mice showed increased acetylation and reduced activity of AKT. *Sirt1* null mice that lived to adulthood were resistant to angiotensin II- and exercise-induced cardiac hypertrophy. These studies arrived at similar results to Oka et al.; however, Planavila's work suggests *Sirt1* null mice are prone to develop HF.[17,27,30]

MI, or heart attack, usually results from an unstable coronary plaque rupture event that prevents blood flow to downstream cardiac muscle.[36,37] Modern-day coronary interventions aim to restore blood flow in a timely manner and, as a result, ischemia/reperfusion (I/R) injury is a clinically relevant phenomenon.[38,39] In order to explore the role of SIRT1 in I/R injury, Hsu et al. employed cardiac-specific OE and cardiac-specific *Sirt1* KO mice.[19] I/R injury suppressed SIRT1 expression in the region of ischemia. *Sirt1* KO mice demonstrated an exacerbated response to I/R injury, whereas SIRT1 OE mice were protected. SIRT1-mediated cardioprotection occurred through upregulation of cardioprotective molecules: manganese SOD (SOD2), thioredoxin, and the antiapoptotic protein Bcl-XL, and reduced caspase 3 cleavage and expression of the proapoptotic protein Bax. SOD2 levels were regulated by SIRT1-mediated regulation of FOXO1 expression, deacetylation, and activity. How Bcl-XL and Bax are regulated in this context is not completely clear, and may occur partially through SIRT1-mediated deacetylation of p53. Tong et al. also explored functions of SIRT1 in the context of ischemic stress and aging.[20] They found that young mice have smaller infarct sizes compared to aged mice, a clinically relevant observation since MI predominantly afflicts middle-aged and older individuals. This was attributed to decreased nuclear SIRT1 levels in aged hearts, a finding that was supported by demonstrating similarly large infarcts in young *Sirt1* heterozygous mice and old wildtype (WT) mice.

Cardiac homeostasis is also tightly regulated by ion channels that define the cardiac action potential and regulate intracellular calcium handling and muscle contraction.[40] Abnormalities in the cardiac ion channel flux can lead to irregular heart rhythms, which in turn can lead to decompensated HF, debilitating symptoms, and/or sudden cardiac death. In a recent study, Vikram et al. explored the role of SIRT1 in regulating the cardiac sodium current.[41] Based on prior work demonstrating that activity of the sodium channel can be influenced by NAD^+,[42] Vikram et al. hypothesized that SIRT1 interacts with SCN5A, a gene that largely mediates the sodium current. SIRT1-deficient mice showed an attenuated sodium current as a result of impaired SIRT1-mediated lysine deacetylation of SCN5A. SCN5A hyperacetylation inhibited the channel from trafficking to the cell membrane, relocalization that is necessary for SCN5A function.

As this discussion has made clear, the roles of SIRT1 in cardiac biology and pathology are complex and somewhat controversial at the moment, and SIRT1 cannot be said to be definitively cardioprotective or deleterious. However, given the interest in roles for sirtuins as prohealth and longevity factors, several studies have explored the potential for pharmacologic sirtuin activation in cardioprotection. Resveratrol, a polyphenol found in red wine, is a highly studied small molecule that was identified as an SIRT1 activator.[43] Importantly, resveratrol affects many cellular targets potentially relevant for cardiac health, notably including activating AMPK via elevated cellular cAMP levels.[44] In mice, resveratrol exerts major effects on energy homeostasis.[45] It extends the lifespan of mice on a high-fat diet, increasing insulin sensitivity, reducing IGF signaling, enhancing AMPK and PGC1α signaling, and elevating mitochondrial content.[46] Several studies have examined relationships between resveratrol, SIRT1 activation, and cardioprotection.[47,48]

Yoshida et al. employed a dilated cardiomyopathy/HFrEF rat model to assess cardioprotection invoked by resveratrol.[49] They showed that resveratrol inhibited the HF phenotype and the inflammatory response. Tanno et al. studied the impact of resveratrol on cardiac failure using a hamster model of dilated cardiomyopathy.[50] They found that resveratrol blunted adverse cardiac remodeling in this model. This was attributed to upregulation of SOD2 and suppression of cell death. In cultured cells, they showed that the resveratrol-mediated enhancement of SOD2 levels and the decrease in oxidative stress were SIRT1-dependent. Using a rat model of diabetic cardiomyopathy, Sulaiman et al. demonstrated that resveratrol antagonized the reduction in levels of SERCA2a (ATP2A2), a calcium ATPase that regulates sarcoplasmic calcium levels and cardiac contractility, observed in streptozocin-mediated diabetic cardiomyopathy.[51] They found that the link between resveratrol and SERCA2a was SIRT1-dependent. Gu et al. studied resveratrol using a rat model of ischemic cardiomyopathy.[52] They found that resveratrol improved adverse cardiac remodeling and cardiac function after coronary artery ligation. They attributed this effect to increased AMPK activity, via a resveratrol—SIRT1-mediated pathway.

Anthracycline-induced cardiomyopathy represents another clinically relevant model of cardiomyopathy. Doxorubicin (DOX) is an anthracycline that is often employed as a component of effective chemotherapeutic regimens. Unfortunately, DOX can exert—in a cumulative, dose-dependent manner—impairment of cardiac function resulting in adverse cardiac remodeling, manifesting as decreased cardiac function and LV chamber dilation.[53,54] Although mechanisms of DOX-induced cardiac dysfunction are incompletely understood, one common theme is that of ROS-mediated mitochondrial injury, with altered energy/ATP metabolism leading to cell death and cardiac dysfunction.[55–57] Several studies have examined the role of SIRT1 in DOX-induced cardiomyopathy, with many using resveratrol as an SIRT1 activator. Danz et al. demonstrated that resveratrol protected neonatal rat ventricular myocytes in culture from DOX-induced oxidative stress, mitochondrial alterations, and cell death.[58] The protective effect of resveratrol was blocked when cells were treated with the sirtuin inhibitor nicotinamide. Zhang et al. explored the role of SIRT1 in mediating DOX-induced HF in a murine model.[59] They found that resveratrol inhibited the proapoptotic effects of DOX. DOX induced p53 acetylation and cytochrome c release from mitochondria, and increased Bax expression. These deleterious effects were attenuated when mice were concomitantly treated with resveratrol. Ruan et al. explored the role of SIRT1 in DOX-induced cardiotoxicity, in both cell culture and murine models.[60] Using cells in culture they found that SIRT1 overexpression and resveratrol were both able to attenuate DOX-induced oxidative stress and apoptosis by inhibiting the p38/MAPK pathway and activation of caspase-3. Sin et al. treated mice with DOX and resveratrol.[61] Consistent with the other studies, they found resveratrol to be cardioprotective. DOX inhibited the deacetylase activity of SIRT1 and altered the proteasomal activity of USP7, a p53 deubiquitinating protein. The protective effects of resveratrol on DOX-induced cardiotoxicity were attenuated when mice were treated with the selective SIRT1 inhibitors, sirtinol and EX527. Cappetta et al. treated rats with doxorubicin and resveratrol and found that resveratrol attenuated DOX-induced cardiac fibrosis.[62] Fibroblasts from resveratrol-treated rats showed upregulation of SIRT1 and decreased expression of TGF-β and SMAD signaling. Sin et al. showed that resveratrol attenuated DOX-induced cardiotoxicity in a murine model by blunting FOXO1-mediated proapoptotic signaling. In summary, multiple studies have found that resveratrol protects from DOX-induced cardiotoxicity, in a SIRT1-associated mechanism.

Another means of modifying SIRT1 activity is by altering levels of the essential sirtuin substrate NAD^+. NAD^+ is synthesized de novo or salvaged, the major route of NAD^+ generation in mammalian cells. The conversion of NAM to NAM mononucleotide (NMN) in the salvage pathway is performed by NAM phosphoribosyltransferase (NAMPT), the rate-limiting step in NAD^+ salvage. Yano et al. demonstrated that during pressure overload, cardiac NAD^+ levels are preserved because monocyte-derived extracellular NAMPT augments cardiac intracellular NAMPT.[63] NAMPT inhibition by FK866 administration attenuated NAD^+ levels, SIRT1 deacetylase activity, and exacerbated the pathologic response to TAC. Conversely, enhancing NAD^+ levels via NMN administration enhanced SIRT1 deacetylase activity and decreased pressure overload-induced cardiac apoptosis. NMN was also able to attenuate apoptosis and adverse cardiac remodeling in monocyte-depleted mice, suggesting that monocytes are an important regulator of cardiac NAD^+ levels.[63]

12.3 **SIRT2**

Few studies have focused on roles of SIRT2 in the heart. SIRT2 participates in cell cycle regulation, and deacetylates histone H4 and tubulin.[64,65] North et al. identified BUBR1, a cell cycle regulator whose levels decline with age, as a target for SIRT2-mediated deacetylation.[22] *BubR1* hypomorphic mice have a reduced lifespan, possibly in part due to cardiac electrophysiological abnormalities and reduced heart size. SIRT2 overexpression enhanced BUBR1 expression and rescued the electrical abnormalities and diminished cardiac size observed in *BubR1* hypomorphic mice. The effects of SIRT2 OE on cardiac function in normal mice have not been assessed in depth. Yuan et al. examined SIRT2 expression in a rat model of diabetic cardiomyopathy.[66] They identified an association between SIRT2 levels and deacetylation of α-tubulin, a modification that stabilizes microtubules. The authors found an association between the two proteins and showed that as SIRT2 levels declined, acetylation of α-tubulin increased.

Recently, Tang et al. utilized *Sirt2* null and cardiac-specific SIRT2 OE mice to study the role of SIRT2 in the development of angiotensin II-induced and age-associated cardiac hypertrophy. SIRT2 levels decreased with the development of hypertrophy. Lack of SIRT2 resulted in more severe cardiac hypertrophy and adverse cardiac remodeling. In contrast, SIRT2 OE mice were protected from angiotensin II-induced cardiac hypertrophy. Mechanistically, SIRT2 interacted with and deacetylated liver kinase B1, a key upstream kinase for AMP-activated protein kinase.[32]

12.4 **SIRT3**

Targeting mitochondrial metabolism may provide new therapeutic opportunities to treat or prevent HF. The heart consumes the most energy of any organ, utilizing roughly 20−30 times its own weight in ATP per day, and is therefore exquisitely sensitive to bioenergetic perturbations.[67] Mitochondria generate the great majority of ATP in cells via oxidative phosphorylation, and perform many other biochemical reactions necessary for organismal function, including fatty acid oxidation (FAO). In young, healthy individuals, FAO, rather than glucose metabolism, represents the primary energy source for the heart. During natural aging and in early HF, major metabolic shifts occur in the heart, leading to decreased FAO and increased glucose use.[68,69] Later in the course of HF, both glucose metabolism and FAO are impaired.[69] Studies in mouse models suggest that these metabolic changes are reversible if the underlying cause of failure is removed.[70] Cardiomyocyte metabolism, and how it changes during aging or pathologic states, is a complex and somewhat controversial area that has been reviewed in depth elsewhere.[33,71]

SIRT3 is a sirtuin found predominantly in the mitochondrial matrix that possesses robust deacetylase activity. Consequently, mice lacking SIRT3 have increased mitochondrial protein acetylation in the myocardium as well as in other tissues.[72,73] In contrast, the other two predominantly mitochondrial sirtuins, SIRT4 and SIRT5, do not function as protein deacetylases at the bulk level. SIRT3-deficient mice showed no significant metabolic abnormalities under basal conditions or short-term fasting. However, when fed a high-fat diet, *Sirt3* KO mice developed worsened obesity, insulin resistance, hyperlipidemia, and steatohepatitis.[74] In humans, polymorphisms that lead to increased SIRT3 levels have been associated with extended lifespan, and another polymorphism that decreased SIRT3 levels was associated with metabolic syndrome.[75,76] Molecularly, SIRT3 has been reported to deacetylate, and in general, activate, numerous proteins involved in ROS management (e.g., SOD2), FAO, energy production, and many other processes.[77]

Mitochondria and ATP levels play critical roles in cardiac homeostasis and stress responses. Ahn et al. showed that in *Sirt3* null mice, cardiac ATP levels are reduced by over 50%.[73] Horton et al. explored a potential role for mitochondrial protein acetylation in both human and murine models of HF.[78] They found that in both mice and humans, HF resulted in a significant increase in mitochondrial protein acetylation. Sundaresan et al. employed *Sirt3* null and SIRT3 OE mice to explore the role of SIRT3 in the development of cardiac hypertrophy and HF.[79] First, they studied SIRT3 levels in response to different hypertrophic stimuli. They found that mild hypertrophy induced by phenylephrine, angiotensin II, or swimming resulted in increased SIRT3 expression. In response to more severe hypertrophic stimuli, TAC, or isoproterenol infusion, they found a downregulation of mitochondrial SIRT3. At baseline, they observed that *Sirt3* KO mice have cardiac hypertrophy and increased fibrosis compared to wildtype controls. When *Sirt3* KO mice were stressed with angiotensin II, isoproterenol, or phenylephrine, they developed an exaggerated molecular and phenotypic signature of cardiac hypertrophy. Others have also reported cardiac hypertrophy, fibrosis, and a decline in cardiac function in *Sirt3* KOs.[80] To study the response to hypertrophic stimuli in SIRT3-overexpressing mice, Sundaresan et al. generated cardiac-specific SIRT3 OE and subjected them to angiotensin II infusion. SIRT3 OE mice exhibited a blunted hypertrophic phenotype, molecular phenotype, and fibrosis when compared to littermate controls. The protective effect was dependent upon the SIRT3-induced activation of FOXO3A, thereby enhancing MnSOD and catalase activity and decreasing cellular ROS levels.

With aging, the incidence of HF increases. In mice, aging results in increased heart size, an increase in cardiac fibrosis, and stress susceptibility. Several studies have explored the role of SIRT3 in the aging heart.[81,82] Sundaresan et al. showed that *Sirt3* KO mice develop hypertension and fibrosis of multiple organs as the mice age to 15 months. This profibrotic response is blunted in SIRT3-overexpressing mice. They concluded that *Sirt3* deletion induced TGF-B1 expression and regulated Smad3 and β-catenin signaling via GSK-3β deacetylation. Hafner et al. also found that *Sirt3* KO mice exhibited age-dependent cardiac phenotypes, including the development of hypertrophy and fibrosis at 13 months by age. They focused on SIRT3-mediated regulation of the mitochondrial permeability transition pore (mPTP). They found that SIRT3 deacetylates cyclophilin D, a key regulator of the mPTP. *Sirt3* KOs showed greatly increased mortality in response to TAC.

Mitochondrial function and ROS are critical aspects of the cardiac response to IR injury.[83] Utilizing *Sirt3* heterozygous mice, Porter et al. employed a Langendorff preparation to explore the impact of SIRT3 on IR injury at 7 months of age, and compared the response to that of older wildtype mice.[84] They found that *Sirt3* heterozygous mice were more susceptible to IR injury. This was associated with impaired mitochondrial complex I function and reduced oxygen consumption when compared to age-matched controls. The impairment observed in 7-month-old *Sirt3* heterozygous mice was similar to that observed in 18-month-old wildtype mice, suggesting that in wildtype mice, age-associated declines in SIRT3 function may play a role in age-related impaired responses to IR injury. A similar hypothesis was evaluated using younger (10-week-old) *Sirt3* null mice, and a working heart model of IR injury by Koentges et al.[80] However, they found no increased susceptibility to ischemia reperfusion injury in *Sirt3* null mice when compared to WT controls.

Given that DOX-induced cardiotoxicity is a result of mitochondrial dysfunction and overproduction of ROS, SIRT3 has been studied within the context of DOX-induced cardiotoxicity. Pillai et al. studied the role of SIRT3 in regulating DOX-induced cardiotoxicity by examining the effects of DOX treatment in SIRT3 overexpressing and *Sirt3* null mice.[85] They found that DOX treatment attenuated SIRT3 levels and resulted in increased acetylation of mitochondrial proteins. *Sirt3* null mice could not tolerate a full dose of DOX, showing an increased incidence of adverse cardiac remodeling and death. Conversely, SIRT3 OE were protected from DOX-induced cardiotoxicity, ROS production, and cell death.

Overall, available data showing cardioprotective roles for SIRT3 are more consistent than for SIRT1. Thus, SIRT3 activators may be of clinical utility to treat or prevent various cardiac pathologies. In this regard, Honokiol is a naturally occurring biphenolic compound that enhances the activity of SIRT3.[86–88] Pilliai et al. administered Honokiol to mice and observed that this agent elevates SIRT3 levels and deacetylase activity.[89] In a murine model of pressure overload, Honokiol increased SIRT3 levels and attenuated cardiac hypertrophy and failure. This protective effect was mediated at least partly by deacetylation of SOD2. The protective effects of Honokiol were abolished in *Sirt3* null mice. Pillai et al. treated mice and cells with Honokiol, and then subjected them to DOX.[90] Honokiol attenuated DOX-induced ROS production and mitochondrial damage in cultured neonatal rat cardiomyocytes. Honokiol-treated mice were also protected from DOX-induced cardiotoxicity, without inhibiting the chemotherapeutic properties of DOX.

The role of augmenting NAD^+ to harness SIRT3's cardioprotective effect has also been investigated. Pillai et al. studied the role of NAD^+ in an angiotensin II- and isoproterenol-mediated cardiac hypertrophy.[91] They found that exogenous NAD^+ attenuated the hypertrophic response. Utilizing *Sirt1* and *Sirt3* null animals, they showed that NAD^+-mediated cardioprotection occurs through activation of SIRT3 but not SIRT1. Recently, Martin et al. also studied the effects of NMN administration in a Friedreich's ataxia (FA) cardiomyopathy murine model.[92] FA is a progressive inherited mitochondrial disease with prominent neurological and cardiac manifestations. Administration of NMN improved cardiac function in a murine FA model, which shows mitochondrial protein hyperacetylation at baseline. Using *Sirt3* null mice, they showed that NMN-mediated cardioprotection is dependent upon SIRT3.

12.5 **SIRT4**

SIRT4 is a mitochondrial sirtuin that has recently been reported to remove a newly discovered class of lysine modifications derived from acyl-CoA products of leucine metabolism.[93,94] SIRT4 regulates aspects of intermediary metabolism relevant for cardiac physiology. In particular, SIRT4 suppresses FAO via deacetylation and inhibition of malonyl CoA-decarboxylase,[95] and inhibits entry of glucose-derived carbon into the TCA cycle via modification of pyruvate dehydrogenase complex.[96]

Luo et al. studied the role of SIRT4 in development of angiotensin II-mediated cardiac hypertrophy using *Sirt4* KO and SIRT4 overexpressing mice. *Sirt4* null mice were resistant to the development of hypertrophy and cardiac fibrosis, whereas SIRT4 OE mice had an exaggerated response to angiotensin II. Mechanistically, SIRT4 prevented SOD2 from interacting with SIRT3, resulting in elevated levels of ROS. Treating SIRT4 OE mice with a SOD mimetic blocked the exaggerated response to angiotensin II observed in the SIRT4 OEs.[97]

Xiao et al. also found that inhibiting SIRT4 attenuated adverse cardiac responses to pressure overload and angiotensin II.[98] They utilized miR-497 to target the 3'-UTR of SIRT4. Overexpression of miR-497 via lentivirus inhibited the hypertrophic response to TAC in vivo and in cardiac cells treated with angiotensin II in vitro. Thus, both studies by Xiao and Luo suggest that SIRT4 can function to sensitize the heart to injury.

12.6 SIRT5

Like SIRT3 and SIRT4, SIRT5 resides primarily in the mitochondrial matrix. However, unlike SIRT3, it possesses weak deacetylase activity and, instead removes succinyl, malonyl, and glutaryl (Ksucc, Kmal, and Kglu) groups from lysines on an array of protein targets, thereby regulating cellular metabolism.[77,99−102] Among these modifications, Ksucc is the best characterized, and the one most closely linked to mitochondrial function. SIRT5 suppresses glucose oxidation, while promoting FAO.[100,101] SIRT5 also suppresses ROS levels, and enhances urea cycle function.[103−106] Sirt5 KO mice are fertile and healthy, with minimal metabolic phenotypes elicited mainly in response to prolonged fasting.[72,100,106,107]

In order to determine which tissues may be regulated by SIRT5 mediated Ksucc, Sadhukan et al. investigated which tissues have the highest levels of succinyl-CoA, the precursor of lysine succinylation.[108] Using mice, they found that succinyl-CoA is the most abundant acyl-CoA in cardiac tissue. Building upon this finding, they identified ECHA, a subunit of mitochondrial trifunctional enzyme that carries out FAO, as a target of SIRT5-mediated lysine desuccinylation and regulation. They found that in the absence of Sirt5, during aging mice develop cardiac hypertrophy, decreased cardiac function, and increased cardiac fibrosis.

Boylston et al. examined differences in the mitochondrial lysine succinylome in Sirt5 null versus WT murine hearts.[109] They found that lysine succinylation differed markedly for proteins that regulate fatty acid β-oxidation, branched chain amino acid catabolism, and respiratory chain proteins. When they subjected Sirt5 null mice to IR injury using a Langendorff perfusion model, they observed that SIRT5-deficient hearts showed an increase in infarct area and decreased recovery after the IR insult. Thus, although SIRT5 appears to exert relatively mild effects on overall metabolism and physiology, it plays protective roles in the myocardium during aging or in response to IR insult.

Recently, Herschberger et al. evaluated the role of SIRT5 in the cardiac response to TAC-induced pressure overload. TAC resulted in increased mortality and severe adverse cardiac remodeling in SIRT5-deficient mice when compared to WT controls. High-resolution MS-based metabolomics and proteomic analysis revealed that Sirt5 KO mice have impaired FAO, glucose oxidation, and attenuated NAD^+/NADH levels, suggesting that SIRT5-mediated acylation is important for regulating cardiac mitochondrial energy production.[110]

12.7 SIRT6

SIRT6 is a chromatin-bound deacetylase predominantly found in the nucleus. SIRT6 deacetylates histone 3 lysine 9 (H3K9) and K56, thereby, serving as a repressor of gene expression.[111−113]

SIRT6 functions as a transcriptional co-repressor of key transcription factors including NF-κB and HIF1α.[111,113] SIRT6 also functions biochemically as an ADP-ribosyltransferase[114,115] and a defatty-acylase.[9] SIRT6 is highly expressed in the developing and the adult heart.[114]

In order to explore the role of SIRT6 in cardiac biology, Sundaresan et al. studied global *Sirt6* KO mice, as well as mice with cardiac-specific manipulation of SIRT6 levels.[116] First, they demonstrated that induction of cardiac stress by angiotensin II, isoproterenol, or TAC results in a decrease in SIRT6 levels. Human HF samples also show decreased SIRT6 levels compared to nonfailing heart samples. They found that global *Sirt6* deletion results in the development of cardiac hypertrophy, chamber dilation, and a decline in cardiac function. Heterozygous *Sirt6* mice also develop cardiac hypertrophy and fibrosis, as do animals with cardiac-specific *Sirt6* deletion. Sundaresan et al. developed cardiac-specific SIRT6 overexpressing mice. SIRT6 overexpressors showed an attenuated response when subjected to TAC or isoproterenol. ChIP assays revealed that SIRT6 localizes to the promotor regions of IGF signaling-related genes. Coimmunoprecipitation experiments revealed that SIRT6 interacts with the stress-responsive transcription factor c-Jun to deacetylate H3 and inhibit expression of these genes. In this regard, c-Jun activation is a key regulator of cardiac hypertrophy and failure.[117−119]

A study by Zhang et al. also highlighted the importance of SIRT6 in the regulation of gene expression via chromatin effects in the context of CV disease.[120] They found that human atherosclerotic plaques show decreased SIRT6 levels compared to normal vascular intima. Next, they bred *Sirt6* heterozygous mice with *ApoE* null mice, an established model for atherosclerosis,[121] and found that SIRT6 heterozygosity results in an increase in atherosclerotic plaque burden and plaque instability compared to *ApoE* null controls. Mechanistically, they found that SIRT6 regulates H3K9 and H3K56 acetylation marks at the regulatory region of NKG2D ligand, a mediator of the innate immune response.

12.8 SIRT7

SIRT7 is predominantly a nucleolar protein originally reported to interact with RNA polymerase I to drive rDNA transcription.[122] SIRT7 functions as a histone deacetylase that is also reported to have histone desuccinylase activity.[123,124] Vakhrusheva et al. found that *Sirt7* null mice develop progressive cardiac hypertrophy and fibrosis, with increased basal cardiomyocyte apoptosis.[125] The authors traced these phenotypes to a role for SIRT7 in deacetylating and regulating p53.

Araki et al. showed that SIRT7 levels increase in murine hearts subjected to MI induced via permanent left anterior descending artery ligation.[126] Using global *Sirt7* KO mice, they showed that the absence of SIRT7 results in an increase in LV myocardial rupture after MI. They found that the increase in LV rupture was a result of impaired TGFBRI (ALK5) expression and downstream signaling.

12.9 SUMMARY

Sirtuins play diverse roles in regulating cardiac homeostasis and stress responses (Fig. 12.1). SIRT1 is the most widely studied sirtuin in CV disease. Unfortunately, it is difficult at the moment to unambiguously discern whether SIRT1 has beneficial or detrimental impacts on CV

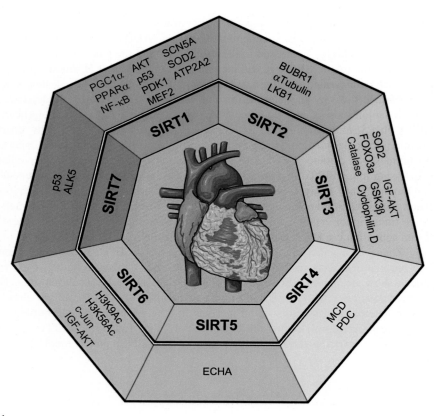

FIGURE 12.1

Sirtuins and their substrates that modulate cardiac health and cardiovascular biology. Mammalian sirtuins (SIRT1-7) play important roles in regulating cardiac biology and cardiovascular pathology by regulating various substrates or interactors as indicated. Some graphics in the figures were obtained and modified from Servier Medical Art from Servier (http://www.servier.com/Powerpoint-image-bank).

pathophysiology, given discrepant reports in this area. In general, many groups have reported that resveratrol exerts protective effects on myocardium. However these effects are unlikely to occur through SIRT1 alone. In contrast, multiple independent reports have identified protective roles for SIRT3 in the heart. Furthermore, SIRT3 activators, such as Honokiol, and NAD^+ supplementation, seem to attenuate CV disease. Recent work on SIRT2 suggests that it has a cardioprotective role. SIRT4 has not been extensively studied in the heart, but current data suggest that SIRT4 inhibition might be a useful therapeutic strategy for cardioprotection, at least in the short term. SIRT5 appears to be necessary for the cardiac response to IR or pressure overload injury, and for maintaining normal cardiac structure and function during aging. It will be of great interest to determine whether supraphysiologic SIRT5 activity is actually beneficial for myocardium. All work on SIRT6 generated to date supports the notion that this protein plays a cardioprotective function. Similarly, SIRT7 also appears to be necessary for normal cardiac homeostasis and stress responses.

Given the interest of sirtuins in cancer biology, targeting sirtuins pharmacologically through small-molecule activators, such as NAD^+ precursors or allosteric activators, and inhibitors is of significant clinical interest. Understanding the role of sirtuins in CV biology will be critical to developing sirtuin-based therapies for CV disease, and to provide insight into potential cardiac side effects when targeting sirtuins for extracardiac disease states.

ACKNOWLEDGMENTS

Support: AS: A. Alfred Taubman Medical Research Institute and NIH P30 AG024824. DL: Glenn Foundation for Medical Research and NIH (R01GM101171, 2R01HL114858, and R21AG053561). AG and WG were supported by T32 awards (T32-GM113900 and T32-HL007853, respectively).

REFERENCES

1. Benjamin EJ, Blaha MJ, Chiuve SE, Cushman M, Das SR, Deo R, et al. Heart disease and stroke statistics-2017 update: a report from the American Heart Association. *Circulation* 2017;**135**(10): e146−603.
2. Writing Group M, Mozaffarian D, Benjamin EJ, Go AS, Arnett DK, Blaha MJ, et al. Heart disease and stroke statistics-2016 update: a report from the American Heart Association. *Circulation* 2016;**133**(4): e38−60.
3. Vigen R, Maddox TM, Allen LA. Aging of the United States population: impact on heart failure. *Curr Heart Fail Rep* 2012;**9**(4):369−74.
4. Bhatt AS, Ambrosy AP, Velazquez EJ. Adverse remodeling and reverse remodeling after myocardial infarction. *Curr Cardiol Rep* 2017;**19**(8):71.
5. Upadhya B, Taffet GE, Cheng CP, Kitzman DW. Heart failure with preserved ejection fraction in the elderly: scope of the problem. *J Mol Cell Cardiol* 2015;**83**:73−87.
6. Metra M, Teerlink JR. Heart failure. *Lancet (London, England)* 2017.
7. Kalogeropoulos AP, Samman-Tahhan A, Hedley JS, McCue AA, Bjork JB, Markham DW, et al. Progression to stage D heart failure among outpatients with stage C heart failure and reduced ejection fraction. *JACC Heart Fail* 2017;**5**(7):528−37.
8. Imai S, Armstrong CM, Kaeberlein M, Guarente L. Transcriptional silencing and longevity protein Sir2 is an NAD-dependent histone deacetylase. *Nature* 2000;**403**(6771):795−800.
9. Jiang H, Khan S, Wang Y, Charron G, He B, Sebastian C, et al. SIRT6 regulates TNF-alpha secretion through hydrolysis of long-chain fatty acyl lysine. *Nature* 2013;**496**(7443):110−13.
10. Lin YY, Lu JY, Zhang J, Walter W, Dang W, Wan J, et al. Protein acetylation microarray reveals that NuA4 controls key metabolic target regulating gluconeogenesis. *Cell* 2009;**136**(6):1073−84.
11. Canto C, Sauve AA, Bai P. Crosstalk between poly(ADP-ribose) polymerase and sirtuin enzymes. *Mol Aspects Med* 2013;**34**(6):1168−201.
12. Kaeberlein M, McVey M, Guarente L. The SIR2/3/4 complex and SIR2 alone promote longevity in Saccharomyces cerevisiae by two different mechanisms. *Genes Dev* 1999;**13**(19):2570−80.
13. Kennedy BK, Austriaco Jr NR, Zhang J, Guarente L. Mutation in the silencing gene SIR4 can delay aging in *S. cerevisiae*. *Cell* 1995;**80**(3):485−96.

14. Kennedy BK, Gotta M, Sinclair DA, Mills K, McNabb DS, Murthy M, et al. Redistribution of silencing proteins from telomeres to the nucleolus is associated with extension of life span in *S. cerevisiae. Cell* 1997;**89**(3):381–91.

15. Giblin W, Skinner ME, Lombard DB. Sirtuins: guardians of mammalian healthspan. *Trends Genetics TIG.* 2014;**30**(7):271–86.

16. Tanno M, Sakamoto J, Miura T, Shimamoto K, Horio Y. Nucleocytoplasmic shuttling of the NAD + -dependent histone deacetylase SIRT1. *J Biol Chem* 2007;**282**(9):6823–32.

17. Alcendor RR, Gao S, Zhai P, Zablocki D, Holle E, Yu X, et al. Sirt1 regulates aging and resistance to oxidative stress in the heart. *Circ Res* 2007;**100**(10):1512–21.

18. Alcendor RR, Kirshenbaum LA, Imai S, Vatner SF, Sadoshima J. Silent information regulator 2alpha, a longevity factor and class III histone deacetylase, is an essential endogenous apoptosis inhibitor in cardiac myocytes. *Circ Res* 2004;**95**(10):971–80.

19. Hsu CP, Zhai P, Yamamoto T, Maejima Y, Matsushima S, Hariharan N, et al. Silent information regulator 1 protects the heart from ischemia/reperfusion. *Circulation* 2010;**122**(21):2170–82.

20. Tong C, Morrison A, Mattison S, Qian S, Bryniarski M, Rankin B, et al. Impaired SIRT1 nucleocytoplasmic shuttling in the senescent heart during ischemic stress. *FASEB J* 2013;**27**(11):4332–42.

21. Gu C, Xing Y, Jiang L, Chen M, Xu M, Yin Y, et al. Impaired cardiac SIRT1 activity by carbonyl stress contributes to aging-related ischemic intolerance. *PLoS One* 2013;**8**(9):e74050.

22. North BJ, Rosenberg MA, Jeganathan KB, Hafner AV, Michan S, Dai J, et al. SIRT2 induces the checkpoint kinase BubR1 to increase lifespan. *EMBO J* 2014;**33**(13):1438–53.

23. Cheng HL, Mostoslavsky R, Saito S, Manis JP, Gu Y, Patel P, et al. Developmental defects and p53 hyperacetylation in Sir2 homolog (SIRT1)-deficient mice. *Proc Natl Acad Sci U S A* 2003;**100**(19):10794–9.

24. McBurney MW, Yang X, Jardine K, Hixon M, Boekelheide K, Webb JR, et al. The mammalian SIR2alpha protein has a role in embryogenesis and gametogenesis. *Mol Cell Biol* 2003;**23**(1):38–54.

25. Kawashima T, Inuzuka Y, Okuda J, Kato T, Niizuma S, Tamaki Y, et al. Constitutive SIRT1 overexpression impairs mitochondria and reduces cardiac function in mice. *J Mol Cell Cardiol* 2011;**51**(6):1026–36.

26. Rockman HA, Ross RS, Harris AN, Knowlton KU, Steinhelper ME, Field LJ, et al. Segregation of atrial-specific and inducible expression of an atrial natriuretic factor transgene in an in vivo murine model of cardiac hypertrophy. *Proc Natl Acad Sci U S A* 1991;**88**(18):8277–81.

27. Sundaresan NR, Pillai VB, Wolfgeher D, Samant S, Vasudevan P, Parekh V, et al. The deacetylase SIRT1 promotes membrane localization and activation of Akt and PDK1 during tumorigenesis and cardiac hypertrophy. *Sci Signal* 2011;**4**(182):ra46.

28. Condorelli G, Drusco A, Stassi G, Bellacosa A, Roncarati R, Iaccarino G, et al. Akt induces enhanced myocardial contractility and cell size in vivo in transgenic mice. *Proc Natl Acad Sci U S A* 2002;**99**(19):12333–8.

29. Shiojima I, Yefremashvili M, Luo Z, Kureishi Y, Takahashi A, Tao J, et al. Akt signaling mediates postnatal heart growth in response to insulin and nutritional status. *J Biol Chem* 2002;**277**(40):37670–7.

30. Planavila A, Dominguez E, Navarro M, Vinciguerra M, Iglesias R, Giralt M, et al. Dilated cardiomyopathy and mitochondrial dysfunction in Sirt1-deficient mice: a role for Sirt1-Mef2 in adult heart. *J Mol Cell Cardiol* 2012;**53**(4):521–31.

31. Ramachandran B, Yu G, Gulick T. Nuclear respiratory factor 1 controls myocyte enhancer factor 2A transcription to provide a mechanism for coordinate expression of respiratory chain subunits. *J Biol Chem* 2008;**283**(18):11935–46.

32. Tang X, Chen XF, Wang NY, Wang XM, Liang ST, Zheng W, et al. SIRT2 acts as a cardioprotective deacetylase in pathological cardiac hypertrophy. *Circulation* 2017;**136**(21):2051–67.

33. De Jong KA, Lopaschuk GD. Complex energy metabolic changes in heart failure with preserved ejection fraction and heart failure with reduced ejection fraction. *Can J Cardiol* 2017;**33**(7):860−71.
34. Madrazo JA, Kelly DP. The PPAR trio: regulators of myocardial energy metabolism in health and disease. *J Mol Cell Cardiol* 2008;**44**(6):968−75.
35. Oka S, Zhai P, Yamamoto T, Ikeda Y, Byun J, Hsu CP, et al. Peroxisome proliferator activated receptor-alpha association with silent information regulator 1 suppresses cardiac fatty acid metabolism in the failing heart. *Circ Heart Fail.* 2015;**8**(6):1123−32.
36. Falk E, Shah PK, Fuster V. Coronary plaque disruption. *Circulation* 1995;**92**(3):657−71.
37. Kristensen SD, Andersen HR, Falk E. What an interventional cardiologist should know about the pathophysiology of acute myocardial infarction. *Semin Interv cardiol* 1999;**4**(1):11−16.
38. Menees DS, Peterson ED, Wang Y, Curtis JP, Messenger JC, Rumsfeld JS, et al. Door-to-balloon time and mortality among patients undergoing primary PCI. *N Engl J Med* 2013;**369**(10):901−9.
39. Nallamothu BK, Normand SL, Wang Y, Hofer TP, Brush Jr JE, et al. Relation between door-to-balloon times and mortality after primary percutaneous coronary intervention over time: a retrospective study. *Lancet (London, England).* 2015;**385**(9973):1114−22.
40. Eisner DA, Caldwell JL, Kistamas K, Trafford AW. Calcium and excitation-contraction coupling in the heart. *Circ Res* 2017;**121**(2):181−95.
41. Vikram A, Lewarchik CM, Yoon JY, Naqvi A, Kumar S, Morgan GM, et al. Sirtuin 1 regulates cardiac electrical activity by deacetylating the cardiac sodium channel. *Nat Med* 2017;**23**(3):361−7.
42. Liu M, Sanyal S, Gao G, Gurung IS, Zhu X, Gaconnet G, et al. Cardiac Na + current regulation by pyridine nucleotides. *Circ Res* 2009;**105**(8):737−45.
43. Baur JA, Ungvari Z, Minor RK, Le Couteur DG, de Cabo R. Are sirtuins viable targets for improving healthspan and lifespan? *Nat Rev Drug Discov* 2012;**11**(6):443−61.
44. Park SJ, Ahmad F, Philp A, Baar K, Williams T, Luo H, et al. Resveratrol ameliorates aging-related metabolic phenotypes by inhibiting cAMP phosphodiesterases. *Cell* 2012;**148**(3):421−33.
45. Lagouge M, Argmann C, Gerhart-Hines Z, Meziane H, Lerin C, Daussin F, et al. Resveratrol improves mitochondrial function and protects against metabolic disease by activating SIRT1 and PGC-1alpha. *Cell* 2006;**127**(6):1109−22.
46. Baur JA, Pearson KJ, Price NL, Jamieson HA, Lerin C, Kalra A, et al. Resveratrol improves health and survival of mice on a high-calorie diet. *Nature* 2006;**444**(7117):337−42.
47. Desquiret-Dumas V, Gueguen N, Leman G, Baron S, Nivet-Antoine V, Chupin S, et al. Resveratrol induces a mitochondrial complex I-dependent increase in NADH oxidation responsible for sirtuin activation in liver cells. *J Biol Chem* 2013;**288**(51):36662−75.
48. Suzuki K, Koike T. Resveratrol abolishes resistance to axonal degeneration in slow Wallerian degeneration (WldS) mice: activation of SIRT2, an NAD-dependent tubulin deacetylase. *Biochem Biophys Res Commun* 2007;**359**(3):665−71.
49. Yoshida Y, Shioi T, Izumi T. Resveratrol ameliorates experimental autoimmune myocarditis. *Circul J* 2007;**71**(3):397−404.
50. Tanno M, Kuno A, Yano T, Miura T, Hisahara S, Ishikawa S, et al. Induction of manganese superoxide dismutase by nuclear translocation and activation of SIRT1 promotes cell survival in chronic heart failure. *J Biol Chem* 2010;**285**(11):8375−82.
51. Sulaiman M, Matta MJ, Sunderesan NR, Gupta MP, Periasamy M, Gupta M. Resveratrol, an activator of SIRT1, upregulates sarcoplasmic calcium ATPase and improves cardiac function in diabetic cardiomyopathy. *Am J Physiol Heart Circ Physiol* 2010;**298**(3):H833−43.
52. Gu XS, Wang ZB, Ye Z, Lei JP, Li L, Su DF, et al. Resveratrol, an activator of SIRT1, upregulates AMPK and improves cardiac function in heart failure. *Gen Mol Res* 2014;**13**(1):323−35.
53. Buja LM, Ferrans VJ, Roberts WC. Drug-induced cardiomyopathies. *Adv Cardiol* 1974;**13**:330−48.

54. Jaenke RS. Delayed and progressive myocardial lesions after adriamycin administration in the rabbit. *Cancer Res* 1976;**36**(8):2958−66.

55. Doroshow JH. Effect of anthracycline antibiotics on oxygen radical formation in rat heart. *Cancer Res* 1983;**43**(2):460−72.

56. Zhang DX, Ma DY, Yao ZQ, Fu CY, Shi YX, Wang QL, et al. ERK1/2/p53 and NF-kappaB dependent-PUMA activation involves in doxorubicin-induced cardiomyocyte apoptosis. *Eur Rev Med Pharmacol Sci* 2016;**20**(11):2435−42.

57. Kotamraju S, Konorev EA, Joseph J, Kalyanaraman B. Doxorubicin-induced apoptosis in endothelial cells and cardiomyocytes is ameliorated by nitrone spin traps and ebselen. Role of reactive oxygen and nitrogen species. *J Biol Chem* 2000;**275**(43):33585−92.

58. Danz ED, Skramsted J, Henry N, Bennett JA, Keller RS. Resveratrol prevents doxorubicin cardiotoxicity through mitochondrial stabilization and the Sirt1 pathway. *Free Radic Biol Med* 2009;**46**(12):1589−97.

59. Zhang C, Feng Y, Qu S, Wei X, Zhu H, Luo Q, et al. Resveratrol attenuates doxorubicin-induced cardiomyocyte apoptosis in mice through SIRT1-mediated deacetylation of p53. *Cardiovasc Res* 2011;**90**(3):538−45.

60. Ruan Y, Dong C, Patel J, Duan C, Wang X, Wu X, et al. SIRT1 suppresses doxorubicin-induced cardiotoxicity by regulating the oxidative stress and p38MAPK pathways. *Cell Physiol Biochem* 2015;**35**(3):1116−24.

61. Sin TK, Tam BT, Yung BY, Yip SP, Chan LW, Wong CS, et al. Resveratrol protects against doxorubicin-induced cardiotoxicity in aged hearts through the SIRT1-USP7 axis. *J Physiol* 2015;**593**(8):1887−99.

62. Cappetta D, Esposito G, Piegari E, Russo R, Ciuffreda LP, Rivellino A, et al. SIRT1 activation attenuates diastolic dysfunction by reducing cardiac fibrosis in a model of anthracycline cardiomyopathy. *Int J Cardiol* 2016;**205**:99−110.

63. Yano M, Akazawa H, Oka T, Yabumoto C, Kudo-Sakamoto Y, Kamo T, et al. Monocyte-derived extracellular Nampt-dependent biosynthesis of NAD(+) protects the heart against pressure overload. *Sci Rep* 2015;**5**:15857.

64. North BJ, Marshall BL, Borra MT, Denu JM, Verdin E. The human Sir2 ortholog, SIRT2, is an NAD + -dependent tubulin deacetylase. *Mol Cell* 2003;**11**(2):437−44.

65. North BJ, Verdin E. Mitotic regulation of SIRT2 by cyclin-dependent kinase 1-dependent phosphorylation. *J Biol Chem* 2007;**282**(27):19546−55.

66. Yuan Q, Zhan L, Zhou QY, Zhang LL, Chen XM, Hu XM, et al. SIRT2 regulates microtubule stabilization in diabetic cardiomyopathy. *Eur J Pharmacol* 2015;**764**:554−61.

67. Neubauer S. The failing heart--an engine out of fuel. *N Engl J Med* 2007;**356**(11):1140−51.

68. Stanley WC, Recchia FA, Lopaschuk GD. Myocardial substrate metabolism in the normal and failing heart. *Physiol Rev* 2005;**85**(3):1093−129.

69. Doehner W, Frenneaux M, Anker SD. Metabolic impairment in heart failure: the myocardial and systemic perspective. *J Am Coll Cardiol* 2014;**64**(13):1388−400.

70. Byrne NJ, Levasseur J, Sung MM, Masson G, Boisvenue J, Young ME, et al. Normalization of cardiac substrate utilization and left ventricular hypertrophy precede functional recovery in heart failure regression. *Cardiovasc Res* 2016;**110**(2):249−57.

71. Lesnefsky EJ, Chen Q, Hoppel CL. Mitochondrial metabolism in aging heart. *Circ Res* 2016;**118**(10):1593−611.

72. Lombard DB, Alt FW, Cheng HL, Bunkenborg J, Streeper RS, Mostoslavsky R, et al. Mammalian Sir2 homolog SIRT3 regulates global mitochondrial lysine acetylation. *Mol Cell Biol* 2007;**27**(24):8807−14.

73. Ahn BH, Kim HS, Song S, Lee IH, Liu J, Vassilopoulos A, et al. A role for the mitochondrial deacetylase Sirt3 in regulating energy homeostasis. *Proc Natl Acad Sci U S A* 2008;**105**(38):14447−52.

74. Hirschey MD, Shimazu T, Jing E, Grueter CA, Collins AM, Aouizerat B, et al. SIRT3 deficiency and mitochondrial protein hyperacetylation accelerate the development of the metabolic syndrome. *Mol Cell* 2011;**44**(2):177−90.

75. Bellizzi D, Rose G, Cavalcante P, Covello G, Dato S, De Rango F, et al. A novel VNTR enhancer within the SIRT3 gene, a human homologue of SIR2, is associated with survival at oldest ages. *Genomics* 2005;**85**(2):258−63.

76. Rose G, Dato S, Altomare K, Bellizzi D, Garasto S, Greco V, et al. Variability of the SIRT3 gene, human silent information regulator Sir2 homologue, and survivorship in the elderly. *Exp Gerontol* 2003;**38**(10):1065−70.

77. Kumar S, Lombard DB. Mitochondrial sirtuins and their relationships with metabolic disease and cancer. *Antioxid Redox Signal* 2015;**22**(12):1060−77.

78. Horton JL, Martin OJ, Lai L, Riley NM, Richards AL, Vega RB, et al. Mitochondrial protein hyperacetylation in the failing heart. *JCI Insight* 2016;**2**(1). pii: e84897.

79. Sundaresan NR, Gupta M, Kim G, Rajamohan SB, Isbatan A, Gupta MP. Sirt3 blocks the cardiac hypertrophic response by augmenting Foxo3a-dependent antioxidant defense mechanisms in mice. *J Clin Invest* 2009;**119**(9):2758−71.

80. Koentges C, Pfeil K, Schnick T, Wiese S, Dahlbock R, Cimolai MC, et al. SIRT3 deficiency impairs mitochondrial and contractile function in the heart. *Basic Res Cardiol* 2015;**110**(4):36.

81. Hafner AV, Dai J, Gomes AP, Xiao CY, Palmeira CM, Rosenzweig A, et al. Regulation of the mPTP by SIRT3-mediated deacetylation of CypD at lysine 166 suppresses age-related cardiac hypertrophy. *Aging (Albany NY)*. 2010;**2**(12):914−23.

82. Sundaresan NR, Bindu S, Pillai VB, Samant S, Pan Y, Huang JY, et al. SIRT3 blocks aging-associated tissue fibrosis in mice by deacetylating and activating glycogen synthase kinase 3beta. *Mol Cell Biol* 2015;**36**(5):678−92.

83. Muntean DM, Sturza A, Danila MD, Borza C, Duicu OM, Mornos C. The role of mitochondrial reactive oxygen species in cardiovascular injury and protective strategies. *Oxid Med Cell Longev* 2016;**2016**:8254942.

84. Porter GA, Urciuoli WR, Brookes PS, Nadtochiy SM. SIRT3 deficiency exacerbates ischemia-reperfusion injury: implication for aged hearts. *Am J Physiol Heart Circ Physiol* 2014;**306**(12):H1602−9.

85. Pillai VB, Bindu S, Sharp W, Fang YH, Kim G, Gupta M, et al. Sirt3 protects mitochondrial DNA damage and blocks the development of doxorubicin-induced cardiomyopathy in mice. *Am J Physiol Heart Circ Physiol* 2016;**310**(8):H962−72.

86. Fried LE, Arbiser JL. Honokiol, a multifunctional antiangiogenic and antitumor agent. *Antioxid Redox Signal* 2009;**11**(5):1139−48.

87. Bai X, Cerimele F, Ushio-Fukai M, Waqas M, Campbell PM, Govindarajan B, et al. Honokiol, a small molecular weight natural product, inhibits angiogenesis in vitro and tumor growth in vivo. *J Biol Chem* 2003;**278**(37):35501−7.

88. Liou KT, Shen YC, Chen CF, Tsao CM, Tsai SK. Honokiol protects rat brain from focal cerebral ischemia-reperfusion injury by inhibiting neutrophil infiltration and reactive oxygen species production. *Brain Res* 2003;**992**(2):159−66.

89. Pillai VB, Samant S, Sundaresan NR, Raghuraman H, Kim G, Bonner MY, et al. Honokiol blocks and reverses cardiac hypertrophy in mice by activating mitochondrial Sirt3. *Nat Commun* 2015;**6**:6656.

90. Pillai VB, Kanwal A, Fang YH, Sharp WW, Samant S, Arbiser J, et al. Honokiol, an activator of Sirtuin-3 (SIRT3) preserves mitochondria and protects the heart from doxorubicin-induced cardiomyopathy in mice. *Oncotarget* 2017;**8**(21):34082−98.

91. Pillai VB, Sundaresan NR, Kim G, Gupta M, Rajamohan SB, Pillai JB, et al. Exogenous NAD blocks cardiac hypertrophic response via activation of the SIRT3-LKB1-AMP-activated kinase pathway. *J Biol Chem* 2010;**285**(5):3133−44.

92. Martin AS, Abraham DM, Hershberger KA, Bhatt DP, Mao L, Cui H, et al. Nicotinamide mononucleotide requires SIRT3 to improve cardiac function and bioenergetics in a Friedreich's ataxia cardiomyopathy model. *JCI Insight* 2017;**2**(14).

93. Anderson KA, Huynh FK, Fisher-Wellman K, Stuart JD, Peterson BS, Douros JD, et al. SIRT4 is a lysine deacylase that controls leucine metabolism and insulin secretion. *Cell Metab* 2017;**25**(4). 838-855.e15.

94. Wagner GR, Bhatt DP, O'Connell TM, Thompson JW, Dubois LG, Backos DS, et al. A class of reactive acyl-CoA species reveals the non-enzymatic origins of protein acylation. *Cell Metab* 2017;**25**(4). 823-837.e8.

95. Laurent G, German NJ, Saha AK, de Boer VC, Davies M, Koves TR, et al. SIRT4 coordinates the balance between lipid synthesis and catabolism by repressing malonyl CoA decarboxylase. *Mol Cell* 2013;**50**(5):686−98.

96. Mathias RA, Greco TM, Oberstein A, Budayeva HG, Chakrabarti R, Rowland EA, et al. Sirtuin 4 is a lipoamidase regulating pyruvate dehydrogenase complex activity. *Cell* 2014;**159**(7):1615−25.

97. Luo YX, Tang X, An XZ, Xie XM, Chen XF, Zhao X, et al. SIRT4 accelerates Ang II-induced pathological cardiac hypertrophy by inhibiting manganese superoxide dismutase activity. *Eur Heart J* 2017;**38**(18):1389−98.

98. Xiao Y, Zhang X, Fan S, Cui G, Shen Z. MicroRNA-497 inhibits cardiac hypertrophy by targeting Sirt4. *PLoS One* 2016;**11**(12):e0168078.

99. Nishida Y, Rardin MJ, Carrico C, He W, Sahu AK, Gut P, et al. SIRT5 regulates both cytosolic and mitochondrial protein malonylation with glycolysis as a major target. *Mol Cell* 2015;**59**(2):321−32.

100. Rardin MJ, He W, Nishida Y, Newman JC, Carrico C, Danielson SR, et al. SIRT5 regulates the mitochondrial lysine succinylome and metabolic networks. *Cell Metab* 2013;**18**(6):920−33.

101. Park J, Chen Y, Tishkoff DX, Peng C, Tan M, Dai L, et al. SIRT5-mediated lysine desuccinylation impacts diverse metabolic pathways. *Mol Cell* 2013;**50**(6):919−30.

102. Tan M, Peng C, Anderson KA, Chhoy P, Xie Z, Dai L, et al. Lysine glutarylation is a protein posttranslational modification regulated by SIRT5. *Cell Metab* 2014;**19**(4):605−17.

103. Zhou L, Wang F, Sun R, Chen X, Zhang M, Xu Q, et al. SIRT5 promotes IDH2 desuccinylation and G6PD deglutarylation to enhance cellular antioxidant defense. *EMBO Rep* 2016;**17**(6):811−22.

104. Liu B, Che W, Zheng C, Liu W, Wen J, Fu H, et al. SIRT5: a safeguard against oxidative stress-induced apoptosis in cardiomyocytes. *Cell Physiol Biochem* 2013;**32**(4):1050−9.

105. Lin ZF, Xu HB, Wang JY, Lin Q, Ruan Z, Liu FB, et al. SIRT5 desuccinylates and activates SOD1 to eliminate ROS. *Biochem Biophys Res Commun* 2013;**441**(1):191−5.

106. Nakagawa T, Lomb DJ, Haigis MC, Guarente L. SIRT5 deacetylates carbamoyl phosphate synthetase 1 and regulates the urea cycle. *Cell* 2009;**137**(3):560−70.

107. Yu J, Sadhukhan S, Noriega LG, Moullan N, He B, Weiss RS, et al. Metabolic characterization of a Sirt5 deficient mouse model. *Sci Rep* 2013;**3**:2806.

108. Sadhukhan S, Liu X, Ryu D, Nelson OD, Stupinski JA, Li Z, et al. Metabolomics-assisted proteomics identifies succinylation and SIRT5 as important regulators of cardiac function. *Proc Natl Acad Sci USA* 2016;**113**(16):4320−5.

109. Boylston JA, Sun J, Chen Y, Gucek M, Sack MN, Murphy E. Characterization of the cardiac succinylome and its role in ischemia-reperfusion injury. *J Mol Cell Cardiol* 2015;**88**:73−81.

110. Hershberger KA, Abraham DM, Martin AS, Mao L, Liu J, Gu H, et al. Sirtuin 5 is required for mouse survival in response to cardiac pressure overload. *J Biol Chem* 2017;**292**(48):19767−81.

111. Kawahara TL, Michishita E, Adler AS, Damian M, Berber E, Lin M, et al. SIRT6 links histone H3 lysine 9 deacetylation to NF-kappaB-dependent gene expression and organismal life span. *Cell* 2009;**136** (1):62−74.
112. Michishita E, McCord RA, Berber E, Kioi M, Padilla-Nash H, Damian M, et al. SIRT6 is a histone H3 lysine 9 deacetylase that modulates telomeric chromatin. *Nature* 2008;**452**(7186):492−6.
113. Zhong L, D'Urso A, Toiber D, Sebastian C, Henry RE, Vadysirisack DD, et al. The histone deacetylase Sirt6 regulates glucose homeostasis via Hif1alpha. *Cell* 2010;**140**(2):280−93.
114. Liszt G, Ford E, Kurtev M, Guarente L. Mouse Sir2 homolog SIRT6 is a nuclear ADP-ribosyltransferase. *J Biol Chem* 2005;**280**(22):21313−20.
115. Mostoslavsky R, Chua KF, Lombard DB, Pang WW, Fischer MR, Gellon L, et al. Genomic instability and aging-like phenotype in the absence of mammalian SIRT6. *Cell* 2006;**124**(2):315−29.
116. Sundaresan NR, Vasudevan P, Zhong L, Kim G, Samant S, Parekh V, et al. The sirtuin SIRT6 blocks IGF-Akt signaling and development of cardiac hypertrophy by targeting c-Jun. *Nat Med* 2012;**18** (11):1643−50.
117. Reiss K, Capasso JM, Huang HE, Meggs LG, Li P, Anversa P. ANG II receptors, c-myc, and c-jun in myocytes after myocardial infarction and ventricular failure. *Am J Physiol* 1993;**264**(3 Pt 2):H760−9.
118. Iwaki K, Sukhatme VP, Shubeita HE, Chien KR. Alpha- and beta-adrenergic stimulation induces distinct patterns of immediate early gene expression in neonatal rat myocardial cells. fos/jun expression is associated with sarcomere assembly; Egr-1 induction is primarily an alpha 1-mediated response. *J Biol Chem* 1990;**265**(23):13809−17.
119. Takemoto Y, Yoshiyama M, Takeuchi K, Omura T, Komatsu R, Izumi Y, et al. Increased JNK, AP-1 and NF-kappa B DNA binding activities in isoproterenol-induced cardiac remodeling. *J Mol Cell Cardiol* 1999;**31**(11):2017−30.
120. Zhang ZQ, Ren SC, Tan Y, Li ZZ, Tang X, Wang TT, et al. Epigenetic regulation of NKG2D ligands is involved in exacerbated atherosclerosis development in Sirt6 heterozygous mice. *Sci Rep* 2016;**6**:23912.
121. Piedrahita JA, Zhang SH, Hagaman JR, Oliver PM, Maeda N. Generation of mice carrying a mutant apolipoprotein E gene inactivated by gene targeting in embryonic stem cells. *Proc Natl Acad Sci U S A* 1992;**89**(10):4471−5.
122. Ford E, Voit R, Liszt G, Magin C, Grummt I, Guarente L. Mammalian Sir2 homolog SIRT7 is an activator of RNA polymerase I transcription. *Genes Dev* 2006;**20**(9):1075−80.
123. Barber MF, Michishita-Kioi E, Xi Y, Tasselli L, Kioi M, Moqtaderi Z, et al. SIRT7 links H3K18 deacetylation to maintenance of oncogenic transformation. *Nature* 2012;**487**(7405):114−18.
124. Tong Z, Wang Y, Zhang X, Kim DD, Sadhukhan S, Hao Q, et al. SIRT7 is activated by DNA and deacetylates histone H3 in the chromatin context. *ACS Chem Biol* 2016;**11**(3):742−7.
125. Vakhrusheva O, Smolka C, Gajawada P, Kostin S, Boettger T, Kubin T, et al. Sirt7 increases stress resistance of cardiomyocytes and prevents apoptosis and inflammatory cardiomyopathy in mice. *Circ Res* 2008;**102**(6):703−10.
126. Araki S, Izumiya Y, Rokutanda T, Ianni A, Hanatani S, Kimura Y, et al. Sirt7 contributes to myocardial tissue repair by maintaining transforming growth factor-beta signaling pathway. *Circulation* 2015;**132** (12):1081−93.

SIRTUINS IN BRAIN AND NEURODEGENERATIVE DISEASE

13

Éva M. Szegő[1], Tiago F. Outeiro[1,2] and Aleksey G. Kazantsev[3]

[1]*University Medical Center Göttingen, Göttingen, Germany* [2]*Max Planck Institute for Experimental Medicine, Göttingen, Germany* [3]*Massachusetts General Hospital and Harvard Medical School, Cambridge, MA, United States*

13.1 INTRODUCTION

Sirtuins, named after the silent information regulator 2 (Sir2) gene, were first discovered in budding yeast. In humans, the sirtuin family is composed of seven members (SIRT1−7).[1,2] Sirtuin-mediated deacetylation is strictly coupled to the cleavage of NAD^+,[3] yielding nicotinamide (NAM) and *O*-acetyl adenosine diphosphate (ADP)-ribose, along with the deacetylated lysine residue within the protein substrate.[4] Deacetylation activity has now been demonstrated for six of seven mammalian SIRTs, with evidence of this activity lacking only for SIRT4.[4−10] The deacetylation activities of SIRTs are executed on a broad spectrum of protein substrates, which leads to modulation of diverse cellular functions.[11] Sirtuin members display different subcellular localizations. SIRT1 and SIRT2 are nuclear and cytosolic, SIRT3−5 are mitochondrial, and SIRT6−7 are nuclear proteins.[11]

Human SIRT1, the closest analog of yeast Sir2, has gained much attention as a mediator of metabolic changes and longevity in several model organisms.[1] SIRT1 activation, either genetically or pharmacologically, or via metabolic conditioning associated with caloric restriction (CR), was found efficacious in model systems of neurodegeneration.[12,13] This raised the broad interest in investigating the biological functions of sirtuins in the brain, and in exploring putative therapeutic strategies targeting sirtuin activities in human age-associated neurodegenerative diseases.

13.2 SIRTUIN ISOFORMS AND EXPRESSION IN THE BRAIN

Alternatively spliced variants have been described and characterized for SIRT1,[14,15] SIRT2,[16] SIRT3,[17] SIRT5,[18] and SIRT6[19] (Fig. 13.1).

SIRT1 lacking exon 8 (isoform SIRT1-D8) retains minimal deacetylase activity, but exhibits distinct stress sensitivity, RNA/protein stability, and protein−protein interactions compared to full-length protein.[15] SIRT1-D8 and p53 are involved in an autoregulatory loop, whereby SIRT1-D8 can regulate p53, and reciprocally p53 can influence SIRT1 splice variation.[15] A distinct regulatory role, opposing the function of full-length protein, has been assigned to splice variant SIRT1-D2/9,

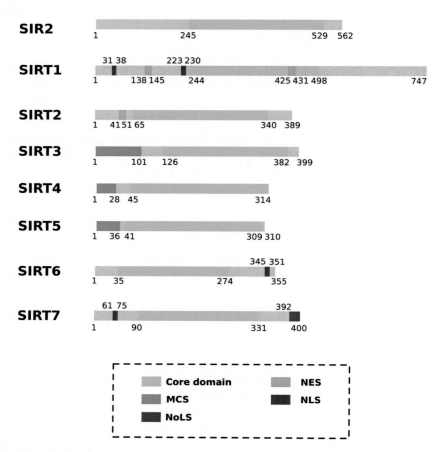

FIGURE 13.1 Sirtuin isoforms.

Schematic representation of the yeast silent information regulator-2 longevity regulator (Sir2) and its seven mammalian homologs, SIRT1−SIRT7.

Core domain, NAD-dependent catalytic domain; *MCS*, mitochondrial targeting sequence; *NES*, nuclear export sequence; *NLS*, nuclear localization sequence; *NoLS*, nucleolar localization sequence; numbers indicate amino acid residues.

lacking exons 2−9, and thereby lacking the catalytic domain.[20] Both full-length SIRT1 and splice variant SIRT1-D2/9 proteins bind to p53, but the former has a negative and the latter a positive influence on stress-response.[20]

An alternatively spliced form of SIRT2 at the N-terminus (SIRT2.2), encodes for a 37-kDa nuclear-cytoplasmic protein with intact catalytic deacetylation domain, and has the highest expression level amongst all sirtuins in the adult central nervous system (CNS).[19] Intriguingly a SIRT2 splice variant, lacking a nuclear export signal, isoform 5, encodes a predominantly nuclear protein with no detectable deacetylase activity, despite the presence of intact catalytic domain.[16]

Two distinct isoforms of the SIRT3 have been identified with the short isoform having no recognizable mitochondrial localization sequence (MLS) and the long isoform having a putative MLS.[17] The long isoform is generated via intra-exon splicing creating a frameshift to expose a novel upstream translation start site and has been shown to be predominantly mitochondrial, with robust deacetylase activity.

SIRT5 isoforms 1 and 2 are different at the C-termini. This determines the protein stability of the former, and is responsible for the predominant mitochondrial localization of the latter, which exists in primates as the preferentially spliced variant.[18]

Although a specific role of each sirtuin isoform is still emerging, it appears evident that alternative splicing is a major mechanism regulating protein activities and contributing to the diversity of sirtuin functions in the cell.

All sirtuins are detected in the adult mammalian brain. However, the magnitudes of RNA and protein expression vary considerably, with the 37-kDa isoform 2 of SIRT2 being the most abundant, and SIRT4 being the least abundant proteins.[19]

Sirtuins display distinct patterns of expression in developing and in adult brains. The levels of SIRT1 and SIRT6 isoform-2 (36 kDa) decrease after brain maturation, while expressions of SIRT2, SIRT3, SIRT5, and SIRT6 isoform-1 (39 kDa) increase.[19] During embryonic development, SIRT1 is also expressed at high levels in the spinal cord and dorsal root ganglia, suggesting the involvement of this deacetylase in neurogenesis.[21] SIRT2 and SIRT3 are nondetectable, or expressed at extremely low levels during development, but are then elevated at the end of brain maturation, and remain at constant high levels throughout the lifespan, suggesting an important role of these deacetylases in the adult brain.[19] Remarkably, high mRNA levels of SIRT2 during brain development are not translated to abundant protein presence, indicating a critical role of stability and turnover in the homeostasis of this deacetylase.[19]

Sirtuin protein expression patterns are nonredundant in different regions of adult CNS. For example, the highest and lowest levels of SIRT1 are observed in cerebellum and spinal cord, respectively. SIRT2 is most abundant in hippocampus, striatum, spinal cord, and brain stem. Elevated levels of SIRT5 isoform-2 (35 kDa) are detected in the cerebellum and brain stem.[19] Such different expression patterns may be viewed as indicative of specific roles of individual sirtuins in specific brain region.

13.3 CELLULAR PATHWAYS AND FUNCTIONS REGULATED BY SIRTUINS

Sirtuins play important roles in several cellular processes, such as regulation of metabolism, mitochondrial functions, cell cycle, or senescence.[22] Since the seven known mammalian sirtuins have partially different cellular localization (Fig. 13.2) and they act on several distinct substrates in different tissues, their function is rather diverse.

13.3.1 SIRT1

SIRT1, the best characterized member of the family, is mainly found in the nucleus, but it can also transiently shuttle to the cytoplasm.[23] SIRT1 regulates the activity of several proteins, including

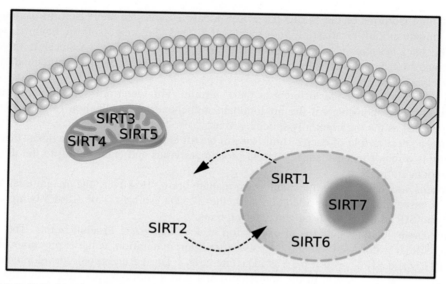

FIGURE 13.2 Cellular localization of sirtuins.

SIRT1, 6, and 7 are predominantly nuclear sirtuins. SIRT2 is mostly cytosolic, but can also shuttle into the nucleus. SIRT3—5 are mitochondrial sirtuins.

histones, transcription factors (TFs), and coactivators. Since lysine acetylation of histones H3 and H4 is critical in histone—DNA binding, SIRT1 activity is responsible for gene silencing. In this context SIRT1 reduces overall transcriptional activity and extends the life span.[3] However, by binding to and deacetylating TFs, SIRT1 modifies the transcription of specific genes. One of the best studied TFs targeted by SIRT1 is the family of forkhead box O proteins (FOXO).[24] SIRT1 deacetylates and modulates the activity of FOXO1, FOXO3, and FOXO4, which, in turn, increases the transcription of stress-response-related genes including growth arrest and DNA damage 45 (GADD45), or superoxide dismutase 1 (SOD1),[25] and increases cellular stress resistance (for review, see Refs.[22,26]). SIRT1 regulates both the activity (deacetylation) and the expression (binding to its promoter region) of p53,[27,28] a protein activated in response to DNA damage and responsible for the upregulation of proapoptotic proteins.[29] Sirtuins also directly increase the activity of enzymes involved in DNA repair mechanisms. For example, SIRT1 deacetylates and activates apurinic/apyrimidinic endonuclease 1 (APE1), xeroderma pigmentosum group A (XPA), or DNA-dependent protein kinase (DNA-PK), poly(ADP-ribose) polymerases (PARPs), therefore facilitating base/nucleotide excision repair or nonhomologous end-joining repair,[30] and attenuating telomere shortening.[31] By inhibiting nuclear factor kappa B (NF-κB) signal transduction (through deacetylation of the p65 subunit), SIRT1 reduces the activation of proinflammatory responses.[32]

Several studies suggest that sirtuins control pathways involved in the regulation of metabolism and calorie restriction.[33,34] By repressing the transcription of mitochondrial uncoupling protein UPC-2 gene, SIRT1 increases glucose-induced insulin secretion from the β-cells of the pancreas.[35] As a next regulation step, it promotes the transcription of gluconeogenic genes (deacetylation of peroxisome proliferator-activated receptor gamma coactivator [PGC1α] and FOXO1), inhibits glycolysis (deacetylation of both glycolytic enzymes, such as phosphoglycerate mutase 1, and their key transcriptional inducer, hypoxia-inducible factor-1α [HIF-1α]), and fat synthesis (sterol regulatory element binding protein), stimulating fatty acid use (PPARα deacetylation and increased expression) in the liver.[33] Interestingly, SIRT1 regulates not only glucose availability, but also controls mitochondrial biogenesis by acting synergistically with AMP-activated kinase to enhance transcriptional activity of PGC1α—a major TF controlling the expression of nuclear-encoded mitochondrial proteins.[35,36] In addition, SIRT1 supervises autophagy, modulating cellular metabolism by protein and organelle degradation via deacetylation through the activation of several Atg proteins.[37]

By controlling a key signaling mechanism, the phosphatidylinositide 3-kinase (PI3K)/Akt pathway, SIRT1 plays an important role in cell viability and differentiation.[38] Under basal conditions, lysine acetylation of Akt suppresses its activity. However, upon growth factor stimulation, SIRT1 deacetylates the pleckstrin homology domain of both Akt and its kinase, phosphoinositide-dependent protein kinase 1. In this stage, the binding of the kinases to phosphatidylinositide 3 increases, resulting in Akt phosphorylation and a 100-fold increase in its activity.[39] By interacting with Akt signaling, SIRT1 may affect cell fate determination[40] and longevity.[41,42]

Sirtuins are also known to be major regulators of inflammation. By deacetylation of histones and TFs, such as NF-κB or activator protein 1 (AP-1), SIRT1 represses the transcription of several genes related to inflammation.[43]

13.3.2 SIRT2

SIRT2 was originally described as a cytosolic protein. However, recent data show that it can also shuttle to the nucleus.[44] Together with histone deacetylase 6 (HDAC6), SIRT2 is a major α-tubulin deacetylase.[36,45] Since lysine acetylation of α-tubulin is a signal for the recruitment of motor complexes (kinesin-1 and dynein/dynactin), and stimulates both anterograde and retrograde transport[46,47] and neurite outgrowth,[48] SIRT2 inhibits growth cone collapse and neurite outgrowth.[49,50]

SIRT2 regulates mitotic exit via tubulin and histone H4 deacetylation.[51] In addition, SIRT2 enhances DNA binding by FOXO proteins,[25] and deactivates and enhances the degradation of p53.[52] Therefore, it is important in tumor suppression and genome maintenance. In addition, by binding to and deacetylating β-catenin, SIRT2 inhibits Wnt signaling and the expression of Wnt target genes in tumors.[53] By deacetylating histone H3 or cyclin-dependent kinase 9 (CDK9), SIRT2 plays an important role in the DNA damage response.[54] Therefore, it is not surprising that SIRT2 knockout mice develop several types of cancer, and SIRT2 expression itself is decreased in liver or breast cancer.[54] Like SIRT1, SIRT2 increases Akt activity and downstream signaling,[55,56] deacetylates NF-κB, and inhibits transcription of proinflammatory genes.[57]

13.3.3 SIRT3

Mitochondrial SIRT3 regulates ATP production, apoptosis, and cell signaling.[9,58] SIRT3 is regulated by nutrient availability,[59] and as a global regulator of mitochondrial protein acetylation.[60] By deacetylating acetyl-CoA (coenzyme A) synthetase, isocitrate dehydrogenase, ATP synthase, cytochromes, SOD2 or catalase,[60] SIRT3 regulates virtually all major mitochondrial functions including oxidative phosphorylation, antioxidant mechanisms, or mitochondrial permeability transition pore opening. By activating tricarboxylic acid cycle enzymes, SIRT3 shifts glycolysis-based metabolism towards a more catabolic metabolism, especially during fasting. In addition to metabolic control, SIRT3 also plays a key role in mitochondrial protein quality control, by deacetylating heat shock proteins 10 and 60.[59,60]

13.3.4 **SIRT4**

SIRT4 has little deacetylase activity, and strong NAD^+-dependent mono-ADP-ribosylation activity.[61] One of the best-known targets of SIRT4 is glutamate dehydrogenase (GDH). Via inhibition of GDH, SIRT4 controls glutamine catabolism and metabolic reprogramming of cancer cells.[62] By repressing GDH activity in pancreatic cells, SIRT4 also opposes the effect of SIRT1 on insulin secretion.[61] Knockdown of SIRT4 increases fatty acid oxidation in a SIRT1-dependent manner.[63] Interestingly, by preventing NF-κB nuclear translocation, SIRT4 has an antiinflammatory function.[64]

SIRT4 also has lipoamidase activity, hydrolyzing lipoamide cofactors from the pyruvate dehydrogenase complex that catalyzes the decarboxylation of pyruvate to generate acetyl CoA, and links glycolysis to the tricarboxylic acid cycle.[65] Thus, by inhibition of the complex, SIRT4 is a regulator of metabolism and fuel switch.

13.3.5 **SIRT5**

The third mitochondrial sirtuin, SIRT5, has mainly desuccinylase and demalonylase activity, and plays a major role in lipid β-oxidation and ketogenesis.[66] By deacetylation of carbamoyl synthetase 1 (CPS1), SIRT5 increases the activity of the mitochondrial urea cycle and ammonia elimination, especially when protein catabolism increases.[67]

13.3.6 **SIRT6**

Together with SIRT7, SIRT6 can be almost exclusively found in the nucleus of fibroblast cells.[68] However, during mitosis or under stress, SIRT6 also localizes in the cytoplasm.[69-71] In the nucleus, SIRT6 associates with telomeres, where it deacetylates histone H3. In addition, SIRT6 promotes DNA end resection through C-terminal binding protein interacting protein (CtIP) deacetylation.[72] Reduced SIRT6 levels result in telomere dysfunction with end-to-end chromosomal fusions, genomic instability, DNA damage and premature cellular senescence.[6,73] By ADP-ribosylation and, therefore, activation of PARP1, SIRT6 stimulates DNA repair after oxidative stress.[74] Interestingly, SIRT6 attenuates NF-κB-dependent signaling, inflammation, and senescence by direct binding to NF-κB and by deacetylation of H3 histone at NF-κB target genes.[75]

SIRT6 also plays an important role in metabolism. Together with SIRT3, SIRT6 reduces glycolysis by repressing HIF-1α activity,[76] reduces triglyceride synthesis[77] and increases fatty acid oxidation.[78] Acting against SIRT1, by activation (deacetylation) of the general control nonrepressed protein 5 (GCN5), SIRT6 decreases PGC-1α activity.[79] SIRT6 deacetylates FoxO1, the key TF that mediates activation of the rate-limiting enzymes of gluconeogenesis, resulting in its transport to the cytoplasm and in a reduction in gluconeogenesis.[80]

13.3.7 SIRT7

SIRT7 is mainly located in the nucleus and in the nucleolus of mitotically active cells, where it regulates proliferation.[23] Recently, SIRT7 was linked to ribosome biogenesis and rRNA transcription and processing.[81] Deacetylation of U3-55k, a core component of the U3–small nucleolar RNP complex, is a prerequisite for pre-rRNA processing. However, under stress, when SIRT7 is released from nucleoli, hyperacetylation of U3-55k leads to attenuation of pre-rRNA processing.[81] In addition, SIRT7 activates both RNA polymerase I and II transcription.[82,83] By direct binding, SIRT7 reduces the protein levels and target gene expression of HIF-1α and HIF-2α.[84] Interestingly, SIRT7 controls mitochondrial homeostasis by the deacetylation of GA binding protein (GABPβ1), thereby enhancing the expression of mitochondrial genes.[85]

13.4 SIRTUINS AND NAD$^+$ HOMEOSTASIS

The strict dependence of sirtuins on NAD$^+$ as a cofactor suggests that modulation of intracellular concentrations of NAD$^+$, estimated as ~ 0.3 mM, can impact on the activity of all sirtuins. In experimental models, manipulation of NAD$^+$ levels changes SIRT2-dependent deacetylation of α-tubulin lysine 40 (K40).[45] It is less clear whether supplementation of NAD$^+$ might be in a range which is translated into a significant change of sirtuin activities in the brain.[86] Nevertheless, NAD$^+$ concentrations in the brain decline with age and, speculatively, might have a negative impact on activities of all sirtuins and other NAD$^+$-dependent enzymes like PARP1.[87] Restoring NAD$^+$ by supplementing its precursors or intermediates can ameliorate these age-associated functional defects.[88]

Oxidation-induced DNA damage activates PARP1, which rapidly cleaves substrate NAD$^+$ molecules to form ADP-ribose polymers on histones and other cellular proteins as a part of the cellular defense and repair program.[89,90] Activated PARP1 acts in a highly sequential, efficient, and damage-dependent manner,[91] but excessive oxidative stress may lead to overactivation of PARP1 and, therefore, to a rapid depletion of NAD$^+$ and ATP and cell-death[92,93] (Fig. 13.3). PARP1 inhibition decreases acetylation and restores activity of SIRT3-dependent enzymes in cells treated with cisplatin, a DNA-damaging agent,[94] and leads to marked improvement of the respiratory chain defect in the brain of Sco2 knockout/knockin mouse, a mitochondrial disease model characterized by impaired cytochrome c oxidase biogenesis.[86] Excessive oxidative stress, associated with pathology in neurodegenerative diseases, may reduce NAD$^+$ brain levels by activated PARP1, which subsequently negatively impacts on sirtuin performance.

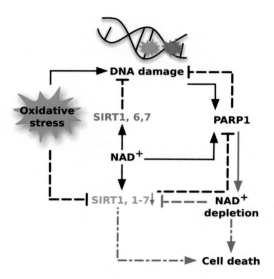

FIGURE 13.3 Influence of oxidative stress on homeostasis of NAD^+-dependent sirtuin and PARP1 enzymes.

Excessive oxidative stress is a common feature associated with neurodegeneration. Oxidation-induced DNA damage activates PARP1 that targets damaged sites. Strong damage (red lines) may lead to PARP1 overactivation resulting in rapid NAD^+ depletion, inhibition of SIRT1−7, and ultimately cell death. Dashed red lines: NAD^+-depletion-induced cell death (mechanism includes, e.g., release of apoptosis-inducing factor, loss of plasma membrane homeostasis, inhibition of glycolysis, and mitochondrial functions).[169] Plain black lines with arrowhead and dashed black lines with bars show activation and inhibition, respectively.

13.5 SIRTUIN FUNCTIONS IN THE BRAIN

Sirtuins play a role in neurogenesis and neuronal differentiation. SIRT1 inhibits neurogenesis probably by repressing the TF Mash1.[95] Similarly, inhibition of SIRT1 leads to neurogenesis and neuronal differentiation via increased signaling through the Wnt and Notch signaling pathways.[96] In contrast, SIRT2 plays a positive role in neuronal differentiation. SIRT2 promotes neuronal differentiation of mesenchymal stem cells via its tubulin deacetylase activity and stimulation of the extracellular signal-regulated kinase (ERK)-cAMP response element-binding protein (CREB) signaling pathway.[50] More specifically, SIRT2 potentiates the differentiation of nigral dopaminergic neurons via the AKT/GSK-3β/β-catenin pathway.[56]

SIRT6 regulates the expression of core pluripotent genes (Oct4, Sox2, and Nanog) by H3 deacetylation, which in turn controls stem cell differentiation and neuroectoderm development.[97]

SIRT1, 2, and 6 are also reported to influence synaptic plasticity. Absence of SIRT1 impairs cognitive abilities and synaptic plasticity, and decreases dendritic branching and complexity in an ERK1/2-dependent manner.[98,99] SIRT1 increases presenilin[100] and brain-derived neurotrophic factor (BDNF)[101] expression, regulates p53 stability, and increases LTP.[102] In addition, via the regulation of microRNA-134, it induces the expression of CREB-binding protein-dependent genes.[103]

SIRT1 promotes neurite outgrowth via downregulation of the mammalian target of rapamycin (mTOR) pathway[104] and GSK-3β, and parallel activation of Akt.[105]

SIRT1 regulates several, mainly metabolism-, food intake-, and circadian rhythm-related functions in the hypothalamus.[106] Depletion of SIRT1 from the hypothalamus prevents calorie restriction regulation of the somatotropic axis,[107] influences food intake during the dark cycle,[108] or improves leptin sensitivity.[109] Via activation of PGC-1α, SIRT1 increases the expression of the circadian clock protein BMAL1, and influences circadian control.[106] Since the SIRT1 brain level decreases with age, it might contribute to altered circadian rhythm in the elderly population.

Activation of SIRT2 and SIRT6 mediates depression-like phenotypes. SIRT2 inhibition has anti-depressant effects via modulation of the glutamate and serotonin systems in the prefrontal cortex.[110] SIRT2 activation, in turn, inhibits learning and decreases neuroblast proliferation in the adult dentate gyrus.[111] SIRT2 also inhibits neurite outgrowth and growth cone collapse.[49] Expression of SIRT6 increases under chronic stress, and decreases pAkt/Akt and pGSK3β/GSK3β ratios in the hippocampus.[112]

SIRT2 is present in neurons and astrocytes[19] and is highly expressed in oligodendroglia, mainly in the myelin sheet.[113] SIRT2 inhibits oligodendroglia differentiation, via regulation of tubulin acetylation, and contributes to glial support of axonal integrity.[114] Interestingly, SIRT2 was also shown to facilitate oligodendroglial differentiation by increasing the expression of myelin basic protein and promoting arborization[115] and remyelination after nerve injury in the peripheral nervous system, by regulation of Par-3 acetylation.[116] In contrast, SIRT1 limits the expansion and proliferation of oligodendroglia precursor cells, and inhibition of SIRT1 ameliorates remyelination and minimizes axonal damage by affecting platelet-derived growth factor receptor α expression.[117]

Sirtuins can also influence the innate immune system. Microglia, the innate immune cells of the brain, react against pathogens and damaged cells. However, sustained inflammation mediated by microglia has been associated with neurodegenerative diseases.[118] SIRT1, SIRT2, and SIRT3 decrease microglia activation and inflammatory responses[119–122] via inhibition of NF-κB signaling or FOXO3 activation. In aging microglia, the levels of SIRT1 decrease, leading to an NF-κB-mediated increase in IL-1β, contributing to cognitive decline.[123]

13.6 SIRTUINS AND NEURODEGENERATIVE DISEASE

Together, age-associated neurodegenerative disorders, including prevalent diseases, such as Alzheimer's (AD) and Parkinson's diseases (PD), as well as rare diseases, such as Huntington's disease (HD) and amyotrophic lateral sclerosis (ALS), affect millions of patients worldwide. The prevalence of these disorders is projected to increase due to an ongoing extension of human life span. Currently, there are no disease-modifying therapies for any neurodegenerative disease, leading to a tremendous socioeconomic impact.

Although the causative factors and clinical manifestations are distinct for each neurodegenerative disease, the molecular pathogenesis shares common pathways including, but not limited to, metabolic and energetic deficiencies, excessive oxidative stress due to mitochondrial impairment, neuroinflammation, disturbances in protein homeostasis (proteostasis), and protein misfolding and aggregation.

13.7 SIRTUINS AND ALZHEIMER'S DISEASE

In mechanistic studies, loss of SIRT1 is associated with accumulation of β-amyloid (Aβ) and tau in the cerebral cortex of AD patients.[124] Via inhibition of the serine/threonine Rho kinase 1 (ROCK1), SIRT1 increases the activity of the nonamyloidogenic α-secretase and attenuates Aβ accumulation in the brain[125,126] (Fig. 13.4A). By modulation of the amyloidogenic pathway, SIRT1 might be the link between metabolic stress and increased risk of AD, as PGC-1α-mediated

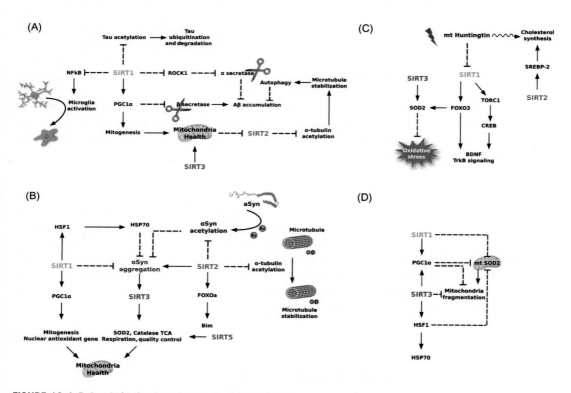

FIGURE 13.4 Role of sirtuins in neurodegenerative diseases.

(A) Alzheimer's disease: SIRT1 decreases Aβ production and Aβ-mediated microglia activation, and facilitates tau degradation, and together with SIRT3, improves mitochondrial functions. In contrast, by deacetylation of α-tubulin and hence modulating microtubule stability, SIRT2 inhibits Aβ clearance via autophagy. (B) Parkinson's disease: SIRT1 inhibits the aggregation of αsyn via induction of HSP70, and activates nuclear antioxidant genes and mitogenesis. SIRT3 and 5 improve mitochondrial health as well. SIRT2 increases αsyn aggregation and apoptotic signaling. (C) Huntington's disease: Via FOXO3 and TORC1 activation, SIRT1 increases BDNF expression and TrkB signaling, and acting synergistically with SIRT3, activates SOD2 and improves ROS elimination. Similar to mutant huntingtin, SIRT2 increases sterol synthesis as well. (D) Amyotrophic lateral sclerosis: SIRT3 and possibly SIRT1 protect against mitochondrial fragmentation and increased oxidative stress.

suppression of BACE1 transcription is SIRT1-dependent.[127] In addition, SIRT1 deacetylates tau, rendering it more prone for ubiquitination and degradation.[128,129]

SIRT1 also inhibits NF-κB signaling in microglia surrounding the Aβ-plaques, and attenuates neuronal cell death.[130] SIRT2 inhibition may have similar effects, reducing Aβ production[131] and Thr231-phospho-tau levels, and improving cognitive performance.[129] The levels of SIRT3 protein decrease in AD mouse models,[132] and exercise induces neuroprotection and increases SIRT3 levels.[133]

13.8 SIRTUINS AND PARKINSON'S DISEASE

The involvement of SIRT1 in the pathology of PD is controversial. SIRT1 suppresses α-synuclein inclusion formation in worms[134] (Fig. 13.4B). In addition, resveratrol has protective effects in the 1-methyl-4-phenyl-1,2,3,6-tetrahydropyridine (MPTP) mouse model[135,136] and in HtrA2 KO mice.[137] Resveratrol activates SIRT1 and, in turn, increases degradation of α-synuclein via LC-3 deacetylation-induced autophagy. In addition, resveratrol activates PGC-1α via SIRT1, and subsequently the transcription of antioxidant genes. However, SIRT1 transgenic mice do not show resistance to the MPTP model.[138]

In contrast, inhibition of SIRT2 leads to protection in different PD models: decreases αsynuclein toxicity in yeast,[139] and reduces the effect of MPTP in aged mice.[140,141] SIRT2 KO mice are protected against MPTP due to increased Foxo3a acetylation and decreased Bim expression.[142] Recently, it was shown that SIRT2 increases α-synuclein toxicity by deacetylating specific lysine residues.[143] Overexpression of the mitochondrial SIRT3 increases cell viability and decreases α-synuclein accumulation in vitro by decreasing ROS generation[144] and acetylation, hence activating citrate synthase and isocitrate dehydrogenase 2.[145] Interestingly, the levels of SOD2 in MPTP-treated SIRT5-deficient mice decrease more than in control animals, leading to exacerbated nigrostriatal pathology.[146]

13.9 SIRTUINS AND HUNTINGTON'S DISEASE

In a transgenic mouse model of HD, overall histone H3 K9/K14 acetylation is decreased,[147] suggesting a role for HDACs in HD pathology. Interestingly, by binding to SIRT1 and inhibiting its deacetylase activity, mutant huntingtin (Htt) inhibits transcription activity of Foxo3 that results in decreased BDNF levels and tyrosine receptor kinase B (TrkB) signaling in vivo[148] (Fig. 13.4C). Overexpression of SIRT1 enhances Foxo3a deacetylation and protects against Htt toxicity. SIRT1 activates the CREB-regulated transcription coactivator 1 (TORC1) as well.[149] Although many studies described protective effects of SIRT1 in HD models, resveratrol failed to improve motor performance or striatal atrophy in mice.[150]

Although SIRT2 is not upregulated in HD models, genetic or pharmacological inhibition of SIRT2 is protective in different HD models.[151,152] SIRT2 induces nuclear trafficking and transcription activation of the sterol response element binding protein 2 (SREBP-2) and increases cholesterol synthesis in invertebrates.[151] In addition, inhibition of SIRT2 improves motor function and reduces neuropathology in a mouse HD model.[152] In contrast, in another mouse model, loss of SIRT2 has no effect either on α-tubulin or H4 acetylation, and it does not improve HD progression,[153] leaving the putative effects of SIRT2 unresolved.

SIRT3 activity and mitochondrial biogenesis are reduced in neurons expressing mutant Htt mice,[154] and ε-viniferin, a resveratrol dimer decreases Htt-induced ROS production via SIRT3-mediated SOD2 deacetylation.[154]

13.10 SIRTUINS AND AMYOTROPHIC LATERAL SCLEROSIS

Expression of SIRT1 and SIRT2 is, in general, reduced in the cortex and spinal cord of ALS patients.[155] However, neuronal-specific levels of SIRT1, 2, and 5 are increased.[155] In an ALS mouse model, SIRT1 is upregulated, but resveratrol-induced activation of SIRT1 still protects against mutant SOD1-mediated neuronal loss.[156] Moreover, intraperitoneal resveratrol improves ALS symptoms in mice.[157] Via deacetylation of HSF1, SIRT1 overexpression increases HSP70i levels in another model, although it does not improve the pathology[158] (Fig. 13.4D).

While SIRT3 protects against mitochondrial fragmentation and neuronal cell death by mutant SOD1,[159] pharmacological inhibition of SIRT2 is not neuroprotective against ALS.[141]

13.11 DRUG DISCOVERY AND DEVELOPMENT OF SIRTUIN MODULATORS

Much of the data from translational research implicates SIRT1 activation as putatively neuroprotective. Therefore, the initial thrust in drug discovery was focused on the identification of sirtuin agonists. The proof of principle has been initially established with the natural compound resveratrol.

Resveratrol, a polyphenol, is an enriched component of red wine, which induces expression and activation of several neuroprotective pathways involving SIRT1 and 5′ AMP-activated protein kinase. Given the well-known absence of resveratrol toxicity even at relatively high doses, much of the efficacy studies have been focused on the examination of this polypharmacological compound in models of neurodegeneration. Resveratrol, administered by IP injection, attenuates oxidative stress and improves behavior in the MPTP mouse model of PD.[160] Chronic IP injection of resveratrol also delayed disease onset and extended survival of the ALS transgenic mouse model overexpressing mutant G93A-SOD1, increasing the number of surviving motor neurons.[161] Long-term resveratrol oral treatment with resveratrol significantly prevents memory loss, reduces the amyloid burden, and increases mitochondrial complex IV protein levels in brain of an AD mouse model.[162] In a mouse model of HD, continuous treatment with resveratrol significantly improves motor coordination and learning, and enhances expression of mitochondrial-encoded electron transport chain genes.[163] A correlation between the observed efficacy and activation of SIRT1 is not always apparent, which makes data interpretation not straightforward, due to known resveratrol polypharmacology. Importantly, the target validation study with selective brain-permeable small-molecule activator of SIRT1 demonstrated that treatment attenuates brain atrophy, improves motor function, and extends survival in an HD mouse model, providing important evidence of neuroprotective effects of pathway activation.[164] Consistently, treatment with high concentrations of NAM, a metabolic product of the deacetylation reaction and pan-inhibitor of sirtuin activities, including SIRT1, blocked mitochondrial-related transcription and worsened the motor phenotype in HD mice.

In contrast, NAM treatment restores cognitive deficits in AD transgenic mice via a mechanism involving sirtuin inhibition.[129] NAM selectively reduces a specific species of phospho-tau (Thr231), associated with microtubule depolymerization, and increases acetylation of α-tubulin, a major SIRT2 substrate associated with enhanced microtubule stability.[129]

Interestingly, new studies identified α-synuclein as a novel substrate of SIRT2 deacetylase,[143] providing a mechanistic link to previously observed efficacy of SIRT2 inhibitors in PD neuronal and fly models of aSyn neurotoxicity.[139] Moreover, SIRT2 inhibitors are also protective in HD neuronal and *Drosophila* models.[151,152] A selective brain-permeable SIRT2 inhibitor, AK-7, was found to be neuroprotective in PD and HD mouse models.[152,141,165] The latest studies showed that the SIRT2 inhibitor rescues age-related induced PD phenotype in mice.[141,166] In AD mice, treatment with selective SIRT2 inhibitors improves cognitive deficits, alongside reduced accumulation of Ab.[131]

The neuroprotective effects of SIRT2 inhibitors across different disease models suggest the involvement of multiple pathways modulated by deacetylation of diverse protein substrates.[167] Strikingly, it appears that SIRT1 activation and SIRT2 inhibition might be redundant in mediating neuroprotection, as it was shown, for example, that resveratrol and a selective SIRT2 antagonist repress reactive gliosis in AD astrocytes.[168]

13.12 FUTURE PERSPECTIVES

Despite significant progress, the biological functions of sirtuins in the CNS still have to be further elucidated. In particular, the identification of biological roles for sirtuin-specific isoforms and the understanding of their interactions with other sirtuin members will be instrumental for our understanding of their role in the brain, a complex and specialized organ with multiple cell types in which sirtuins emerge as important players. Additionally, it remains imperative to define a specific pathway(s) and mechanism(s) modulating the impact of sirtuins on aging and neurodegeneration.

The discovery and development of potent and selective brain-permeable modulators of all sirtuin members remains a challenge, due to our limited understanding of their specific functions and localization. Nevertheless, the discovery of SIRT1, SIRT3, and SIRT5 activators and SIRT2 inhibitors, and the systematic evaluation of these small molecules in mouse models of neurodegenerative diseases are essential.

In conclusion, basic and translational research on sirtuins holds promise for the development of sirtuin modulators for clinical use and treatment of neurodegenerative conditions in the near future and, in addition, for the development of novel tool compounds with which to more effectively probe sirtuin biology.

ACKNOWLEDGMENTS

We thank Dr. Diana F. Lázaro for excellent assistance with the preparation of figures. ÉMS is supported by the Dorothea Schlözer program. TFO is supported by the DFG Center for Nanoscale Microscopy and Molecular Physiology of the Brain (CNMPB).

REFERENCES

1. Michan S, Sinclair D. Sirtuins in mammals: insights into their biological function. *Biochem J* 2007;**404**:1−13.
2. Shore D, Squire M, Nasmyth KA. Characterization of two genes required for the position-effect control of yeast mating-type genes. *EMBO J* 1984;**3**:2817−23.
3. Imai S, Armstrong CM, Kaeberlein M, et al. Transcriptional silencing and longevity protein Sir2 is an NAD-dependent histone deacetylase. *Nature* 2000;**403**:795−800.
4. Landry J, Sutton A, Tafrov ST, et al. The silencing protein SIR2 and its homologs are NAD-dependent protein deacetylases. *Proc Natl Acad Sci U S A* 2000;**97**:5807−11.
5. North BJ, Marshall BL, Borra MT, et al. The human Sir2 ortholog, SIRT2, Is an NAD + -dependent tubulin deacetylase. *Mol Cell* 2003;**11**:437−44.
6. Michishita E, McCord RA, Berber E, et al. SIRT6 is a histone H3 lysine 9 deacetylase that modulates telomeric chromatin. *Nature* 2008;**452**:492−6.
7. Barber MF, Michishita-Kioi E, Xi Y, et al. SIRT7 links H3K18 deacetylation to maintenance of oncogenic transformation. *Nature* 2012;**487**:114−18.
8. Du J, Zhou Y, Su X, et al. Sirt5 is a NAD-dependent protein lysine demalonylase and desuccinylase. *Science* 2011;**334**:806−9.
9. Onyango P, Celic I, McCaffery JM, et al. SIRT3, a human SIR2 homologue, is an NAD-dependent deacetylase localized to mitochondria. *Proc Natl Acad Sci U S A* 2002;**99**:13653−8.
10. Ahuja N, Schwer B, Carobbio S, et al. Regulation of insulin secretion by SIRT4, a mitochondrial ADP-ribosyltransferase. *J Biol Chem* 2007;**282**:33583−92.
11. Taylor DM, Maxwell MM, Luthi-Carter R, et al. Biological and potential therapeutic roles of sirtuin deacetylases. *Cell Mol Life Sci* 2008;**65**:4000−18.
12. Bishop NA, Guarente L. Genetic links between diet and lifespan: shared mechanisms from yeast to humans. *Nat Rev Genet* 2007;**8**:835−44.
13. Cohen HY, Miller C, Bitterman KJ, et al. Calorie restriction promotes mammalian cell survival by inducing the SIRT1 deacetylase. *Science* 2004;**305**:390−2.
14. Lynch CJ, Milner J. Loss of one p53 allele results in four-fold reduction of p53 mRNA and protein: a basis for p53 haplo-insufficiency. *Oncogene* 2006;**25**:3463−70.
15. Lynch CJ, Shah ZH, Allison SJ, et al. SIRT1 undergoes alternative splicing in a novel auto-regulatory loop with p53. *PLoS One* 2010;**5**:e13502.
16. Rack JGM, VanLinden MR, Lutter T, et al. Constitutive nuclear localization of an alternatively spliced sirtuin-2 isoform. *J Mol Biol* 2014;**426**:1677−91.
17. Bao J, Lu Z, Joseph JJ, et al. Characterization of the murine SIRT3 mitochondrial localization sequence and comparison of mitochondrial enrichment and deacetylase activity of long and short SIRT3 isoforms. *J Cell Biochem* 2010;**110**:238−47.
18. Matsushita N, Yonashiro R, Ogata Y, et al. Distinct regulation of mitochondrial localization and stability of two human Sirt5 isoforms. *Genes Cells* 2011;**16**:190−202.
19. Sidorova-Darmos E, Wither RG, Shulyakova N, et al. Differential expression of sirtuin family members in the developing, adult, and aged rat brain. *Front Aging Neurosci* 2014;**6**:333.
20. Shah ZH, Ahmed SU, Ford JR, et al. A deacetylase-deficient SIRT1 variant opposes full-length sirt1 in regulating tumor suppressor p53 and governs expression of cancer-related genes. *Mol Cell Biol* 2012;**32**:704−16.
21. Sakamoto J, Miura T, Shimamoto K, et al. Predominant expression of Sir2alpha, an NAD-dependent histone deacetylase, in the embryonic mouse heart and brain. *FEBS Lett* 2004;**556**:281−6.

22. Kupis W, Pałyga J, Tomal E, et al. The role of sirtuins in cellular homeostasis. *J Physiol Biochem* 2016;**72**:371−80.
23. Haigis MC, Guarente LP. Mammalian sirtuins--emerging roles in physiology, aging, and calorie restriction. *Genes Dev* 2006;**20**:2913−21.
24. Xiong S, Salazar G, Patrushev N, et al. FoxO1 mediates an autofeedback loop regulating SIRT1 expression. *J Biol Chem* 2011;**286**:5289−99.
25. Martins R, Lithgow GJ, Link W. Long live FOXO: unraveling the role of FOXO proteins in aging and longevity. *Aging Cell* 2016;**15**:196−207.
26. Tanno M, Sakamoto J, Miura T, et al. Nucleocytoplasmic shuttling of the NAD + -dependent histone deacetylase SIRT1. *J Biol Chem* 2007;**282**:6823−32.
27. Feng Y, Liu T, Dong S-Y, et al. Rotenone affects p53 transcriptional activity and apoptosis via targeting SIRT1 and H3K9 acetylation in SH-SY5Y cells. *J Neurochem* 2015;**134**:668−76.
28. Fusco S, Maulucci G, Pani G. Sirt1: def-eating senescence? *Cell Cycle* 2012;**11**:4135−46.
29. Hao Q, Cho W. Battle against cancer: an everlasting saga of p53. *Int J Mol Sci* 2014;**15**:22109−27.
30. Wu X, Cao N, Fenech M, et al. Role of sirtuins in maintenance of genomic stability: relevance to cancer and healthy aging. *DNA Cell Biol* 2016;**35**:542−75.
31. Palacios JA, Herranz D, De Bonis ML, et al. SIRT1 contributes to telomere maintenance and augments global homologous recombination. *J Cell Biol* 2010;**191**:1299−313.
32. Kauppinen A, Suuronen T, Ojala J, et al. Antagonistic crosstalk between NF-κB and SIRT1 in the regulation of inflammation and metabolic disorders. *Cell Signal* 2013;**25**:1939−48.
33. Chang H-C, Guarente L. SIRT1 and other sirtuins in metabolism. *Trends Endocrinol Metab* 2014;**25**:138−45.
34. Guarente L. Sir2 links chromatin silencing, metabolism, and aging. *Genes Dev* 2000;**14**:1021−6.
35. Rodgers JT, Lerin C, Haas W, et al. Nutrient control of glucose homeostasis through a complex of PGC-1α and SIRT1. *Nature* 2005;**434**:113−18.
36. Guedes-Dias P, Oliveira JMA. Lysine deacetylases and mitochondrial dynamics in neurodegeneration. *Biochim Biophys Acta − Mol Basis Dis* 1832;**2013**:1345−59.
37. Lee IH, Cao L, Mostoslavsky R, et al. A role for the NAD-dependent deacetylase Sirt1 in the regulation of autophagy. *Proc Natl Acad Sci U S A* 2008;**105**:3374−9.
38. Pillai VB, Sundaresan NR, Gupta MP. Regulation of Akt signaling by sirtuins: its implication in cardiac hypertrophy and aging. *Circ Res* 2014;**114**:368−78.
39. Sundaresan NR, Pillai VB, Wolfgeher D, et al. The deacetylase SIRT1 promotes membrane localization and activation of Akt and PDK1 during tumorigenesis and cardiac hypertrophy. *Sci Signal* 2011;**4**:ra46.
40. Yu JSL, Cui W. Proliferation, survival and metabolism: the role of PI3K/AKT/mTOR signalling in pluripotency and cell fate determination. *Development* 2016;**143**:3050−60.
41. Lai C-H, Ho T-J, Kuo W-W, et al. Exercise training enhanced SIRT1 longevity signaling replaces the IGF1 survival pathway to attenuate aging-induced rat heart apoptosis. *Age (Omaha)* 2014;**36**:9706.
42. Sharples AP, Hughes DC, Deane CS, et al. Longevity and skeletal muscle mass: the role of IGF signalling, the sirtuins, dietary restriction and protein intake. *Aging Cell* 2015;**14**:511−23.
43. Vachharajani VT, Liu T, Wang X, et al. Sirtuins link inflammation and metabolism. *J Immunol Res* 2016;8167273.
44. Jęśko H, Strosznajder RP. Sirtuins and their interactions with transcription factors and poly(ADP-ribose) polymerases. *Folia Neuropathol* 2016;**3**:212−33.
45. Skoge RH, Dölle C, Ziegler M. Regulation of SIRT2-dependent α-tubulin deacetylation by cellular NAD levels. *DNA Repair (Amst)* 2014;**23**:33−8.
46. Reed NA, Cai D, Blasius TL, et al. Microtubule acetylation promotes kinesin-1 binding and transport. *Curr Biol* 2006;**16**:2166−72.

47. Li L, Yang X-J. Tubulin acetylation: responsible enzymes, biological functions and human diseases. *Cell Mol Life Sci* 2015;**72**:4237–55.
48. Creppe C, Malinouskaya L, Volvert M-L, et al. Elongator controls the migration and differentiation of cortical neurons through acetylation of alpha-tubulin. *Cell* 2009;**136**:551–64.
49. Pandithage R, Lilischkis R, Harting K, et al. The regulation of SIRT2 function by cyclin-dependent kinases affects cell motility. *J Cell Biol* 2008;**180**:915–29.
50. Jeong S-G, Cho G-W. The tubulin deacetylase sirtuin-2 regulates neuronal differentiation through the ERK/CREB signaling pathway. *Biochem Biophys Res Commun* 2017;**482**:182–7.
51. Inoue T, Hiratsuka M, Osaki M, et al. The molecular biology of mammalian SIRT proteins: SIRT2 in cell cycle regulation. *Cell Cycle* 2007;**6**:1011–18.
52. van Leeuwen IMM, Higgins M, Campbell J, et al. Modulation of p53 C-Terminal Acetylation by mdm2, p14ARF, and Cytoplasmic SirT2. *Mol Cancer Ther* 2013;**12**:471–80.
53. Nguyen P, Lee S, Lorang-Leins D, et al. SIRT2 interacts with β-catenin to inhibit Wnt signaling output in response to radiation-induced stress. *Mol Cancer Res* 2014;**12**:1244–53.
54. Zhang H, Head PE, Yu DS. SIRT2 orchestrates the DNA damage response. *Cell Cycle* 2016;**15**:2089–90.
55. Ramakrishnan G, Davaakhuu G, Kaplun L, et al. Sirt2 deacetylase is a novel AKT Binding partner critical for AKT activation by insulin. *J Biol Chem* 2014;**289**:6054–66.
56. Szegő ÉM, Gerhardt E, Outeiro TF. Sirtuin 2 enhances dopaminergic differentiation via the AKT/GSK-3β/β-catenin pathway. *Neurobiol Aging* 2017;**56**:7–16.
57. Rothgiesser KM, Erener S, Waibel S, et al. SIRT2 regulates NF-κB dependent gene expression through deacetylation of p65 Lys310. *J Cell Sci* 2010;**123**:4251–8.
58. Verdin E, Hirschey MD, Finley LWS, et al. Sirtuin regulation of mitochondria: energy production, apoptosis, and signaling. *Trends Biochem Sci* 2010;**35**:669–75.
59. Osborne B, Bentley NL, Montgomery MK, et al. The role of mitochondrial sirtuins in health and disease. *Free Radic Biol Med* 2016;**100**:164–74.
60. Rardin MJ, Newman JC, Held JM, et al. Label-free quantitative proteomics of the lysine acetylome in mitochondria identifies substrates of SIRT3 in metabolic pathways. *Proc Natl Acad Sci U S A* 2013;**110**:6601–6.
61. Haigis MC, Mostoslavsky R, Haigis KM, et al. SIRT4 inhibits glutamate dehydrogenase and opposes the effects of calorie restriction in pancreatic beta cells. *Cell* 2006;**126**:941–54.
62. Csibi A, Fendt S-M, Li C, et al. The mTORC1 pathway stimulates glutamine metabolism and cell proliferation by repressing SIRT4. *Cell* 2013;**153**:840–54.
63. Nasrin N, Wu X, Fortier E, et al. SIRT4 regulates fatty acid oxidation and mitochondrial gene expression in liver and muscle cells. *J Biol Chem* 2010;**285**:31995–2002.
64. Tao Y, Huang C, Huang Y, et al. SIRT4 suppresses inflammatory responses in human umbilical vein endothelial cells. *Cardiovasc Toxicol* 2015;**15**:217–23.
65. Mathias RA, Greco TM, Oberstein A, et al. Sirtuin 4 is a lipoamidase regulating pyruvate dehydrogenase complex activity. *Cell* 2014;**159**:1615–25.
66. Rardin MJ, He W, Nishida Y, et al. SIRT5 regulates the mitochondrial lysine succinylome and metabolic networks. *Cell Metab* 2013;**18**:920–33.
67. Nakagawa T, Lomb DJ, Haigis MC, et al. SIRT5 deacetylates carbamoyl phosphate synthetase 1 and regulates the urea cycle. *Cell* 2009;**137**:560–70.
68. Michishita E, Park JY, Burneskis JM, et al. Evolutionarily conserved and nonconserved cellular localizations and functions of human SIRT proteins. *Mol Biol Cell* 2005;**16**:4623–35.
69. Ardestani PM, Liang F. Sub-cellular localization, expression and functions of Sirt6 during the cell cycle in HeLa cells. *Nucleus* 2012;**3**:442–51.
70. Bhardwaj A, Das S. SIRT6 deacetylates PKM2 to suppress its nuclear localization and oncogenic functions. *Proc Natl Acad Sci U S A* 2016;**113**:E538–47.

71. Jedrusik-Bode M, Studencka M, Smolka C, et al. The sirtuin SIRT6 regulates stress granule formation in *C. elegans* and mammals. *J Cell Sci* 2013;**126**:5166−77.

72. Kaidi A, Weinert BT, Choudhary C, et al. Human SIRT6 promotes DNA end resection through CtIP deacetylation. *Science* 2010;**329**:1348−53.

73. Mostoslavsky R, Chua KF, Lombard DB, et al. Genomic instability and aging-like phenotype in the absence of mammalian SIRT6. *Cell* 2006;**124**:315−29.

74. Kugel S, Mostoslavsky R. Chromatin and beyond: the multitasking roles for SIRT6. *Trends Biochem Sci* 2014;**39**:72−81.

75. Kawahara TLA, Michishita E, Adler AS, et al. SIRT6 links histone H3 lysine 9 deacetylation to NF-kappaB-dependent gene expression and organismal life span. *Cell* 2009;**136**:62−74.

76. Zhong L, D'Urso A, Toiber D, et al. The histone deacetylase Sirt6 regulates glucose homeostasis via Hif1alpha. *Cell* 2010;**140**:280−93.

77. Kim H-S, Xiao C, Wang R-H, et al. Hepatic-specific disruption of SIRT6 in mice results in fatty liver formation due to enhanced glycolysis and triglyceride synthesis. *Cell Metab* 2010;**12**:224−36.

78. Liu TF, Vachharajani VT, Yoza BK, et al. NAD + -dependent sirtuin 1 and 6 proteins coordinate a switch from glucose to fatty acid oxidation during the acute inflammatory response. *J Biol Chem* 2012;**287**:25758−69.

79. Dominy JE, Lee Y, Jedrychowski MP, et al. The deacetylase Sirt6 activates the acetyltransferase GCN5 and suppresses hepatic gluconeogenesis. *Mol Cell* 2012;**48**:900−13.

80. Zhang P, Tu B, Wang H, et al. Tumor suppressor p53 cooperates with SIRT6 to regulate gluconeogenesis by promoting FoxO1 nuclear exclusion. *Proc Natl Acad Sci U S A* 2014;**111**:10684−9.

81. Chen S, Blank MF, Iyer A, et al. SIRT7-dependent deacetylation of the U3-55k protein controls pre-rRNA processing. *Nat Commun* 2016;**7**:10734.

82. Chen S, Seiler J, Santiago-Reichelt M, et al. Repression of RNA Polymerase I upon Stress Is Caused by Inhibition of RNA-Dependent Deacetylation of PAF53 by SIRT7. *Mol Cell* 2013;**52**:303−13.

83. Blank MF, Chen S, Poetz F, et al. SIRT7-dependent deacetylation of CDK9 activates RNA polymerase II transcription. *Nucleic Acids Res* 2017;**45**:2675−86.

84. Hubbi ME, Hu H, Kshitiz, et al. Sirtuin-7 inhibits the activity of hypoxia-inducible factors. J Biol Chem 2013;**288**:20768−75.

85. Ryu D, Jo YS, Lo Sasso G, et al. A SIRT7-dependent acetylation switch of GABPβ1 controls mitochondrial function. *Cell Metab* 2014;**20**:856−69.

86. Cerutti R, Pirinen E, Lamperti C, et al. NAD(+)-dependent activation of Sirt1 corrects the phenotype in a mouse model of mitochondrial disease. *Cell Metab* 2014;**19**:1042−9.

87. Zhu X, Ma X, Hu Y. PARP1: a promising target for the development of PARP1-based candidates for anticancer intervention. *Curr Med Chem* 2016;**23**:1756−74.

88. Imai S, Guarente L. NAD + and sirtuins in aging and disease. *Trends Cell Biol* 2014;**24**:464−71.

89. Virág L. Structure and function of poly(ADP-ribose) polymerase-1: role in oxidative stress-related pathologies. *Curr Vasc Pharmacol* 2005;**3**:209−14.

90. Petermann E, Keil C, Oei SL. Poly-ADP-ribosylation in health and disease. *C Cell Mol Life Sci* 2005;**62**:731−8.

91. Erdélyi K, Bakondi E, Gergely P, et al. Pathophysiologic role of oxidative stress-induced poly (ADP-ribose) polymerase-1 activation: focus on cell death and transcriptional regulation. *Cell Mol Life Sci* 2005;**62**:751−9.

92. Szabó C. Poly(ADP-ribose) polymerase activation by reactive nitrogen species--relevance for the pathogenesis of inflammation. *Nitric Oxide* 2006;**14**:169−79.

93. Cosi C, Marien M. Implication of poly (ADP-ribose) polymerase (PARP) in neurodegeneration and brain energy metabolism. Decreases in mouse brain NAD + and ATP caused by MPTP are prevented by the PARP inhibitor benzamide. *Ann N Y Acad Sci* 1999;**890**:227−39.

94. Yoon SP, Kim J. Poly(ADP-ribose) polymerase 1 contributes to oxidative stress through downregulation of sirtuin 3 during cisplatin nephrotoxicity. *Anat Cell Biol* 2016;**49**:165−76.
95. Prozorovski T, Schulze-Topphoff U, Glumm R, et al. Sirt1 contributes critically to the redox-dependent fate of neural progenitors. *Nat Cell Biol* 2008;**10**:385−94.
96. Kim BS, Lee C-H, Chang G-E, et al. A potent and selective small molecule inhibitor of sirtuin 1 promotes differentiation of pluripotent P19 cells into functional neurons. *Sci Rep* 2016;**6**:34324.
97. Etchegaray J-P, Chavez L, Huang Y, et al. The histone deacetylase SIRT6 controls embryonic stem cell fate via TET-mediated production of 5-hydroxymethylcytosine. *Nat Cell Biol* 2015;**17**:545−57.
98. Michan S, Li Y, Chou MM-H, et al. SIRT1 is essential for normal cognitive function and synaptic plasticity. J Neurosci 2010;**30**: 9695−707.
99. Abe-Higuchi N, Uchida S, Yamagata H, et al. Hippocampal sirtuin 1 signaling mediates depression-like behavior. *Biol Psychiatry* 2016;**80**:815−26.
100. Torres G, Dileo JN, Hallas BH, et al. Silent information regulator 1 mediates hippocampal plasticity through presenilin1. *Neuroscience* 2011;**179**:32−40.
101. Zocchi L, Sassone-Corsi P. SIRT1-mediated deacetylation of MeCP2 contributes to BDNF expression. *Epigenetics* 2012;**7**:695−700.
102. Lisachev PD, Pustylnyak VO, Shtark MB. Sirt1 regulates p53 stability and expression of its target S100B during long-term potentiation in rat hippocampus. *Bull Exp Biol Med* 2016;**160**(432−4).
103. Gao J, Wang W-Y, Mao Y-W, et al. A novel pathway regulates memory and plasticity via SIRT1 and miR-134. *Nature* 2010;**466**:1105−9.
104. Guo W, Qian L, Zhang J, et al. Sirt1 overexpression in neurons promotes neurite outgrowth and cell survival through inhibition of the mTOR signaling. *J Neurosci Res* 2011;**89**:1723−36.
105. Li X, Chen C, Tu Y, et al. Sirt1 promotes axonogenesis by deacetylation of Akt and inactivation of GSK3. *Mol Neurobiol* 2013;**48**:490−9.
106. Guarente L. Calorie restriction and sirtuins revisited. *Genes Dev* 2013;**27**:2072−85.
107. Cohen DE, Supinski AM, Bonkowski MS, et al. Neuronal SIRT1 regulates endocrine and behavioral responses to calorie restriction. *Genes Dev* 2009;**23**:2812−17.
108. Dietrich MO, Antunes C, Geliang G, et al. Agrp neurons mediate Sirt1's action on the melanocortin system and energy balance: roles for Sirt1 in neuronal firing and synaptic plasticity. *J Neurosci* 2010;**30**:11815−25.
109. Sasaki T, Kikuchi O, Shimpuku M, et al. Hypothalamic SIRT1 prevents age-associated weight gain by improving leptin sensitivity in mice. *Diabetologia* 2014;**57**:819−31.
110. Erburu M, Muñoz-Cobo I, Diaz-Perdigon T, et al. SIRT2 inhibition modulate glutamate and serotonin systems in the prefrontal cortex and induces antidepressant-like action. *Neuropharmacology* 2017;**117**:195−208.
111. Yoo DY, Kim DW, Kim MJ, et al. Sodium butyrate, a histone deacetylase Inhibitor, ameliorates SIRT2-induced memory impairment, reduction of cell proliferation, and neuroblast differentiation in the dentate gyrus. *Neurol Res* 2015;**37**:69−76.
112. Mao Q, Gong X, Zhou C, et al. Up-regulation of SIRT6 in the hippocampus induced rats with depression-like behavior via the block Akt/GSK3β signaling pathway. *Behav Brain Res* 2017;**323**:38−46.
113. Li W, Zhang B, Tang J, et al. Sirtuin 2, a mammalian homolog of yeast silent information regulator-2 longevity regulator, is an oligodendroglial protein that decelerates cell differentiation through deacetylating alpha-tubulin. *J Neurosci* 2007;**27**:2606−16.
114. Werner HB, Kuhlmann K, Shen S, et al. Proteolipid protein is required for transport of sirtuin 2 into CNS myelin. *J Neurosci* 2007;**27**:7717−30.

115. Ji S, Doucette JR, Nazarali AJ. Sirt2 is a novel in vivo downstream target of Nkx2.2 and enhances oligodendroglial cell differentiation. *J Mol Cell Biol* 2011;**3**:351−9.

116. Beirowski B, Gustin J, Armour SM, et al. Sir-two-homolog 2 (Sirt2) modulates peripheral myelination through polarity protein Par-3/atypical protein kinase C (aPKC) signaling. *Proc Natl Acad Sci* 2011;**108**: E952−61.

117. Rafalski VA, Ho PP, Brett JO, et al. Expansion of oligodendrocyte progenitor cells following SIRT1 inactivation in the adult brain. *Nat Cell Biol* 2013;**15**:614−24.

118. Plaza-Zabala A, Sierra-Torre V, Sierra A. Autophagy and microglia: novel partners in neurodegeneration and aging. Int J Mol Sci 2017;**18**: 598.

119. Jiang D-Q, Wang Y, Li M-X, et al. SIRT3 in neural stem cells attenuates microglia activation-induced oxidative stress injury through mitochondrial pathway. *Front Cell Neurosci* 2017;**11**:7.

120. Rangarajan P, Karthikeyan A, Lu J, et al. Sirtuin 3 regulates Foxo3a-mediated antioxidant pathway in microglia. *Neuroscience* 2015;**311**:398−414.

121. Li L, Sun Q, Li Y, et al. Overexpression of SIRT1 induced by resveratrol and inhibitor of miR-204 suppresses activation and proliferation of microglia. *J Mol Neurosci* 2015;**56**:858−67.

122. Pais TF, Szegő ÉM, Marques O, et al. The NAD-dependent deacetylase sirtuin 2 is a suppressor of microglial activation and brain inflammation. *EMBO J* 2013;**32**:2603−16.

123. Cho S-H, Chen JA, Sayed F, et al. SIRT1 deficiency in microglia contributes to cognitive decline in aging and neurodegeneration via epigenetic regulation of IL-1. *J Neurosci* 2015;**35**:807−18.

124. Julien C, Tremblay C, Emond V, et al. Sirtuin 1 reduction parallels the accumulation of tau in Alzheimer disease. *J Neuropathol Exp Neurol* 2009;**68**:48−58.

125. Qin W, Yang T, Ho L, et al. Neuronal SIRT1 activation as a novel mechanism underlying the prevention of Alzheimer disease amyloid neuropathology by valorie restriction. *J Biol Chem* 2006;**281**:21745−54.

126. Qin W, Chachich M, Lane M, et al. Calorie restriction attenuates Alzheimer's disease type brain amyloidosis in Squirrel monkeys (Saimiri sciureus). *J Alzheimers Dis* 2006;**10**:417−22.

127. Wang R, Li JJ, Diao S, et al. Metabolic stress modulates Alzheimer's β-secretase gene transcription via SIRT1-PPARγ-PGC-1 in neurons. *Cell Metab* 2013;**17**:685−94.

128. Min S-W, Cho S-H, Zhou Y, et al. Acetylation of tau inhibits its degradation and contributes to tauopathy. *Neuron* 2010;**67**:953−66.

129. Green KN, Steffan JS, Martinez-Coria H, et al. Nicotinamide restores cognition in Alzheimer's disease transgenic mice via a mechanism involving sirtuin inhibition and selective reduction of Thr231-phosphotau. *J Neurosci* 2008;**28**:11500−10.

130. Chen J, Zhou Y, Mueller-Steiner S, et al. SIRT1 protects against microglia-dependent amyloid-beta toxicity through inhibiting NF-kB signaling. *J Biol Chem* 2005;**280**:40364−74.

131. Biella G, Fusco F, Nardo E, et al. Sirtuin 2 inhibition improves cognitive performance and acts on amyloid-β protein precursor processing in two Alzheimer's disease mouse models. *J Alzheimers Dis* 2016;**53**:1193−207.

132. Yang W, Zou Y, Zhang M, et al. Mitochondrial Sirt3 expression is decreased in APP/PS1 double transgenic mouse model of Alzheimer's disease. *Neurochem Res* 2015;**40**:1576−82.

133. Bo H, Kang W, Jiang N, et al. Exercise-induced neuroprotection of hippocampus in APP/PS1 transgenic mice via upregulation of mitochondrial 8-oxoguanine DNA glycosylase. *Oxid Med Cell Longev* 2014;**2014**:1−14.

134. van Ham TJ, Thijssen KL, Breitling R, et al. C. elegans model identifies genetic modifiers of alpha-synuclein inclusion formation during aging. *PLoS Genet* 2008;**4**:e1000027.

135. Guo Y-J, Dong S-Y, Cui X-X, et al. Resveratrol alleviates MPTP-induced motor impairments and pathological changes by autophagic degradation of α-synuclein via SIRT1-deacetylated LC3. *Mol Nutr Food Res* 2016;**60**:2161−75.

136. Mudò G, Mäkelä J, Di Liberto V, et al. Transgenic expression and activation of PGC-1α protect dopaminergic neurons in the MPTP mouse model of Parkinson's disease. *Cell Mol Life Sci* 2012;**69**:1153−65.

137. Gerhardt E, Gräber S, Szegő ÉM, et al. Idebenone and resveratrol extend lifespan and improve motor function of HtrA2 knockout mice. *PLoS One* 2011;**6**:e28855.

138. Kitao Y, Ageta-Ishihara N, Takahashi R, et al. Transgenic supplementation of SIRT1 fails to alleviate acute loss of nigrostriatal dopamine neurons and gliosis in a mouse model of MPTP-induced parkinsonism. *F1000Research* 2015;**4**:130.

139. Outeiro TF, Kontopoulos E, Altmann SM, et al. Sirtuin 2 inhibitors rescue alpha-synuclein-mediated toxicity in models of Parkinson's disease. *Science* 2007;**317**:516−19.

140. Guan Q, Wang M, Chen H, et al. Aging-related 1-methyl-4-phenyl-1,2,3,6-tetrahydropyridine-induced neurochemial and behavioral deficits and redox dysfunction: improvement by AK-7. *Exp Gerontol* 2016;**82**:19−29.

141. Chen X, Wales P, Quinti L, et al. The sirtuin-2 inhibitor AK7 is neuroprotective in models of Parkinson's disease but not amyotrophic lateral sclerosis and cerebral ischemia. *PLoS One* 2015;**10**:e0116919.

142. Liu L, Arun A, Ellis L, et al. SIRT2 enhances 1-methyl-4-phenyl-1,2,3,6-tetrahydropyridine (MPTP)-induced nigrostriatal damage via apoptotic pathway. *Front Aging Neurosci* 2014;**6**:184.

143. de Oliveira RM, Vicente Miranda H, Francelle L, et al. The mechanism of sirtuin 2−mediated exacerbation of alpha-synuclein toxicity in models of Parkinson disease. *PLOS Biol* 2017;**15**:e2000374.

144. Zhang J-Y, Deng Y-N, Zhang M, et al. SIRT3 acts as a neuroprotective agent in rotenone-induced Parkinson cell model. *Neurochem Res* 2016;**41**:1761−73.

145. Cui X-X, Li X, Dong S-Y, et al. SIRT3 deacetylated and increased citrate synthase activity in PD model. *Biochem Biophys Res Commun* 2017;**484**:767−73.

146. Liu L, Peritore C, Ginsberg J, et al. Protective role of SIRT5 against motor deficit and dopaminergic degeneration in MPTP-induced mice model of Parkinson's disease. *Behav Brain Res* 2015;**281**:215−21.

147. McFarland KN, Das S, Sun TT, et al. Genome-wide histone acetylation is altered in a transgenic mouse model of Huntington's disease. *PLoS One* 2012;**7**:e41423.

148. Jiang M, Wang J, Fu J, et al. Neuroprotective role of Sirt1 in mammalian models of Huntington's disease through activation of multiple Sirt1 targets. *Nat Med* 2011;**18**:153−8.

149. Jeong H, Cohen DE, Cui L, et al. Sirt1 mediates neuroprotection from mutant huntingtin by activation of the TORC1 and CREB transcriptional pathway. *Nat Med* 2011;**18**:159−65.

150. Ho DJ, Calingasan NY, Wille E, et al. Resveratrol protects against peripheral deficits in a mouse model of Huntington's disease. *Exp Neurol* 2010;**225**:74−84.

151. Luthi-Carter R, Taylor DM, Pallos J, et al. SIRT2 inhibition achieves neuroprotection by decreasing sterol biosynthesis. *Proc Natl Acad Sci U S A* 2010;**107**:7927−32.

152. Chopra V, Quinti L, Kim J, et al. The Sirtuin 2 inhibitor AK-7 is neuroprotective in Huntington's disease mouse models. *Cell Rep* 2012;**2**:1492−7.

153. Bobrowska A, Donmez G, Weiss A, et al. SIRT2 ablation has no effect on tubulin acetylation in brain, cholesterol biosynthesis or the progression of Huntington's disease phenotypes in vivo. *PLoS One* 2012;**7**:e34805.

154. Fu J, Jin J, Cichewicz RH, et al. trans-(−)-ε-Viniferin increases mitochondrial sirtuin 3 (SIRT3), activates AMP-activated protein kinase (AMPK), and protects cells in models of Huntington disease. *J Biol Chem* 2012;**287**:24460−72.

155. Körner S, Böselt S, Thau N, et al. Differential sirtuin expression patterns in amyotrophic lateral sclerosis (ALS) postmortem tissue: neuroprotective or neurotoxic properties of sirtuins in ALS? *Neurodegener Dis* 2013;**11**:141–52.

156. Kim D, Nguyen MD, Dobbin MM, et al. SIRT1 deacetylase protects against neurodegeneration in models for Alzheimer's disease and amyotrophic lateral sclerosis. *EMBO J* 2007;**26**:3169–79.

157. Han S, Choi J-R, Soon Shin K, et al. Resveratrol upregulated heat shock proteins and extended the survival of G93A-SOD1 mice. *Brain Res* 2012;**1483**:112–17.

158. Watanabe S, Ageta-Ishihara N, Nagatsu S, et al. SIRT1 overexpression ameliorates a mouse model of SOD1-linked amyotrophic lateral sclerosis via HSF1/HSP70i chaperone system. *Mol Brain* 2014;**7**:62.

159. Song W, Song Y, Kincaid B, et al. Mutant SOD1G93A triggers mitochondrial fragmentation in spinal cord motor neurons: neuroprotection by SIRT3 and PGC-1α. *Neurobiol Dis* 2013;**51**:72–81.

160. Anandhan A, Tamilselvam K, Vijayraja D, et al. Resveratrol attenuates oxidative stress and improves behaviour in 1-methyl-4-phenyl-1,2,3,6-tetrahydropyridine (MPTP) challenged mice. *Ann Neurosci* 2010;**17**:113–19.

161. Han S, Choi J-R, Soon Shin K, et al. Resveratrol upregulated heat shock proteins and extended the survival of G93A-SOD1 mice. *Brain Res* 2012;**1483**:112–17.

162. Porquet D, Griñán-Ferré C, Ferrer I, et al. Neuroprotective role of trans-resveratrol in a murine model of familial Alzheimer's disease. *J Alzheimers Dis* 2014;**42**:1209–20.

163. Naia L, Rosenstock TR, Oliveira AM, et al. Comparative mitochondrial-based protective effects of resveratrol and nicotinamide in Huntington's disease models. *Mol Neurobiol* 2017;**54**:5385–99.

164. Jiang M, Zheng J, Peng Q, et al. Sirtuin 1 activator SRT2104 protects Huntington's disease mice. *Ann Clin Transl Neurol* 2014;**1**:1047–52.

165. Taylor DM, Balabadra U, Xiang Z, et al. A brain-permeable small molecule reduces neuronal cholesterol by inhibiting activity of sirtuin 2 deacetylase. *ACS Chem Biol* 2011;**6**:540–6.

166. Guan Q, Wang M, Chen H, et al. Aging-related 1-methyl-4-phenyl-1,2,3,6-tetrahydropyridine-induced neurochemial and behavioral deficits and redox dysfunction: improvement by AK-7. *Exp Gerontol* 2016;**82**:19–29.

167. Rauh D, Fischer F, Gertz M, et al. An acetylome peptide microarray reveals specificities and deacetylation substrates for all human sirtuin isoforms. *Nat Commun* 2013;**4**:2327.

168. Scuderi C, Stecca C, Bronzuoli MR, et al. Sirtuin modulators control reactive gliosis in an in vitro model of Alzheimer's disease. *Front Pharmacol* 2014;**5**:89.

169. Alano CC, Garnier P, Ying W, et al. NAD + depletion is necessary and sufficient for poly(ADP-ribose) polymerase-1-mediated neuronal death. *J Neurosci* 2010;**30**:2967–78.

Index

Printed in the United States
by Bookmasters